**Prentice Hall Advanced Reference Series**

*Polymer Science and Engineering*

**PRENTICE HALL**
**Polymer Science and Engineering**
James E. Mark, Series Editor

ROE, ED. *Computer Simulation of Polymers*
VERGNAUD *Liquid Transport Processes in Polymeric Materials: Modeling and Industrial Applications*

*FORTHCOMING BOOKS IN THIS SERIES (tentative titles)*

BALLAUF *Stiff Chain Polymers*
CLARSON AND SEMLYEN, EDS. *Siloxane Polymers*
FRIED *Polymer Science and Technology*
GALIATSATOS, BOUE, AND SIESLER *Molecular Characterization of Networks*
MARK, WEST, AND ALLCOCK *Inorganic Polymers: An Introduction*
SAIZ AND RIANDE *Dipole Moments and Birefringence of Polymers*
VILGIS AND BOUE *Random Fractals*

# LIQUID TRANSPORT PROCESSES IN POLYMERIC MATERIALS

## Modeling and Industrial Applications

J. M. Vergnaud

*University of Saint-Etienne*

Prentice Hall, Englewood Cliffs, New Jersey 07632

Library of Congress Cataloging-in-Publication Data

Vergnaud, J. M.
Liquid transport processes in polymeric materials : modeling and industrial applications / J.M. Vergnaud.
p. cm.
Includes bibliographical references and index.
ISBN 0-13-538315-3
1. Polymers--Permeability. 2. Diffusion--Mathematical models.
I. Title.
QD381.9.P45V47 1991
620.1'9292--dc20 90-43100
CIP

Editorial/production supervision
and interior design: *Brendan M. Stewart*
Cover design: *Wanda Lubelska*
Manufacturing buyer: *Kelly Behr* and *Patrice Fraccio*
**Prentice Hall Advanced Reference Series**
**Prentice Hall Polymer Science and Engineering Series**


A Division of Simon & Schuster
Englewood Cliffs, New Jersey 07632

This book can be made available to businesses and organizations at a special discount when ordered in large quantities. For more information contact:

Prentice-Hall, Inc.
Special Sales and Markets
College Division
Englewood Cliffs, N.J. 07632



Printed in the United States of America
10 9 8 7 6 5 4 3 2 1

ISBN 0-13-538315-3

Prentice-Hall International (UK) Limited, *London*
Prentice-Hall of Australia Pty. Limited, *Sydney*
Prentice-Hall Canada Inc., *Toronto*
Prentice-Hall Hispanoamericana, S.A., *Mexico*
Prentice-Hall of India Private Limited, *New Delhi*
Prentice-Hall of Japan, Inc., *Tokyo*
Simon & Schuster Asia Pte. Ltd., *Singapore*
Editora Prentice-Hall do Brasil, Ltda., *Rio de Janeiro*

# Contents

# Preface

This book, devoted to the modeling of liquid transport through various polymers, has several purposes.

First and foremost, the material presented here is concerned with the study of the processes of liquid transports which take place in various solids, and especially in polymers. The focus is to provide the reader with pertinent information on these processes. Through this, it is hoped that a better understanding of the process will be gained, before developing industrial applications in various fields. For example, applications of matter transports are considered in some cases:

- the decrease in the transport of liquid and plasticizer in the case of plasticized PVC in contact with various liquids.
- the control of the rate of delivery of a drug from dosage forms in the patient's stomach, or the rate of release of active agent in water in agricultural applications.

Special consideration is also given to the modeling of the process and to the lastest applications of computer simulation, frequently the only means of solving fundamental problems concerning matter transports through polymers. Emphasis is placed on the method used to build up numerical models, beginning with experiments carried out by using short tests and long real tests, and progressing through numerical analysis and computerization.

Experiments are necessary first to obtain a precise knowledge of the process, then to determine data such as diffusivities, the concentration dependency of diffusivities, and the capacity of absorption, by using short tests, and finally, to test the validity of the model with long real tests performed by varying the operational conditions.

Modeling generally obliges the user to carry out experiments in a more precise way and with more attention, because one has to write down the process and draw its various steps at the end of these experiments. When experimental results do not coincide well with the theoretical results, the model has to be improved, and sometimes the experiments have to be performed in a more precise way.

Numerical methods are built up by writing down the various steps of the process on the flowsheet, and by dealing with the mathematical treatment and numerical analysis. Numerical analysis is sometimes considered the art of making the assumptions and approximations which are necessary to transform differential equations into simple equations in common use by computers. It is then understandable that the best person to make these assumptions is the researcher who is responsible for the development of the process, because with a good knowledge of the process one will be obliged to focus in greater detail on these assumptions, and will try to find the best ones.

Numerical analysis has to be used when mathematical treatment is not capable of finding an analytical solution for the problem. In matter of diffusion, the mathematical treatment is powerful in only simple cases:

- when the diffusivity is constant.
- when the initial conditions are simple, for example, when the concentration of the diffusing substance is uniform in the solid.
- when the material is isotropic.
- when the boundary conditions are simple.

Numerical methods can be used in all cases because all the facts can be taken into account in the model: not only the various initial and boundary conditions, but also the concentration dependency of the diffusivity, as well as the various properties of the materials, isotropic or not.

Moreover, numerical models are of help for the user, bringing many advantages. As it is necessary to draw the process on the flowsheet with the various steps, parameters, and assumptions, researchers must be clear in their ideas, and this is of help for planning the work. After the validity of the model has been tested, a simple coincidence with something wrong in the process may appear when theoretical results coincide well with experiments. More experiements should be conducted under various conditions to make sure that the model describes the process properly in all cases.

As each parameter appears clearly in the model, the role played by all parameters can be determined by simulating the process. The optimization of the process in finding the best values for these parameters is then possible.

Modeling and simulation are often used in industrial applications. For instance, from experiments done on a laboratory scale, results in a full-scale plant process can be evaluated by calculation. In the case of polymeric devices, the dimensions of the device as well as the properties desirable for the polymers can be predicted by calculation in terms of the purpose.

Simulation of the process with the help of the model can help in predicting the behavior of large industrial samples without having to carry out tedious, costly, and time-consuming experiments on the real sample, and without detriment to the material, by using the data determined by short tests performed on small aliquot samples. Modeling is generally of help to reduce the number of experiments, and especially to determine the plan of the experiments necessary for the research.

The book is divided into three parts, and each part consists of many chapters.

The first part presents a general overview of the state of the mathematical treatment of diffusion. It is necessary for researchers to have good background knowledge of the mathematical treatment of diffusion equations to know when analytical solutions are available, and the conditions in which they are obtained, and to know how to determine the values of parameters in proper conditions. For instance, at least two ways can be used for determining the value of the diffusivity: (1) by using the square root dependency of the amount of diffusing substance with time, when this amount is low enough and thus for short times, and (2) by drawing the amount of diffusing substance in logarithmic form in terms of time, when this amount is high enough and thus for long times. Special attention is given to the assumptions made so as to obtain an analytical solution for the problem, because it is necessary to know exactly the conditions upon which the solution is given.

Three chapters are considered in the first part:

- A plane sheet of solid with a diffusing substance in contact with the two faces, the diffusion being conducted under transient or stationary conditions.
- A cylinder, either of infinite length or of finite length. The case of the hollow cylinder is also considered.
- The sphere is also studied, this shape being of interest for various applications.

Two cases of interest are described:

- When the solid is in perfect contact with the liquid by considering a large volume or a finite volume of liquid. Absorption and desorption of the liquid are studied.
- When the solid is in contact with the diffusing substance in a vapor state or with atmosphere, enabling the condensation or evaporation of the substance on the surface which diffuses through the solid.

The second part looks at the concept of numerical analysis in order to accustom researchers to this new way of working. Explicit numerical methods are essentially developed, because they are easily used with microcomputers.

Chapter 4 presents examples of numerical analysis in the case of plane sheets, when diffusion proceeds within the solid with constant or concentration-dependent diffusivities, by using various conditions on the surface.

Chapter 5 describes the numerical treatment for cylinders of infinite and finite length, as well as for hollow cylinders of infinite and finite length (tubing, annulus), when the diffusivity is constant or concentration dependent.

Chapter 6 traces the development of numerical analysis for solid and hollow spheres, with constant and concentration-dependent diffusivities.

Chapter 7 deals with the numerical analysis in the case of three-dimensional transports of matter with various main diffusivities along the three main directions of diffusion.

In the second part, two cases of interest are emphasized:

- When the solid is in perfect contact with the diffusing substance in liquid state, and then the concentration of the liquid on the surface reaches equilibrium as soon as the process starts.
- When the diffusing substance evaporates or condenses from or on the surface, the rate of transport being proportional to the difference between the concentration on the surface and the concentration at equilibrium.

The third part examines various approaches of studies of matter transports controlled by diffusion with different materials and liquids. Chapters 8 through 13 discuss each of the various problems, moving through the difficulties of experiments and calculation, and finally showing industrial applications.

Chapter 8 shows the complexity of the matter transport taking place between plasticized PVC and various liquids, and how powerful the numerical methods are to provide good knowledge of the process. Two matter transports are observed, with the liquid entering the PVC enabling the plasticizer to leave the polymer. Moreover, the kinetics of absorption of the liquid exhibit a maximum. And finally, the diffusivities for each of these two liquids are expressed in terms of the concentrations of the plasticizer and liquid, in such a way that these diffusivities are connected with each other. The knowledge obtained from experiments and models enables one to describe a method of preparation of flexible plasticized PVC with low matter transports.

Chapter 9 deals with the problems of absorption and desorption of liquids with rubber samples of various shapes: sheets, cylinders, tubings, and annuli. From experiments performed with thin sheets made of the same material, the data obtained such as the diffusivity and capacity of absorption (or desorption) are successfully used with the help of models to predict the behavior of large real rubber devices of various shapes.

Chapter 10 focuses on the problems of drug delivery for curing the patient, and the preparation and study of dosage forms able to deliver a drug in the stomach at a constant rate. Simple plane and spherical devices obtained by dispersing the drug in biocompatible polymers are studied, the result being that the liquid enters the polymer, dissolves the drug, and then enables the drug to diffuse through the liquid located in the polymer. More complex devices, consisting of a core and shell with defined properties, are built in order to obtain constant rates of delivery.

Chapter 11 discusses the problem of release of an active agent (insecticide, for example) in water, with application in agriculture. As constant rate of release is desirable for the agent, devices made of a core and shell with special properties are prepared and studied. As for the previous chapter with drug delivery, all parameters characterizing the core and shell (dimensions, nature, and properties of the materials) are determined with the help of the model.

Chapter 12 examines the drying of coatings, which is controlled by diffusion of the solvent through the coating and evaporation on the surface. The problem is not simple, because of the concentration-dependent diffusivity and of the presence of various layers. Nevertheless, the models are able to describe the process and to predict the kinetics of drying for coatings made of one and many layers.

Chapter 13 studies the absorption and desorption of moisture by wood samples of various sizes with one-, two-, and three-dimensional transports. All data, such as the three main diffusivities and their dependency with moisture concentration, and the capacity of absorption (amount of moisture absorbed at equilibrium), are determined in a rather fast way by using thin sheets of wood cut along the main axes of diffusion. The models tested are then able to describe the process of absorption and desorption of moisture, whether the conditions of the surrounding atmosphere are constant or programmed relative humidity, and whatever the shape of the wood samples.

As this book is essentially of practical use, no long literature survey is developed concerning the various theories elaborated on the diffusion of liquids through polymers or wood. In the case of rubber and rubbery materials, the essential results from these theories are summarized, as far as they are of practical use.

# Acknowledgments

A number of people have both influenced my work and supported my efforts. Prominent among them is my good friend and research colleague, Jean Bouzon, Professor at the University of St. Etienne. And most especially, I appreciate the kind help of my wife for showing me round the world of pharmacy with the problems of drug delivery.

I am grateful for the collaboration of my students and colleagues, D. Messadi, J. L. Taverdet, A. Grigorakakis, A. Droin, A. Eddine, M. P. Petitjean, I. Malley, Y. Khatir, P. Magron, F. Magnard, H. David, A. Aboutaybi.

Deep gratitude is extended to all the colleagues of the Galenic laboratory of Lyon I, with Prof. M. Rollet, Mrs. J. Bardon, and C. Chaumat.

My deepest appreciation is given to my students, A. Aboutaybi and A. El Brouzi, for their drawings which are presented throughout the book. Special thanks to my colleague Norton who read the initial drafts. Many thanks to Miss C. Cervantes for her competent typing of the manuscript.

# Chapter One

# Diffusion in a Plane Sheet

## 1-1 INTRODUCTION

The plane sheet of material is so thin that all the diffusing fluid (liquid, vapor, or gas) enters or leaves through the parallel faces, with a negligible amount through the edges.

One-dimensional diffusion in the material bounded by the two parallel planes at $x = 0$ and $x = L$, is considered along the $x$-axis. The diffusivity is constant.

## 1-2 NON-STEADY STATE WITH CONSTANT DIFFUSIVITY

All the solutions of the diffusion equation are obtained either by the Laplace transform or by the method of separation of the variables.

### 1-2-1 Initial Distribution is $f(x)$ in the Material. Concentration is Constant on Surfaces.

Assumptions: the diffusivity $D$ is constant.

$$\text{Initial} \quad t = 0 \quad 0 < x < L \quad C = f(x) \tag{1-1}$$

$$\begin{aligned} \text{Boundary} \quad t \geq 0 \quad & x = 0 \quad C = C_1 \\ & x = L \quad C = C_2 \end{aligned} \tag{1-2}$$

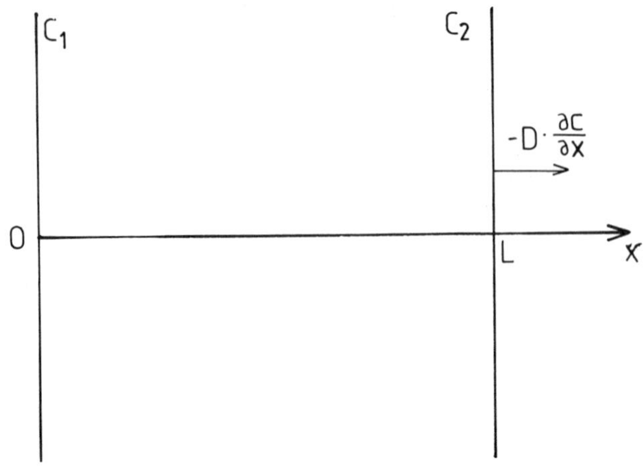


**Figure 1–1** Scheme of the plane sheet of thickness $L$, with constant concentration on the faces.

The equation for diffusion in one dimension when $D$ is constant

$$\frac{\partial C}{\partial t} = D \cdot \frac{\partial^2 C}{\partial x^2} \tag{1-3}$$

has a solution in the form of a trigonometrical series, obtained by the method of separation of the variables (Figure 1–1):

$$C_{x,t} = C_1 + (C_2 - C_1)\frac{x}{L} + \frac{2}{\pi} \cdot \sum_{n=1}^{\infty} \frac{1}{n}(C_2 \cos n\pi - C_1)\sin\frac{n\pi x}{L} \cdot \exp\left(-\frac{n^2\pi^2}{L^2}Dt\right) + \frac{2}{L}\sum_{n=1}^{\infty} \sin\frac{n\pi x}{L} \cdot \exp\left(-\frac{n^2\pi^2}{L^2}Dt\right) \cdot \int_o^L f(x') \cdot \sin\frac{n\pi x'}{L} \cdot \exp\left(-\frac{n^2\pi^2}{L^2}Dt\right) \tag{1-4}$$

When the problem is symmetrical about the central plane of the sheet, the boundary conditions are:

$$t \geq 0 \qquad x = 0 \quad \text{and} \quad x = L \qquad C = C_1\ (C_2 = C_1) \tag{1-5}$$

$$C_{x,t} = C_1 - \frac{4C_1}{\pi} \cdot \sum_{n=0}^{\infty} \frac{1}{2n+1} \cdot \sin\frac{(2n+1)\pi x}{L} \cdot \exp\left(\frac{-(2n+1)^2\pi^2}{L^2}Dt\right) \tag{1-6}$$

### 1-2-2 Initial Distribution is Uniform in the Material ($C_i$). Surface Concentrations Are Equal($C_1$).

This case is very frequent: the material $0 < x < L$ is initially at a uniform concentration $C_i$, and the surfaces are kept at a constant concentration $C_1$ during the whole process.

$$\text{Initial} \qquad t = 0 \qquad 0 < x < L \qquad C = C_i \tag{1-7}$$

$$\text{Boundary} \qquad t \geq 0 \qquad x = 0 \quad x = L \qquad C = C_1 \tag{1-8}$$

The solution is for the concentration $C_{x,t}$ at time $t$ and position $x$ (Figure 1–2):

$$\frac{C_1 - C_{xt}}{C_1 - C_i} = \frac{4}{\pi} \cdot \sum_{n=0}^{\infty} \frac{1}{2n + 1} \cdot \sin\left(\frac{(2n + 1)\pi x}{L}\right) \cdot \exp\left(-\frac{(2n + 1)^2\pi^2}{L^2}\right) Dt \tag{1-9}$$

The total amount of diffusing substance which has entered (or left) the plane sheet at time $t$, $M_t$, is obtained in terms of the amount of diffusing substance transported after infinite time:

$$\frac{M_\infty - M_t}{M_\infty} = \frac{8}{\pi^2} \sum_{n=0}^{\infty} \frac{1}{(2n + 1)^2} \cdot \exp\left(-\frac{(2n + 1)^2\pi^2}{L^2} Dt\right) \tag{1-10}$$

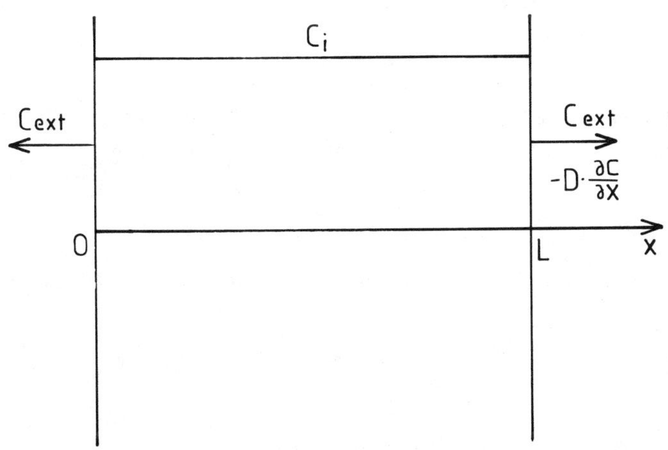


**Figure 1–2** Scheme of the plane sheet of thickness $L$, with uniform initial concentration in the sheet, and equal constant concentrations on the faces.

Another expression for the concentration $C_{x,t}$ is:

$$\frac{C_{x,t} - C_i}{C_1 - C_i} = \sum_{n=0}^{\infty} (-1)^n erfc \frac{(2n + 1)L - 2x}{4(Dt)^{0.5}} + \sum_{n=0}^{\infty} (-1)^n erfc \frac{(2n + 1)L + 2x}{4(Dt)^{0.5}} \tag{1-11}$$

and for the total amount of diffusing substance transported at time $t$, $M_t$:

$$\frac{M_t}{M_\infty} = \frac{4}{L} (Dt)^{0.5} \cdot \left\{ \pi^{-0.5} + 2 \sum_{n=1}^{\infty} (-1)^n \cdot ierfc \frac{nL}{2(Dt)^{0.5}} \right\} \tag{1-12}$$

***Remark. Partition Factor.*** The concentration of diffusing substance on the surface of the solid, $C_1$, can be considered as being at equilibrium with the constant concentration of diffusing substance in the medium outside the sheet, $C_{ext}$.

$K$ being the partition factor between the solid and the medium outside the sheet, there is the relationship: $C_1 = K \cdot C_{ext}$.

***Small Times.*** For small times, when $M_t/M_\infty$ is small enough, equation (1-12) is reduced to the well-known relation:

$$\frac{M_t}{M_\infty} = \frac{4}{L}\left(\frac{Dt}{\pi}\right)^{0.5} \tag{1-13}$$

The relative error made on $M_t/M_\infty$ by using equation (1-13) instead of equation (1-10) is:

1/1000 when $M_t/M_\infty \leq 0.5$

1/100 when $M_t/M_\infty \leq 0.63$

***Long Times.*** For long times, when $M_t/M_\infty$ is sufficiently large, equation (1-10) is reduced to the first term with $n = 0$.

$$\frac{M_\infty - M_t}{M_\infty} = \frac{8}{\pi^2} \cdot \exp\left(-\frac{\pi^2 Dt}{L^2}\right) \tag{1-14}$$

The relative error made on $M_t/M_\infty$ by using equation (1-14) instead of equation (1-10) is:

1/1000 when $M_t/M_\infty \geq 0.55$

### 1-2-3 Case of the Membrane

**Initial distribution is uniform in the material. Surface concentrations are different and kept constant.** One face of the membrane ($x = 0$) is kept at

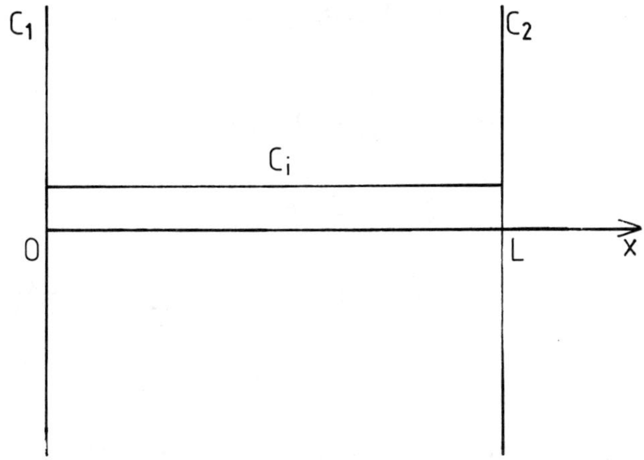


**Figure 1–3** Scheme of the plane sheet of thickness $L$, with uniform initial concentration in the sheet, and constant concentrations on the faces.

a constant concentration $C_1$ and the other face ($x = L$) at concentration $C_2$. The membrane is initially at a uniform concentration $C_i$. (Figure 1–3).

| | | | | |
|---|---|---|---|---|
| Initial | $t = 0$ | $0 < x < L$ | $C = C_i$ | (1-15) |
| Boundary | $t > 0$ | $x = 0$ | $C = C_1$ | (1-16) |
| | | $x = L$ | $C = C_2$ | |

The solution of the diffusion equation

$$\frac{\partial C}{\partial t} = D \cdot \frac{\partial^2 C}{\partial x^2} \tag{1-3}$$

is obtained by the method of separation of the variables:

$$\begin{aligned} C_{x,t} &= C_1 + (C_2 - C_1)\frac{x}{L} \\ &+ \frac{2}{\pi} \cdot \sum_{n=1}^{\infty} \frac{1}{n}(C_2 \cos n\pi - C_1) \cdot \sin\frac{n\pi x}{L} \cdot \exp\left(\frac{-n^2\pi^2}{L^2} Dt\right) \\ &+ \frac{4C_i}{\pi} \cdot \sum_{m=0}^{\infty} \frac{1}{2m+1} \cdot \sin\frac{(2m+1)\pi x}{L} \cdot \exp\left(-\frac{(2m+1)^2\pi^2}{L^2} Dt\right) \end{aligned} \tag{1-17}$$

The amount of diffusing substance, $M_t$, which passes through the membrane during time $t$ per unit area is obtained by integrating Fick's first law with respect to time, by calculating the gradient of concentration at $x = L$.

$$M_t = -\int_o^t D\cdot\left(\frac{\partial C}{\partial x}\right)_{x=L}\cdot dt \tag{1-18}$$

$$M_t = \frac{D(C_1 - C_2)}{L}t + \frac{2L}{\pi^2}\cdot\sum_{n=1}^{\infty}\frac{1}{n^2}(C_1\cos n\pi - C_2)\cdot\left\{1 - \exp\left(-\frac{n^2\pi^2}{L^2}\right)Dt\right\} + \frac{4C_iL}{\pi^2}\cdot\sum_{m=0}^{\infty}\frac{1}{(2m+1)^2}\cdot\left\{1 - \exp\left(-\frac{(2m+1)^2L^2}{L^2}Dt\right)\right\} \tag{1-19}$$

The amount of diffusing substance which enters the membrane per unit area during time $t$ is also calculated by integrating equation (1-18) with respect to time, by considering the gradient of concentration at $x = 0$.

$$M_t' = -\int_o^t D\cdot\left(\frac{\partial C}{\partial x}\right)_{x=0}\cdot dt \tag{1-20}$$

**Initial concentration is zero in the material. Surface concentrations are different and kept constant, one of them being zero.** Very often, the initial concentration in the membrane is zero, and the concentration at the face through which the diffusing substance emerges is also zero. (Figure 1–4).

| | | | | |
|---|---|---|---|---|
| Initial | $t = 0$ | $0 < x < L$ | $C = 0$ | (1-21) |
| Boundary | $t > 0$ | $x = 0$ | $C = C_1$ | (1-22) |
| | | $x = L$ | $C = 0$ | |

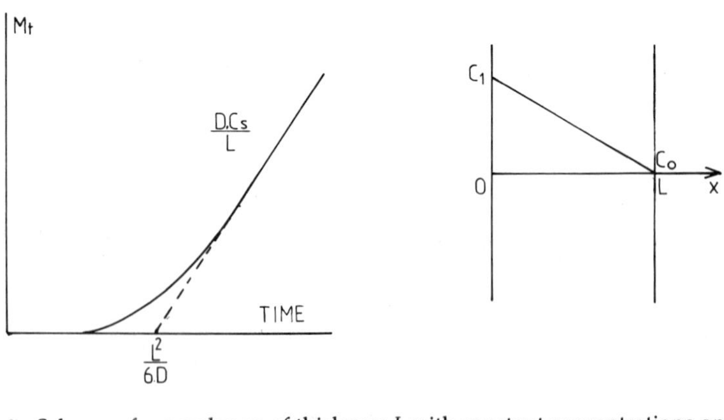


**Figure 1–4** Scheme of a membrane of thickness $L$ with constant concentrations on the faces. Amount of matter transported as a function of time.

The solution of Fick's second law, equation (1-3), is easily obtained by the method of separation of the variables, or by using equation (1-17) written in the general case.

$$C_{x,t} = C_1 \left(1 - \frac{x}{L}\right) - \frac{2C_1}{\pi} \cdot \sum_{n=1}^{\infty} \frac{1}{n} \cdot \sin \frac{n\pi x}{L} \cdot \exp\left(-\frac{n^2\pi^2}{L^2} Dt\right) \quad (1\text{-}23)$$

The amount of diffusing substance which has passed through the membrane per unit area in time $t$ is calculated by integrating Fick's first law with respect to time, by considering the gradient of concentration at $x = L$.

$$M_t = -\int_o^t D \cdot \left(\frac{\partial C}{\partial x}\right)_{x=L} \cdot dt \quad (1\text{-}18)$$

$$M_t = \frac{DC_1}{L} \cdot \left\{t + \frac{2L^2}{\pi^2 D} \cdot \sum_{n=1}^{\infty} \frac{(-1)^n}{n^2} \left[1 - \exp\left(\frac{-n^2\pi^2}{L^2} Dt\right)\right]\right\} \quad (1\text{-}24)$$

When time is large enough, the exponential term in equation (1-24) tends to zero, and this equation is reduced to:

$$M_t = \frac{DC_1}{L}\left(t - \frac{L^2}{6D}\right) \quad (1\text{-}25)$$

Then, after a transient period determined by

$$t = \frac{L^2}{6D} \quad (1\text{-}26)$$

a steady state is reached, in which the concentration remains constant at all points of the membrane, with $C_1$ at $x = 0$ and $C = 0$ at $x = L$.

The matter transport under steady state is especially studied in the following section 1-3 on steady state.

### 1-2-4 Plane Sheet in Contact with a Stirred Solution of Limited Volume

**The diffusing substance enters the plane sheet.** From the experimental point of view, it is useful to have a limited volume of solution. The rate of uptake of diffusing matter by the plane sheet is deduced from the measurement of the concentration in the solution. This concentration in the solution is uniform, when the solution is well stirred.

Two facts have to be considered:

1. The concentration of diffusing substance in the solution falls as this substance enters the sheet (or increases as this substance leaves the sheet).

2. The total amount of diffusing substance in the solution and in the sheet is constant as diffusion proceeds.

The infinite sheet of thickness $2L$ is placed in the solution of finite volume of thickness $2a$. The sheet occupies the place $-L \leq x \leq L$, while the solution occupies the place $-a - L \leq x \leq -L$ and $L \leq x \leq L + a$. Initial condition ($t = 0$) when the substance enters the sheet

$$-L \leq x \leq L \qquad C = 0 \qquad \text{sheet} \tag{1-27}$$

$$\begin{aligned} -L - a \leq x \leq -L \qquad & C = C_0 \qquad \text{liquid} \\ L \leq x \leq L + a \qquad & \end{aligned} \tag{1-28}$$

The boundary condition ($t > 0$) is obtained by writing that the rate at which the diffusing substance leaves the solution is always equal to the rate at which this substance enters the sheet.

$$a \cdot \frac{\partial C}{\partial t} = \pm D \cdot \frac{\partial C}{\partial x} \qquad x = \pm L \tag{1-29}$$

Equation (1-29) is written when the concentration of diffusing substance on the surface of the sheet is the same as that in the solution.

Very often, there is a partition factor $K$, expressing that the concentration of diffusing substance on the sheet surface is $K$ times that in the solution. Then a modified thickness of solution $a/K$ must be considered in place of $a$. Equation (1-29) is rewritten as follows:

$$\frac{a}{K} \cdot \frac{\partial C}{\partial t} = \pm D \cdot \frac{\partial C}{\partial x} \qquad x = \pm L \tag{1-30}$$

The solution is obtained by the use of the Laplace transform.

The concentration within the sheet, $C_{x,t}$, is given in terms of the concentration in the sheet at infinite time, $C_\infty$, by the expression:

$$\frac{C_{x,t}}{C_\infty} = 1 + \sum_{n=1}^{\infty} \frac{2(1 + \alpha) \cdot \exp(-Dq_n^2 t/L^2)}{1 + \alpha + \alpha^2 q_n^2} \cdot \frac{\cos(q_n x/L)}{\cos q_n} \tag{1-31}$$

The amount of diffusing substance which has entered the sheet at time $t$, $M_t$, is expressed as a fraction of the corresponding quantity entered after infinite time, $M_\infty$:

$$\frac{M_\infty - M_t}{M_\infty} = \sum_{n=1}^{\infty} \frac{2\alpha(1 + \alpha)}{1 + \alpha + \alpha^2 q_n^2} \cdot \exp\left(-\frac{Dq_n^2 t}{L^2}\right) \tag{1-32}$$

where the $q_n$s are the nonzero positive roots of

$$\tan q_n = -\alpha \cdot q_n \tag{1-33}$$

**Table 1–1**

Roots of Tan $q_n = -\alpha q_n$ (equation 1-33)

| $\alpha$ | $q_1$ | $q_2$ | $q_3$ | $q_4$ | $q_5$ | $q_6$ | $q_7$ | $q_8$ | $q_9$ | $q_{10}$ |
|---|---|---|---|---|---|---|---|---|---|---|
| 9.0000 | 1.6385 | 4.7359 | 7.8681 | 11.0057 | 14.1451 | 17.2852 | 20.4258 | 23.5667 | 26.7077 | 29.8489 |
| 4.0000 | 1.7155 | 4.7648 | 7.8857 | 11.0183 | 14.1549 | 17.2933 | 20.4326 | 23.5726 | 26.7129 | 29.8535 |
| 2.3333 | 1.8040 | 4.8014 | 7.9081 | 11.0344 | 14.1674 | 17.3036 | 20.4413 | 23.5801 | 26.7196 | 29.8595 |
| 1.5000 | 1.9071 | 4.8490 | 7.9378 | 11.0558 | 14.1841 | 17.3173 | 20.4529 | 23.5902 | 26.7285 | 29.8674 |
| 1.0000 | 2.0288 | 4.9132 | 7.9787 | 11.0856 | 14.2075 | 17.3364 | 20.4692 | 23.6043 | 26.7409 | 29.8786 |
| 0.6667 | 2.1746 | 5.0037 | 8.0385 | 11.1296 | 14.2421 | 17.3649 | 20.4934 | 23.6253 | 26.7595 | 29.8953 |
| 0.4286 | 2.3521 | 5.1386 | 8.1334 | 11.2010 | 14.2990 | 17.4119 | 20.5335 | 23.6602 | 26.7904 | 29.9229 |
| 0.2500 | 2.5704 | 5.3540 | 8.3029 | 11.3349 | 14.4080 | 17.5034 | 20.6120 | 23.7289 | 26.8514 | 29.9778 |
| 0.1111 | 2.8363 | 5.7172 | 8.6587 | 11.6532 | 14.6870 | 17.7481 | 20.8283 | 23.9218 | 27.0250 | 30.1354 |
| 0.0000 | 3.1416 | 6.2832 | 9.4248 | 12.5664 | 15.7080 | 18.8496 | 21.9912 | 25.1328 | 28.2744 | 31.4160 |

and $\alpha$ is the ratio of the volumes of solution and sheet (written without and with the partition coefficient $K$).

$$\alpha = \frac{a}{L} \quad \text{or} \quad \alpha = \frac{a}{K \cdot L} \tag{1-34}$$

Some roots of equation (1-33) are given in Table 1–1, for various values of the ratio $\alpha$ corresponding to various values of final fractional uptake. The fractional uptake of the sheet is defined by the ratio:

$$\text{Fractional uptake} = \frac{M_\infty}{2aC_0} \tag{1-35}$$

since the thickness of the solution is $2a$.

By writing that the amount of diffusing substance initially contained in the solution ($2aC_o$) is equal to the total amount of diffusing substance in solution and sheet in the final equilibrium state:

$$2aC_0 = \frac{2aC_\infty}{K} + 2L \cdot C_\infty \tag{1-36}$$

it is easy to find the expression of the fractional uptake in terms of the ratio $\alpha$.

$$\text{Fractional uptake} = \frac{1}{1 + \alpha} \tag{1-37}$$

***Remarks.*** The value of $\alpha$ is given for various values of the fractional uptake.

| Fractional Uptake (%) | Ratio $\alpha$ |
|---|---|
| 100 | 0 |
| 50 | 1 |
| 10 | 9 |
| 0 | $\infty$ |

In the case of an infinite amount of solute, the fractional uptake tends to zero, and $\alpha$ becomes infinite. The roots of equation $\tan q_n = -\alpha \cdot q_n$ are then:

$$q_n = \left(\frac{2n + 1}{2}\right)\pi \tag{1-38}$$

and equation (1-32) expressing the amount of diffusing substance in the sheet, $M_t$, becomes:

$$\frac{M_\infty - M_t}{M_\infty} = \frac{8}{\pi^2} \cdot \sum_{n=0}^{\infty} \frac{1}{(2n + 1)^2} \cdot \exp\left(-\frac{(2n + 1)^2\pi^2}{4L^2} Dt\right) \tag{1-39}$$

Equation (1-39) is the same as equation (1-10) obtained in the case of a constant concentration $C_0$ at the surface of the sheet and of a very large volume of liquid (since the thickness of the sheet is $L$ in section 1-2-2 and $2L$ in section 1-2-4).

**The diffusing substance leaves the plane sheet.** All the solute is initially uniformly distributed through the sheet, and the substance diffuses out into the solution. Initial conditions ($t = 0$) are:

$$-L \le x \le L \qquad C = C_0 \qquad \text{sheet} \tag{1-40}$$

$$\left.\begin{array}{l} -L - a \le x \le -L \\ L \le x \le L + a \end{array}\right. \qquad C = 0 \qquad \text{liquid} \tag{1-41}$$

Boundary conditions ($t > 0$) are the same as in the previous case (equation 1-30).

$$\frac{a}{K} \cdot \frac{\partial C}{\partial t} = \pm D \cdot \frac{\partial C}{\partial x} \qquad x = \pm L \tag{1-30}$$

The concentration within the sheet, $C_{x,t}$, is given in terms of the concentration in the sheet at $t = 0$, $C_0$, and at infinite time, $C_\infty$.

$$\frac{C_0 - C_{x,t}}{C_0 - C_\infty} = 1 + \sum_{n=1}^{\infty} \frac{2(1 + \alpha) \cdot \exp(-Dq_n^2 t/L^2)}{1 + \alpha + \alpha^2 q_n^2} \cdot \frac{\cos(q_n x/L)}{\cos q_n} \tag{1-42}$$

The amount of diffusing substance leaving the sheet up to time $t$, $M_t$, is expressed in terms of the corresponding amount after infinite time, $M_\infty$.

$$\frac{M_\infty - M_t}{M_\infty} = \sum_{n=1}^{\infty} \frac{2\alpha(1 + \alpha)}{1 + \alpha + \alpha^2 q_n^2} \cdot \exp\left(-\frac{Dq_n^2 t}{L^2}\right) \tag{1-32}$$

This is the same equation as in the case of absorption of diffusing substance by the sheet.

The amount of diffusing substance initially contained in the sheet ($2LC_0$) is equal to the total amount of diffusing substance in solution and sheet at the end of the process, at equilibrium:

$$2LC_o = \frac{2aC_\infty}{K} + 2LC_\infty \tag{1-43}$$

Then the fractional uptake of the sheet is given by:

$$\frac{M_\infty}{2LC_o} = \frac{\alpha}{1 + \alpha} \tag{1-44}$$

where $\alpha$ is always the limit ratio of the amounts of solute in solution and sheet when equilibrium is reached.

***Remark 1.*** When $t \rightarrow \infty$, or rather when $M_t/M_\infty$ is sufficiently large (larger than 0.55), equation (1-32) is reduced to the first term of the series:

$$\frac{M_\infty - M_t}{M_\infty} \rightarrow \frac{2\alpha(1 + \alpha)}{1 + \alpha + \alpha^2 q_1^2} \cdot \exp\left(-\frac{q_1^2}{L^2} Dt\right) \tag{1-32'}$$

The diffusivity $D$ can be calculated from the slope of the straight line obtained by plotting $ln(M_\infty - M_t/M_\infty)$ as a function of time.

### 1-2-5 Transport with Diffusion and Surface Evaporation

In this case, the transport of substance is obtained either by conduction within the polymer or by evaporation on the sheet surface. Two problems are of importance, the one concerned with the drying of woods, polymers, and elastomers, the other with the absorption of vapor or moisture by the material (see Fig. 1–5).

Generally, the best assumption is that the rate of transport by diffusion next to the surface is equal to the rate of transport by evaporation. The rate of evaporation is proportional to the difference between the actual concentration on the surface, $C_o$, and the concentration $C_{ext}$ which corresponds with the vapor pressure in the atmosphere near the surface.

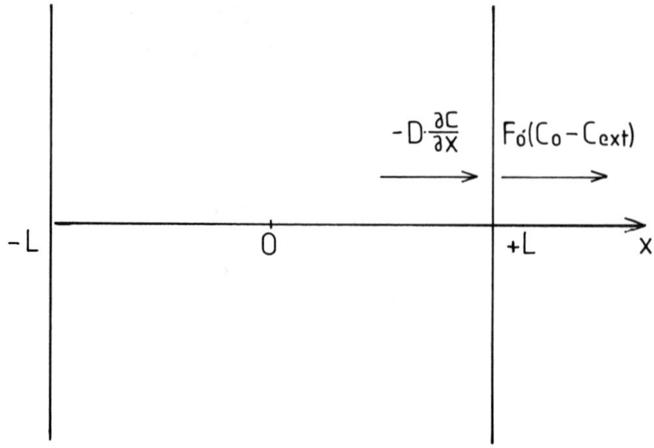


**Figure 1–5** Scheme of the plane sheet with diffusion within the solid and evaporation on the surface.

The rate of loss (or gain) of diffusing substance by evaporation from the surface of the sheet is:

$$-D\cdot\left(\frac{\partial C}{\partial x}\right)_0 = F_o\cdot(C_o - C_{ext}) \tag{1-45}$$

The initial conditions are for a sheet of thickness $2L$:

$$t = 0 \qquad -L < x < L \qquad C = C_i \tag{1-46}$$

The solution of Fick's second equation

$$\frac{\partial C}{\partial t} = D\cdot\frac{\partial^2 C}{\partial x^2} \tag{1-3}$$

with a constant diffusivity is:

$$\frac{C_{x,t} - C_i}{C_o - C_i} = 1 - \sum_{n=1}^{\infty}\frac{2M\cdot\cos\frac{\beta_n x}{L}\cdot\exp\left(-\frac{\beta_n^2}{L^2}Dt\right)}{(\beta_n^2 + M^2 + M)\cdot\cos\beta_n} \tag{1-47}$$

where the $\beta_n$s are the positive roots of

$$\beta\cdot\tan\beta = M \qquad \text{(given in Table 1–2)} \tag{1-48}$$

with the dimensionless number $M$:

$$M = \frac{L\cdot F_0}{D} \tag{1-49}$$

The rate of evaporation of the pure liquid is $F_o$, and it is obtained when $C_o = 1$.

**Table 1–2**
Roots of $\beta \tan \beta = M$ (equation 1-48)

| M | $\beta_1$ | $\beta_2$ | $\beta_3$ | $\beta_4$ | $\beta_5$ | $\beta_6$ | $\beta_7$ | $\beta_8$ | $\beta_9$ | $\beta_{10}$ |
|---|---|---|---|---|---|---|---|---|---|---|
| 0.00 | 0.0000 | 3.1416 | 6.2832 | 9.4248 | 12.5664 | 15.7080 | 18.8496 | 21.9911 | 25.1327 | 28.2743 |
| 0.01 | 0.0998 | 3.1448 | 6.2848 | 9.4258 | 12.5672 | 15.7086 | 18.8501 | 21.9916 | 25.1331 | 28.2747 |
| 0.10 | 0.3111 | 3.1731 | 6.2991 | 9.4354 | 12.5743 | 15.7143 | 18.8549 | 21.9957 | 25.1367 | 28.2779 |
| 0.20 | 0.4328 | 3.2039 | 6.3148 | 9.4459 | 12.5823 | 15.7207 | 18.8602 | 22.0002 | 25.1407 | 28.2814 |
| 0.50 | 0.6533 | 3.2923 | 6.3616 | 9.4775 | 12.6060 | 15.7397 | 18.8760 | 22.0139 | 25.1526 | 28.2920 |
| 1.00 | 0.8603 | 3.4256 | 6.4373 | 9.5293 | 12.6453 | 15.7713 | 18.9024 | 22.0365 | 25.1724 | 28.3096 |
| 2.00 | 1.0769 | 3.6436 | 6.5783 | 9.6296 | 12.7223 | 15.8336 | 18.9547 | 22.0815 | 25.2119 | 28.3448 |
| 5.00 | 1.3138 | 4.0336 | 6.9096 | 9.8928 | 12.9352 | 16.0107 | 19.1055 | 22.2126 | 25.3276 | 28.4483 |
| 10.00 | 1.4289 | 4.3058 | 7.2281 | 10.2003 | 13.2142 | 16.2594 | 19.3270 | 22.4108 | 25.5064 | 28.6106 |
| 100.0 | 1.5552 | 4.6658 | 7.7764 | 10.8871 | 13.9981 | 17.1093 | 20.2208 | 23.3327 | 26.4450 | 29.5577 |
| 900.0 | 1.5691 | 4.7072 | 7.8453 | 10.9834 | 14.1215 | 17.2596 | 20.3977 | 23.5358 | 26.6739 | 29.8120 |

The amount of substance entering or levaing the sheet up to time $t$, $M_t$, is expressed as a fraction of the corresponding quantity after infinite time, $M_\infty$, by the equation:

$$\frac{M_\infty - M_t}{M_\infty} = \sum_{n=1}^{\infty} \frac{2M^2 \cdot \exp\left(-\frac{\beta_n^2}{L^2} Dt\right)}{\beta_n^2 \cdot (\beta_n^2 + M^2 + M)} \tag{1-50}$$

***Remark 1.*** Of course, the substance enters or leaves the sheet, depending on whether $C_{ext}$ is greater or less than $C_o$.

***Remark 2.*** When the atmosphere is well stirred, the concentration $C_{ext}$ may be very low, and then the rate of evaporation is proportional to the actual concentration on the surface $C_0$.

$$-D \cdot \frac{\partial C}{\partial x} = F_0 \cdot C_0 \tag{1-45'}$$

***Remark 3.*** For long times, when $M_t/M_\infty > 0.6$, the first term of the series is preponderant, and equation (1-50) becomes:

$$\frac{M_\infty - M_t}{M_\infty} = \frac{2 \cdot M^2}{\beta_1^2 \cdot (\beta_1^2 + M^2 + M)} \cdot \exp\left(-\frac{\beta_1^2}{L^2} Dt\right) \tag{1-50'}$$

It is thus easy to obtain the diffusivity from the slope of the straight line obtained by plotting $ln(M_\infty - M_t/M_\infty)$ as a function of time.

***Remark 4.*** At the beginning of the process, for $t \rightarrow 0$, the initial rate of evaporation is:

$$\left(\frac{dM}{dt}\right)_{t=0} = F_o \cdot (C_o - C_{ext}) \tag{1-51}$$

This equation is useful for determining the rate of evaporation of the substance, when the concentration $C_o$ on the surface is known.

## 1-3 STEADY STATE CONDITIONS

This is the case of transport of diffusing substance through the plane sheet (very often a membrane) whose surfaces are maintained at constant concentrations.

This problem has been studied in section 1-2-3 (with surfaces kept at constant concentrations different from zero, and with surfaces kept at constant concentrations with one of them being zero) under transient conditions. After a transient period determined by equation (1-26),

$$t = \frac{L^2}{6D} \tag{1-26}$$

the steady state is reached, the concentrations at all points of the membrane remaining constant.

The diffusion equation (Fick's second law) in one dimension becomes:

$$\frac{\partial C}{\partial t} = \frac{\partial}{\partial x}\left(D \cdot \frac{\partial C}{\partial x}\right) = 0 \tag{1-3'}$$

The boundary conditions are as follows, with a membrane of thickness $L$.

$$\begin{aligned} t \geq 0 \qquad x &= 0 \qquad C = C_1 \\ x &= L \qquad C = C_2 \end{aligned} \tag{1-52}$$

### 1-3-1 Transport under Steady State with Constant Diffusivity

Fick's law is reduced to $\partial^2 C/\partial x^2 = 0$. After a first integration with respect to space $x$

$$\frac{\partial C}{\partial x} = \text{constant} \tag{1-53}$$

and a further integration by introducing the boundary conditions, we have:

$$\frac{C_x - C_1}{C_2 - C_1} = \frac{x}{L} \tag{1-54}$$

The rate of transport of diffusing substance is given by Fick's first law.

$$\text{Flux} = -D \cdot \frac{\partial C}{\partial x} \tag{1-55}$$

and becomes, by using the constant value of the gradient of concentration:

$$\text{Flux} = \frac{D}{L}(C_1 - C_2) \tag{1-55'}$$

***Remark 1. Determination of Diffusivity D.*** Equation (1-55′) can be used for determining the diffusivity, from the value of the flux measured, with constant concentrations $C_1$ and $C_2$ on both surfaces of the plane sheet.

***Remark 2. Case of Composite Membrane.*** The composite membrane is composed of $n$ sheets of thicknesses $L_1, L_2, \ldots, L_n$ and diffusivity $D_1, D_2, \ldots, D_n$. The rate of transport is the same across each sheet, and it can be written in the following way:

$$\text{Flux} = \frac{\Delta C_1}{\dfrac{L_1}{D_1}} = \frac{\Delta C_2}{\dfrac{L_2}{D_2}} \cdots \frac{\Delta C_n}{\dfrac{L_n}{D_n}} = \frac{\Delta C}{\sum \dfrac{L_i}{D_i}} \tag{1-56}$$

where $\Delta C_i$ is the drop in concentration for the $i$th sheet and $\Delta C$ is the total drop in concentration for the composite membrane, where $L_i/D_i$ represents the resistance to diffusion of the $i$th sheet, and $\sum (L_i/D_i)$ is the resistance to diffusion of the whole membrane.

Of course, the resistance to diffusion of the whole membrane is the sum of the resistance of the separate sheets. This equation can be used only if there is no barrier to diffusion between every sheet.

***Remark 3. One Surface is Kept at Constant Concentration and Evaporation Takes Place on the Other Surface.*** The boundary conditions are as follows:

$$\begin{array}{lll} t > 0 & x = 0 & C = C_1 \\ & x = L & C = C_2 \\ & x > L & C = C_{ext} \end{array} \tag{1-57}$$

The rate at which solute is evaporated from the surface $L$ is always equal to the rate at which solute is brought by diffusion to this surface.

$$-D \cdot \left(\frac{\partial C}{\partial x}\right)_{x=L} = F_o \cdot (C_2 - C_{ext}) \tag{1-58}$$

where $C_{ext}$ is the concentration in the atmosphere; $F_o$ is the rate of evaporation of the pure diffusing substance; and $C_2$ is the concentration on the surface at equilibrium.

The concentration at abscissa $x$ in the membrane is easily calculated.

$$C_x = C_1 + (C_2 - C_1)\cdot \frac{F_o\cdot x}{D + F_o\cdot L} \tag{1-59}$$

And the rate of transport is:

$$\text{Flux} = \frac{D\cdot F_o}{D + F_o\cdot L}(C_1 - C_2) \tag{1-60}$$

***Remark 4. Diffusion of a Gas and Vapor, with Constant Permeability.*** In the case of transport of a gas and vapor through the membrane, vapor pressure is of better use than concentration. The pressure of gas being kept constant on each face of the membrane, the rate of transport of gas in the steady state is given by the well-known equation:

$$\text{Flux} = \frac{\Delta P}{L}\,Per = \frac{P_1 - P_2}{L}\cdot Per \tag{1-61}$$

where the flux is proportional to the gradient of pressure $\Delta P/L$ constant through the membrane and the permeability *Per*.

If the sorption isotherm for the vapor is linear at equilibrium, the vapor concentration in the membrane is proportional to the external vapor pressure, $S$ being the solubility of the vapor in the material of the membrane.

$$C = S\cdot P \tag{1-62}$$

and a comparison between equations (1-55) and (1-61) expressing the rate of transport either in terms of the concentration or the pressure (following Fick's first law with $D$ and *Per* constant), shows that:

$$Per = D\cdot S \tag{1-63}$$

### 1-3-2 Transport Under Steady State with a Concentration-Dependent Diffusivity

In the case of concentration-dependent diffusivity, the equation describing Fick's second law under steady state is written as follows:

$$\frac{\partial}{\partial x}\left(D\cdot \frac{\partial C}{\partial x}\right) = 0 \tag{1-64}$$

and Fick's first law becomes, after integration

$$\text{Flux} = -D\cdot \frac{\partial C}{\partial x} = \text{constant} \tag{1-55}$$

On integrating between the two surface concentrations, $C_1$ and $C_2$, the following is:

$$\text{Flux} = -\frac{1}{L}\int_{C_1}^{C_2} D \cdot dC \qquad (1\text{-}65)$$

Two ways can be used, following the concentration dependence of diffusivity.

***Remark 1.*** When diffusivity varies little, with the concentration ranging from $C_1$ to $C_2$, the mean value of diffusivity $D_m$ can be used:

$$D_m = \frac{1}{C_1 - C_2} \cdot \int_{C_2}^{C_1} D \cdot dC \qquad (1\text{-}66)$$

And the flux is obtained by:

$$\text{Flux} = \frac{D_m}{L}(C_1 - C_2) \qquad (1\text{-}67)$$

***Remark 2. Concentration-Dependent Diffusivity.*** Analytical solutions can be obtained for equation (1-66) for simple cases. In other more complicated cases, numerical solutions have to be used.

# Chapter Two

# Diffusion in a Cylinder

## 2-1 INTRODUCTION

Three cases are of interest:

1. The long solid cylinder in which diffusion is radial everywhere.
2. The solid cylinder of finite length in which diffusion is radial and longitudinal.
3. The hollow cylinder in which diffusion is radial through the cylinder wall. This is the case of tubings. This problem is considered either under transient or under steady conditions.

In all cases, the diffusivity is constant. In the case of concentration-dependent diffusivity, a numerical method with finite differences must be used.

## 2-2 SOLID CYLINDER OF INFINITE LENGTH UNDER NONSTEADY STATE CONDITIONS

### 2-2-1 Solid Cylinder of Infinite Length with a Uniform Distribution Under Nonsteady State Conditions

The concentration of diffusing substance is a function of radius $r$ and time $t$, following Fick's second law.

$$\frac{\partial C}{\partial t} = \frac{1}{r} \cdot \frac{\partial}{\partial r}\left(rD \cdot \frac{\partial C}{\partial r}\right) \tag{2-1}$$

With constant diffusivity and radial diffusion, Fick's equation can be written

$$\frac{\partial C}{\partial t} = D \cdot \left[\frac{\partial^2 C}{\partial r_2} + \frac{1}{r} \cdot \frac{\partial C}{\partial r}\right] \tag{2-2}$$

By using the method of separation of the variables, with

$$C_{r,t} = U_r \cdot T_t \tag{2-3}$$

it is found for the function $T$

$$T = \exp(-\alpha^2 Dt) \tag{2-4}$$

where $\alpha$ is a constant, and $U$, which is a function of $r$ only, is a solution of:

$$\frac{d^2U}{dr^2} + \frac{1}{r} \cdot \frac{dU}{dr} + \alpha^2 \cdot U = 0 \tag{2-5}$$

The preceding equation is Bessel's equation of order zero. The general solution is:

$$C_{r,t} = \sum_{n=1}^{\infty} A_n \cdot J_0(\alpha_n r) \cdot \exp(-\alpha_n^2 Dt) \tag{2-6}$$

where $J_o$ is the Bessel function of the first kind of order zero. Roots are given in tables of Bessel functions.

The initial and boundary conditions are:

| | | | | |
|---|---|---|---|---|
| Initial | $t = 0$ | $0 < r < R$ | $C = C_i$ | (2-7) |
| Boundary | $t > 0$ | $r = R$ | $C = 0$ | (2-8) |

The boundary condition is satisfied by equation (2-6), provided that

$$J_o(\alpha_n R) = 0 \tag{2-9}$$

and the $\alpha_n$s are roots of equation (2-9)

The initial condition is given by equation (2-10).

$$C_i = \sum_{n=1}^{\infty} A_n \cdot J_o(\alpha_n r) \tag{2-10}$$

and $A_n$ is determined by multiplying both sides of equation (2-10) by $r \cdot J_o(\alpha_n r)$ and integrating from 0 to $R$.

By considering that

$$\int_0^R r \cdot J_0(\alpha r) \cdot dr = \frac{R}{\alpha} \cdot J_1(\alpha R) \tag{2-11}$$

and

$$\int_0^R r\,J_0^2(\alpha r)\cdot dr = \frac{R^2}{2}\cdot J_1^2(\alpha R) \tag{2-12}$$

where $J_1(\alpha R)$ is the Bessel function of the first order, the concentration $C_{r,t}$ is given by:

$$\frac{C_{r,t} - C_i}{C_{ext} - C_i} = 1 - \frac{2}{R}\cdot\sum_{n=1}^{\infty}\frac{J_0(\alpha_n r)}{\alpha_n\cdot J_1(\alpha_n R)}\cdot\exp(-\alpha_n^2 Dt) \tag{2-13}$$

and the amount of diffusing substance which has entered (or left) the cylinder in time $t$, $M_t$, is expressed in terms of the corresponding quantity after infinite time, $M_\infty$, by:

$$\frac{M_\infty - M_t}{M_\infty} = \sum_{n=1}^{\infty}\frac{4}{R^2\alpha_n^2}\exp(-\alpha_n^2 Dt) \tag{2-14}$$

Another expression is also given for the concentration (when $r/R$ is not small) $C_{r,t}$:

$$\frac{C_{r,t} - C_i}{C_{ext} - C_i} = \left(\frac{R}{r}\right)^{0.5}\cdot erfc\,\frac{R - r}{2(Dt)^{0.5}} + \frac{(R - r)(RDt)^{0.5}}{4R\cdot r^{1.5}}\cdot ierfc\,\frac{R - r}{2(Dt)^{0.5}} + \frac{(9R^2 - 7r^2 - 2Rr)Dt}{32\cdot R^{1.5}\cdot r^{2.5}}\cdot i^2erfc\,\frac{R - r}{2(Dt)^{0.5}} + \ldots \tag{2-15}$$

and for the amount of diffusing substance which has entered (or left) the cylinder in time, $M_t$:

$$\frac{M_t}{M_\infty} = \frac{4}{\pi^{0.5}}\left(\frac{Dt}{R^2}\right)^{0.5} - \frac{Dt}{R^2} - \frac{1}{3\pi^{0.5}}\left(\frac{Dt}{R^2}\right)^{1.5} \tag{2-16}$$

***Remark 1. Case of Long Time.*** In the case of long time, equation (2-14) can be reduced to the first term of the series: Equation (2-17) is of interest for determination of diffusivity, from the slope of the line obtained by plotting $\log(M_\infty - M_t/M_\infty)$ as a function of time.

$$\frac{M_\infty - M_t}{M_\infty} = \frac{4}{R^2\cdot\alpha_1^2}\cdot\exp(-\alpha_1^2 Dt) \tag{2-17}$$

***Remark 2. Case of Short Time.*** In the case of short time, equation (2-16) can be reduced to:

$$\frac{M_t}{M_\infty} = \frac{4}{R}\left(\frac{Dt}{\pi}\right)^{0.5} \tag{2-18}$$

expressing a relationship between the amount of matter transported $M_t$ and the square root of time.

### 2-2-2 Solid Cylinder of Infinite Length in a Stirred Solution of Limited Volume

**Case of absorption of diffusing substance by the cylinder.** The cylinder of infinite length, initially free from diffusing substance, is immersed in the solution in which the concentration of substance is $C_o$.

The volume per unit length of the cylinder and solution are $\pi R_2$ and $V$, respectively.

The initial and boundary conditions are:

$$\text{Initial} \quad t = 0 \quad r < R \quad C = 0 \quad \text{cylinder}$$
$$V \quad C = C_i \quad \text{solution} \tag{2-19}$$

$$\text{Boundary} \quad t > 0 \quad -D\left(\frac{\partial C}{\partial r}\right)_{r=R} = \delta \cdot V \cdot \frac{\partial C}{\partial t} \tag{2-20}$$

The concentration of diffusion substance $C_{r,t}$ in the cylinder is:

$$\frac{C_{r,t} - C_\infty}{C_\infty} = \sum_{n=1}^{\infty} \frac{J_0\left(\frac{q_n r}{R}\right)}{J_0(q_n)} \cdot \frac{4(\alpha + 1)}{4 + 4\alpha + \alpha^2 q_n^2} \cdot \exp\left(-\frac{q_n^2}{R^2} Dt\right) \tag{2-21}$$

and the amount of substance in the cylinder after time $t$, $M_t$, is given by:

$$\frac{M_\infty - M_t}{M_\infty} = \sum_{n=1}^{\infty} \frac{4\alpha(1 + \alpha)}{4 + 4\alpha + \alpha^2 q_n^2} \cdot \exp\left(-\frac{q_n^2}{R^2} Dt\right) \tag{2-22}$$

where the $q_n$s are the positive and nonzero roots of:

$$\alpha q_n \cdot J_0(q_n) + 2J_1(q_n) = 0 \quad \text{(given in Table 2–1)} \tag{2-23}$$

and the ratio $\alpha$ of volumes of liquid and cylinder per unit length:

$$\alpha = \frac{V}{\pi R^2} \tag{2-24}$$

This ratio $\alpha$ becomes

$$\alpha = \frac{V}{K\pi R^2} \tag{2-24'}$$

when there is a partition coefficient $K$ for diffusing substance at equilibrium between the cylinder and the solution.

The final fractional uptake of diffusing substance by the cylinder is given by:

$$\text{Fractional uptake} = \frac{M_\infty}{V \cdot C_0} = \frac{1}{1 + \alpha} \tag{2-25}$$

Table 2–1
Roots of $\alpha q_n J_o(q_n) + 2J_1(q_n) = 0$ (equation 2-23)

| Fractional uptake | $\alpha$ | $q_1$ | $q_2$ | $q_3$ | $q_4$ | $q_5$ | $q_6$ |
|---|---|---|---|---|---|---|---|
| 0 | $\infty$ | 2.4048 | 5.5201 | 8.6537 | 11.7915 | 14.9309 | 18.0711 |
| 0.1 | 9.0000 | 2.4922 | 5.5599 | 8.6793 | 11.8103 | 14.9458 | 18.0833 |
| 0.2 | 4.0000 | 2.5888 | 5.6083 | 8.7109 | 11.8337 | 14.9643 | 18.0986 |
| 0.3 | 2.3333 | 2.6962 | 5.6682 | 8.7508 | 11.8634 | 14.9879 | 18.1183 |
| 0.4 | 1.5000 | 2.8159 | 5.7438 | 8.8028 | 11.9026 | 15.0192 | 18.1443 |
| 0.5 | 1.0000 | 2.9496 | 5.8411 | 8.8727 | 11.9561 | 15.0623 | 18.1803 |
| 0.6 | 0.6667 | 3.0989 | 5.9692 | 8.9709 | 12.0334 | 15.1255 | 18.2334 |
| 0.7 | 0.4286 | 3.2645 | 6.1407 | 9.1156 | 12.1529 | 15.2255 | 18.3188 |
| 0.8 | 0.2500 | 3.4455 | 6.3710 | 9.3397 | 12.3543 | 15.4031 | 18.4754 |
| 0.9 | 0.1111 | 3.6374 | 6.6694 | 9.6907 | 12.7210 | 15.7646 | 18.8215 |
| 1.0 | 0 | 3.8317 | 7.0156 | 10.1735 | 13.3237 | 16.4706 | 19.6159 |

**Case of desorption of diffusing substance from the cylinder.** The concentration of diffusing substance is initially uniform in the cylinder, $C_i$, and the solution is free from this substance. The initial conditions are:

$$\text{Initial} \quad t = 0 \quad \begin{array}{lll} r < R & C = C_i & \text{cylinder} \\ V & C = 0 & \text{solution} \end{array} \tag{2-26}$$

The concentration within the cylinder, $C_{rt}$, is then given by:

$$\frac{C_\infty - C_{r,t}}{C_i - C_\infty} = \sum_{n=1}^{\infty} \frac{J_0\left(\frac{q_n r}{R}\right)}{J_0(q_n)} \cdot \frac{4(\alpha + 1)}{4 + 4\alpha + \alpha^2 q_n^2} \cdot \exp\left(-\frac{q_n^2}{R^2} Dt\right) \tag{2-27}$$

and the amount of diffusing substance leaving the cylinder in time $t$ is given by equation (2-22).

The final fractional uptake of diffusing substance by the solution becomes

$$\text{Fractional uptake} = \frac{M_\infty}{\pi R^2 \cdot C_0} = \frac{\alpha}{1 + \alpha} \tag{2-28}$$

### 2-2-3 Evaporation of the Diffusing Substance on the Surface of the Cylinder

The cylinder is initially at the uniform concentration $C_i$. The substance diffuses through the cylinder and evaporates on the surface of cylinder (Figure 2–1).

The initial and boundary conditions are:

$$\text{Initial} \quad t = 0 \quad r < R \quad C = C_i \quad \text{cylinder} \qquad r > 0 \quad C = C_{ext} \quad \text{atmosphere} \tag{2-29}$$

Boundary: The surface condition expresses that the rate of diffusion near the surface is equal to the rate of evaporation.

$$-D\cdot\left(\frac{\partial C}{\partial r}\right)_{r=R} = F_0\cdot(C_{R,t} - C_{ext}) \tag{2-30}$$

where $C_{ext}$ is the concentration in the atmosphere, $C_{R,t}$ is the concentration on the surface, and $F_0$ is the rate of evaporation of the pure substance.

The concentration of substance within the cylinder, $C_{r,t}$, is expressed by:

$$\frac{C_{ext} - C_{r,t}}{C_{ext} - C_i} = 2L\cdot\sum_{n=1}^{\infty}\frac{J_0\left(\frac{\beta_n r}{R}\right)}{(\beta_n^2 + L^2)\cdot J_0(\beta_n)}\cdot\exp\left(-\frac{\beta_n^2}{R^2}Dt\right) \tag{2-31}$$

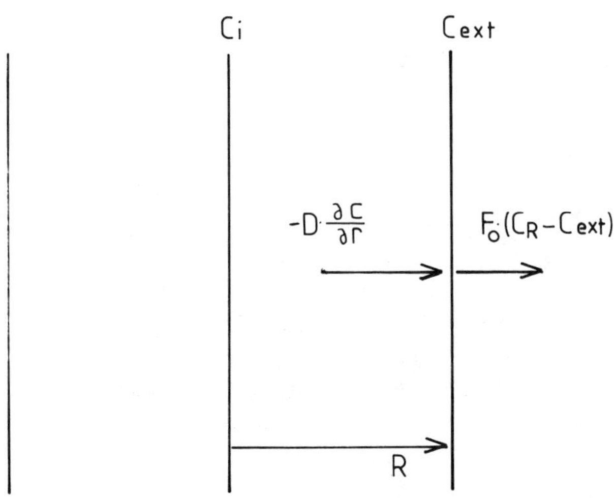


**Figure 2–1** Scheme of a cylinder with diffusion in the solid and evaporation on the surface.

where the $\beta_n$s are the roots of

$$\beta \cdot J_1(\beta) = L \cdot J_o(\beta) \quad \text{(given in Table 2–2).} \tag{2-32}$$

and the dimensionless number $L$ is the ratio of:

$$L = \frac{F_0 R}{D} \tag{2-33}$$

The total amount of diffusing substance leaving the cylinder in time $t$, $M_t$, is expressed as a fraction of the corresponding amount after infinite time, $M_\infty$:

$$\frac{M_\infty - M_t}{M_\infty} = 4L^2 \cdot \sum_{n=1}^{\infty} \frac{\exp\left(-\frac{\beta_n^2}{R^2} Dt\right)}{\beta_n^2 \cdot (\beta_n^2 + L^2)} \tag{2-34}$$

***Remark 1. Evaporation or Condensation of Vapor.*** Evaporation or condensation of vapor takes place, according to the positive or negative value of $(C_{R,t} - C_{ext})$.

$$\begin{aligned} C_{R,t} - C_{ext} > 0 \quad & \text{evaporation} \\ C_{R,t} - C_{ext} < 0 \quad & \text{condensation} \end{aligned} \tag{2-35}$$

***Remark 2. Long Times.*** For long times, the first term in the series (equation 2-34) becomes preponderant. Then the following relationship

**Table 2–2**

Roots of $\beta J_1(\beta) - LJ_0(\beta) = 0$ (equation 2-32)

| L | $\beta_1$ | $\beta_2$ | $\beta_3$ | $\beta_4$ | $\beta_5$ | $\beta_6$ |
|---|---|---|---|---|---|---|
| 0 | 0 | 3.8137 | 7.0156 | 10.1735 | 13.3237 | 16.4706 |
| 0.01 | 0.1412 | 3.8343 | 7.0170 | 10.1745 | 13.3244 | 16.4712 |
| 0.1 | 0.4417 | 3.8577 | 7.0298 | 10.1833 | 13.3312 | 16.4767 |
| 0.2 | 0.6170 | 3.8835 | 7.0440 | 10.1931 | 13.3387 | 16.4828 |
| 0.5 | 0.9408 | 3.9594 | 7.0864 | 10.2225 | 13.3611 | 16.5010 |
| 1.0 | 1.2558 | 4.0795 | 7.1558 | 10.2710 | 13.3984 | 16.5312 |
| 2.0 | 1.5994 | 4.2910 | 7.2884 | 10.3658 | 13.4719 | 16.5910 |
| 5.0 | 1.9898 | 4.7131 | 7.6177 | 10.6223 | 13.6786 | 16.7630 |
| 10.0 | 2.1795 | 5.0332 | 7.9569 | 10.9363 | 13.9580 | 17.0099 |
| 100.0 | 2.3809 | 5.4652 | 8.5678 | 11.6747 | 14.7834 | 17.8931 |
| ∞ | 2.4048 | 5.5201 | 8.6537 | 11.7915 | 14.9309 | 18.0711 |

enables one to obtain the value of diffusivity from the slope of the straight line between Log ($M_\infty - M_t/M_\infty$) and time,

$$\mathrm{Ln}\left(\frac{M_\infty - M_t}{M_\infty}\right) = \mathrm{Ln}\left(\frac{4L^2}{\beta_1^2(\beta_1^2 + L^2)}\right) - \frac{\beta_1^2}{R^2}Dt \tag{2-36}$$

### 2-2-4 Hollow Cylinder and Long Tubing

The most interesting case is that of long tubing for which the internal and external surfaces are maintained at constant concentrations. The initial and boundary conditions are:

| | | | | | |
|---|---|---|---|---|---|
| Initial | $t = 0$ | $R_i \leq r \leq R_e$ | $C = C_0$ | tubing region | (2-37) |
| Boundary | $t > 0$ | $r = R_i$ | $C = C_i$ | internal region | (2-38) |
| | | $r = R_e$ | $C = C_e$ | external region | |

The concentration in the tubing at time $t$ is given by:

$$C_{r,t} = \frac{C_i \cdot Ln\dfrac{R_e}{r} + C_e \cdot Ln\dfrac{r}{R_i}}{Ln\dfrac{R_e}{R_i}} + \pi C_0 \cdot \sum_{n=1}^{\infty} \frac{J_0(\alpha_n R_i)\cdot U_0(\alpha_n r)}{J_0(\alpha_n R_i) + J_0(\alpha_n R_e)} \cdot \exp(-\alpha_n^2 Dt)$$

$$- \pi \cdot \sum_{n=1}^{\infty} \frac{\{C_e \cdot J_0(\alpha_n R_i) - C_i \cdot J_0(\alpha_n R_e)\} J_0(\alpha_n R_i) \cdot U_0(\alpha_n r)}{J_0^2(\alpha_n R_i) - J_0^2(\alpha_n R_e)} \cdot \exp(-\alpha_n^2 Dt) \tag{2-39}$$

where the $\alpha_n$s are roots of

$$U_o(\alpha_n R_i) = 0 \quad \text{(given in Table 2–3).} \tag{2-40}$$

and

$$U_0(\alpha_n r) = J_0(\alpha_n r)\cdot Y_0(\alpha_n R_e) - J_0(\alpha_n R_e)\cdot Y_0(\alpha_n r) \tag{2-41}$$

$J_0$ and $Y_0$ are Bessel functions of the first and second kind, respectively, or order zero.

The amount of diffusing substance leaving the external surface ($r = R_e$) is $M_t$, per unit length of the tube. It is calculated by integrating Fick's first law with respect to time:

$$M_t = -2\pi R_e D \cdot \int_0^R \left(\frac{\partial C}{\partial r}\right)_{R_e} dt \tag{2-42}$$

$$\frac{M_t}{\pi C_i} = \frac{2D(t - t_0)}{Ln\dfrac{R_e}{R_i}} - 4\sum_{n=1}^{\infty} \frac{J_0(\alpha_n R_i) - J_0(\alpha_n R_e)}{\alpha_n^2 \cdot \{J_0^2(\alpha_n R_i) + J_0^2(\alpha_n R_e)} \cdot \exp(-\alpha_n^2 Dt) \tag{2-43}$$

**Table 2–3**

Roots of $U_0(\alpha_n r) = J_0(R_e\alpha_n)Y_0(R_e\alpha_n) - J_0(R_e\alpha_n)Y_0(R_i\alpha_n)$ (equation 2-41)

| $\frac{R_e}{R_i}$ | $R\alpha_1$ | $R\alpha_2$ | $R\alpha_3$ | $R\alpha_4$ | $R\alpha_5$ |
|---|---|---|---|---|---|
| 1.2 | 15.7014 | 31.4126 | 47.1217 | 62.8302 | 78.5385 |
| 1.5 | 6.2702 | 12.5598 | 18.8451 | 25.1294 | 31.4133 |
| 2.0 | 3.1230 | 6.2734 | 9.4182 | 12.5614 | 15.7040 |
| 2.5 | 2.0732 | 4.1773 | 6.2754 | 8.3717 | 10.4672 |
| 3.0 | 1.5485 | 3.1291 | 4.7038 | 6.2767 | 7.8487 |
| 3.5 | 1.2339 | 2.5002 | 3.7608 | 5.0196 | 6.2776 |
| 4.0 | 1.0244 | 2.0809 | 3.1322 | 4.1816 | 5.2301 |

$t_0$ being the intercept on the time-axis:

$$t_0 = \frac{R_i^2 - R_e^2 + (R_i^2 + R_e^2)\cdot Ln\,\frac{R_e}{R_i}}{4\cdot Ln\,\frac{R_e}{R_i}} \tag{2-44}$$

## 2-3 SOLID CYLINDER OF FINITE LENGTH UNDER NONSTEADY STATE CONDITIONS

The equation of diffusion is:

$$\frac{\partial C}{\partial t} = D\cdot\left[\frac{\partial^2 C}{\partial z^2} + \frac{\partial^2 C}{\partial r^2} + \frac{1}{r}\cdot\frac{\partial C}{\partial r}\right] \tag{2-45}$$

with constant diffusivity, taking into account the radial transport and the longitudinal transport (z-axis).

In the case of a finite cylinder initially full of diffusing substance ($C_i$), with the surface kept at concentration zero, the initial and boundary conditions are (Figure 2–2):

$$\text{Initial} \quad t = 0 \quad -L \le z \le L \quad C = C_i \quad \text{cylinder} \qquad 0 \le r \le R \tag{2-46}$$

$$\text{Boundary} \quad t > 0 \quad z = \pm L \quad C = 0 \qquad r = R \tag{2-47}$$

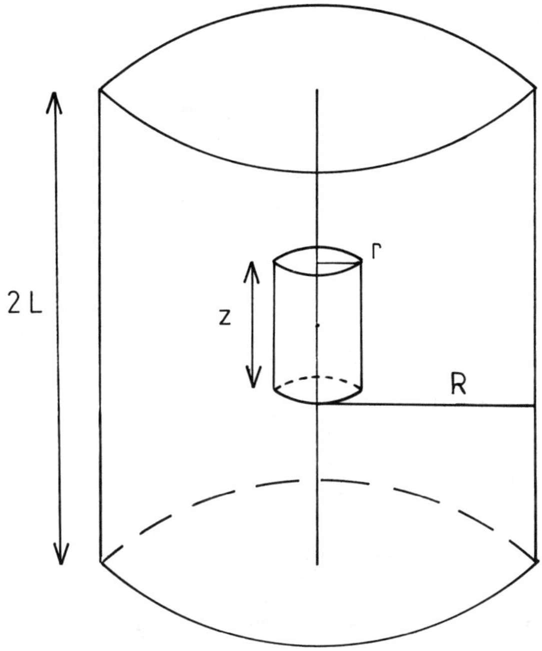


**Figure 2–2** Scheme of a cylinder with finite length, of radius $R$ and length $2L$.

By using the method of separation of the variables

$$C(r,z,t) = R(r) \cdot Z(z) \cdot T(t) \tag{2-48}$$

the equation of diffusion becomes

$$D \cdot \left[ \frac{R''}{R} + \frac{1}{r}\frac{R'}{R} + \frac{Z''}{Z} \right] = \frac{T'}{T} \tag{2-49}$$

The solutions of this equation are:

$$\begin{aligned} R(r) &= K \cdot J_0(\alpha r) \\ Z(z) &= A \cdot \sin \nu z + B \cdot \cos \nu z \\ T(t) &= K' \cdot \exp[-(\alpha^2 + \nu^2)Dt] \end{aligned} \tag{2-50}$$

The general solution of the equation of diffusion is then

$$\begin{aligned} C(r,z,t) = \sum_{n=1}^{\infty} \sum_{m=1}^{\infty} K_m \cdot J_0(\alpha_m r) \\ \cdot [A_n \cdot \sin \nu_n z + B_n \cdot \cos \nu_n z] \cdot \exp[-(\alpha_m^2 + \nu_n^2)Dt] \end{aligned} \tag{2-51}$$

where $\alpha_m$s are the roots of

$$J_o(\alpha_m R) = 0 \tag{2-52}$$

By introducing the boundary conditions for $z = \pm L$, the values of $\nu_n$ are determined, and equation (2-51) becomes:

$$C(r,z,t) = \sum_{n=1}^{\infty} \sum_{m=1}^{\infty} A_m \cdot J_0(\alpha_m r) \cdot \sin\left[\frac{n\pi(L+z)}{2L}\right] \cdot \exp\left[-\left(\alpha_m^2 + \frac{n^2\pi^2}{4L^2}\right)Dt\right] \tag{2-53}$$

The coefficient $A_m$ is determined by introducing the initial conditions, and the solution is then

$$C(r,z,t) = \frac{8C_i}{\pi R} \cdot \sum_{n=0}^{\infty} \sum_{m=1}^{\infty} \frac{J_0(\alpha_m r)}{(2n+1)\,\alpha_m \cdot J_1(\alpha_n r)} \cdot \sin\left[\frac{(2n+1)\pi(z+L)}{2L}\right] \cdot \exp\left[-\left(\alpha_m^2 + \frac{2(n+1)^2\pi^2}{4L^2}\right)Dt\right] \tag{2-54}$$

The amount of diffusing substance remaining in the cylinder of finite length is given by the integral,

$$M_t = \int_0^R 2\pi r dr \cdot \int_{-L}^{+L} C \cdot dz \tag{2-55}$$

which becomes:

$$\frac{M_t}{M_\infty} = \frac{32}{\pi^2} \cdot \sum_{m=1}^{\infty} \frac{1}{(\alpha_m R)^2} \cdot \exp\left[-\frac{(\alpha_m R)^2}{R^2} Dt\right] \cdot \sum_{n=0}^{\infty} \frac{1}{(2n+1)^2} \cdot \exp\left[-\frac{(2n+1)^2\pi^2}{4L^2} Dt\right] \tag{2-56}$$

where $M_\infty$ is the total amount of diffusing substance initially in the cylinder $(2\pi R^2 L C_i)$.

***Remark 1. Case of Short Time.*** In the case of short time, some simplifications can be made in equation (2-56). The roots of equation $J_0(\alpha_m R) = 0$ are approximately equal to:

$$\alpha_m R = \pi(m - 0.25) \tag{2-57}$$

And then, as $(Dt/R)^{0.5}$ is small, the value of the series tends to:

$$\sum_{m=1}^{\infty} \frac{1}{(\alpha_m R)^2} \cdot \exp\left[-\frac{(\alpha_m R)^2}{R^2} Dt\right] = \frac{1}{4} - \frac{1}{R}\left(\frac{Dt}{\pi}\right)^{0.5} \tag{2-58}$$

In the same way, as $Dt/4L^2$ is small, the value of the second series in equation (2-56) tends to:

$$\sum_{n=0}^{\infty} \frac{1}{(2n+1)^2} \cdot \exp\left[-\frac{(2n+1)^2\pi^2}{4L^2} Dt\right] = \frac{\pi^2}{8}\left[1 - \frac{2}{L}\left(\frac{Dt}{\pi}\right)^{0.5}\right] \tag{2-59}$$

Then, for small time, equation (2-56) is reduced to:

$$\frac{M_t}{M_\infty} = 1 - 4\left(\frac{Dt}{\pi}\right)^{0.5} \cdot \left(\frac{1}{R} + \frac{1}{2L}\right) \tag{2-60}$$

This equation is of interest for determining the value of diffusivity from the slope obtained by plotting the amount of diffusing substance $M_t$ remaining in the cylinder as a function of the square root of time.

***Remark 2. Case of Long Time.*** In the case of long time, the first term of each series can be considered in equation (2-56), with the first root of $J_o(\alpha_m R) = 0$ being equal to 2.404825.

## 2-4 HOLLOW CYLINDER OF INFINITE LENGTH UNDER STEADY STATE CONDITIONS

### 2-4-1 Tubing of Infinite Length Under Steady State with Constant Concentration on Each Surface

The hollow cylinder of infinite length under steady state is of high interest because it is the case of long tubing separating two fluids. The surfaces of inner and outer radii $R_i$ and $R_e$ respectively are at constant concentrations (Figure 2–3).

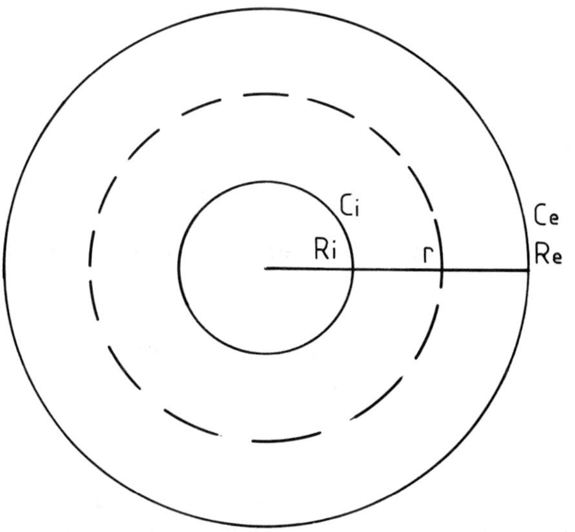


**Figure 2–3** Scheme of the circular cross section of a hollow cylinder of infinite length, with constant concentrations on the surfaces.

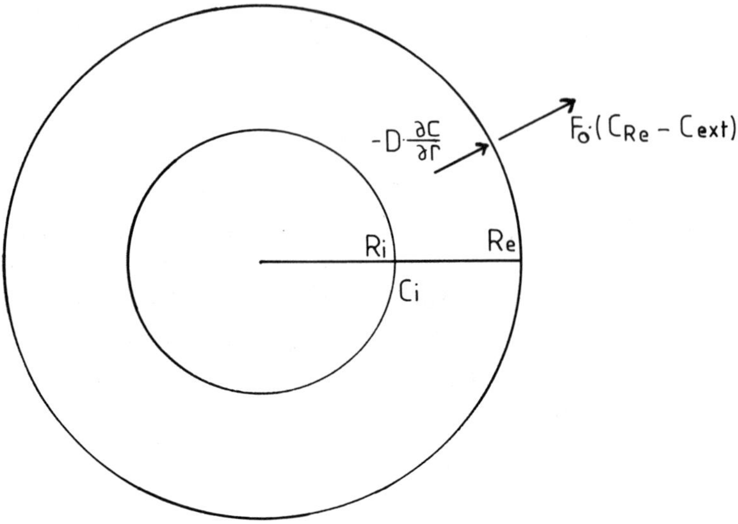


**Figure 2–4** Scheme of the circular cross section of a hollow cylinder of infinite length, with constant concentration on the internal surface and evaporation on the external surface.

The boundary conditions are:

$$\text{Boundary} \quad t > 0 \quad r = R_i \quad C = C_i \qquad r = R_e \quad C = C_e \tag{2-61}$$

The equation describing the steady-state condition

$$\frac{d}{dr}\left(r \cdot \frac{dC}{dr}\right) = 0 \tag{2-62}$$

has a general solution:

$$C = A + B \cdot \text{Ln } r \tag{2-63}$$

The constants $A$ and $B$ are determined from the preceding boundary conditions:

$$C_r = \frac{C_i \cdot Ln \dfrac{R_e}{r} + C_e \cdot Ln \dfrac{r}{R_i}}{Ln \dfrac{R_e}{R_i}} \tag{2-64}$$

The amount of diffusing substance, $M_t$, which diffuses through the unit length of the tubing in time $t$ is easily calculated by integrating Fick's first law with respect to time:

$$M_t = -\int_0^t D\cdot\left(\frac{\partial C}{\partial r}\right)_{r=R_e}\cdot dt \tag{2-65}$$

$$M_t = \frac{2\pi(C_e - C_i)}{Ln\,\dfrac{R_e}{r_i}}\,Dt \tag{2-66}$$

Following the values of concentrations $C_e$ and $C_i$, the diffusing substance leaves or enters the tubing.

***Remark 1. Case of Concentration-Dependent Diffusivity.*** Two ways of calculations can be followed. When diffusivity varies little, the mean value of diffusivity $D_m$ can be considered:

$$D_m = \frac{1}{C_e - C_i}\cdot\int_{C_i}^{C_e} D\cdot dC \tag{2-67}$$

When diffusivity varies significantly, numerical methods can be used.

### 2-4-2 Tubing of Infinite Length with a Constant Concentration at the Inner Surface and Evaporation at the Outer Surface

This case is of very high interest: The liquid in contact with the inner surface diffuses through the tubing and evaporates on the outer surface (Figure 2–4).

The boundary conditions are:

$$\text{Boundary}\quad t>0\quad r=R_i\quad C=C_i$$

$$r=R_e\quad -D\left(\frac{\partial C}{\partial r}\right) = F_0\cdot(C_e - C_{ext}) \tag{2-68}$$

The concentration within the tubing at position $r(R_i < r < R_e)$ is:

$$C_r = \frac{C_i\left(1+\dfrac{F_0}{D}R_e Ln\dfrac{R_e}{r}\right) + C_e\cdot\dfrac{F_0}{D}\cdot R_e\cdot Ln\dfrac{r}{R_i}}{1+\dfrac{F_0}{D}R_e\cdot Ln\dfrac{r}{R_i}} \tag{2-69}$$

The amount of substance diffusing per unit length of the tubing is obtained by integrating Fick's first law. This amount on the outer surface is:

$$M_t = \frac{2\pi(C_e - C_i)\cdot F_0\cdot R_e}{D + F_0\cdot R_e\cdot Ln\dfrac{R_e}{R_i}}\cdot Dt \tag{2-70}$$

# Chapter Three

# Diffusion in a Sphere

## 3-1 INTRODUCTION

When the diffusion is radial, Fick's second law expressing the diffusion equation under transient conditions is in the general form.

$$\frac{\partial C}{\partial t} = \frac{1}{r^2} \cdot \frac{\partial}{\partial r}\left[D \cdot r^2 \cdot \frac{\partial C}{\partial r}\right] \tag{3-1}$$

where C is the concentration at time $t$ and at a distance $r$ from the center of the sphere.

When the diffusivity $D$ is constant, the diffusion equation takes the form of:

$$\frac{\partial C}{\partial t} = D \cdot \left[\frac{\partial^2 C}{\partial r^2} + \frac{2}{r} \cdot \frac{\partial C}{\partial r}\right] \tag{3-2}$$

Analytical solutions can be obtained when the diffusivity is constant. Problems with a concentration-dependent diffusivity need numerical methods.

By putting

$$U = C \cdot r \tag{3-3}$$

equation (3-2) becomes

$$\frac{\partial U}{\partial t} = D \cdot \frac{\partial^2 U}{\partial r^2} \tag{3-4}$$

Equation (3-4) is similar to the equation obtained for diffusion in one dimension through the plane sheet.

## 3-2 DIFFUSION UNDER NONSTEADY STATE

The solutions of some problems of radial diffusion in the sphere can be obtained from the solutions of the problems with diffusion through the plane sheet.

### 3-2-1 Initial Distribution in the Sphere is $f(r)$. Surface Concentration Constant: $C_0$

After the substitution $U = C \cdot r$, the equation of diffusion is:

$$\frac{\partial U}{\partial t} = D \cdot \frac{\partial^2 U}{\partial r^2} \tag{3-4}$$

| | | | | |
|---|---|---|---|---|
| Initial | $t = 0$ | $0 < r < R$ | $u = r \cdot f(r)$ | (3-5) |
| Boundary | $t > 0$ | $r = 0$ | $u = 0$ | (3-6) |
| | | $r = R$ | $u = R \cdot C_0$ | |

where $C_0$ is the constant concentration at the surface of the sphere, and $R$ is the radius of the sphere.

The equation of diffusion (3-4) has been resolved in the case of a plane sheet of thickness $R$, with the initial and boundary conditions shown in equations (3-5) and (3-6).

The analytical solution of this general problem is given by:

$$C_{r,t} = C_0 \cdot \frac{r}{R} + \frac{2C_0}{\pi} \cdot \sum \frac{(-1)^n}{n} \cdot \sin \frac{n\pi r}{R} \exp\left(-\frac{n^2\pi^2}{R^2} Dt\right)$$
$$+ \frac{2}{R^2} \cdot \sum_{n=1}^{\infty} \sin \frac{n\pi r}{R} \cdot \exp\left(-\frac{n^2\pi^2}{R^2} Dt\right) \cdot \int_0^R r' \cdot f(r') \cdot \sin \frac{n\pi r'}{R} \, dr' \tag{3-7}$$

### 3-2-2 Initial Distribution in the Sphere is Constant: $C_i$. Surface Concentration Constant: $C_0$

Very often, the initial distribution of diffusing substance in the sphere is constant (Figure 3–1).

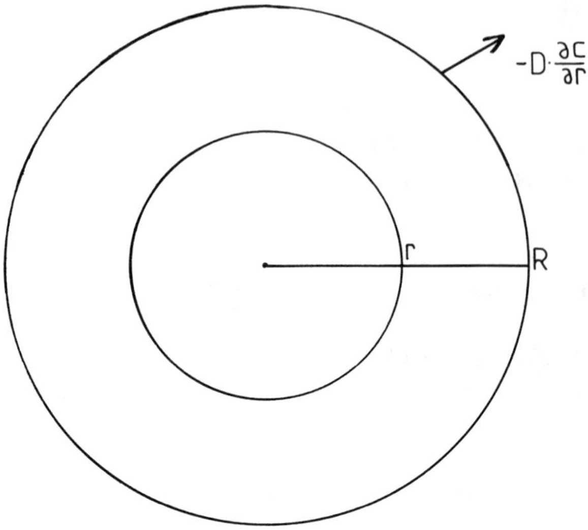


**Figure 3–1** Scheme of the circular cross section of a sphere of radius $R$.

In this case,

| | | | | |
|---|---|---|---|---|
| Initial | $t = 0$ | $0 < r < R$ | $u = C_i$ | (3-8) |
| Boundary | $t > 0$ | $r = 0$ | $u = 0$ | (3-9) |
| | | $r = R$ | $u = R \cdot C_0$ | |

the solution of the equation of diffusion becomes:

$$\frac{C_{r,t} - C_i}{C_0 - C_i} = 1 + \frac{2R}{\pi r} \cdot \sum \frac{(-1)^n}{n} \cdot \sin \frac{n\pi r}{R} \cdot \exp\left(-\frac{n^2\pi^2}{R^2} Dt\right) \tag{3-10}$$

This solution in this form cannot be used for the center of the sphere ($r = 0$). As the limit of $\sin x/x \rightarrow 1$ when $x \rightarrow 0$, equation (3-10) for the center of the sphere is reduced to:

$$\frac{C_{r,t} - C_i}{C_0 - C_i} = 1 + 2 \cdot \sum_{n=1}^{\infty} (-1)^n \cdot \exp\left(-\frac{n^2\pi^2}{R^2} Dt\right) \tag{3-11}$$

The total amount of diffusing substance entering or leaving the sphere is given by integrating Fick's first law with respect to time.

$$M_t = -\int_0^t D \cdot \left(\frac{\partial C}{\partial r}\right)_{r=R} \cdot dt \tag{3-12}$$

By considering that

$$\sum_{n=1}^{\infty} \frac{1}{n^2} = \frac{\pi^2}{6}$$

and

$$M_\infty = \frac{4}{3}\pi R^3 \cdot C_0$$

it is obtained:

$$\frac{M_\infty - M_t}{M_\infty} = \frac{6}{\pi^2} \cdot \sum_{n=1}^{\infty} \frac{1}{n^2} \cdot \exp\left(-\frac{n^2\pi^2}{R^2} Dt\right) \tag{3-13}$$

Another expression of the solution of the equation of diffusion in the sphere (equation 3-4) is given by:

$$\frac{C_{r,t} - C_i}{C_0 - C_i} = \frac{R}{r} \cdot \sum_{n=0}^{\infty} \left\{ erfc \frac{(2n + 1)R - r}{2(Dt)^{0.5}} - erfc \frac{(2n + 1)R + r}{2(Dt)^{0.5}} \right\} \tag{3-14}$$

The kinetics for the matter transported are:

$$\frac{M_t}{M_\infty} = 6 \left(\frac{Dt}{R^2}\right)^{0.5} \cdot \left\{ \pi^{-0.5} + 2 \cdot \sum_{n=1}^{\infty} ierfc \frac{nR}{(Dt)^{0.5}} \right\} - 3 \frac{Dt}{R^2} \tag{3-15}$$

***Remark 1. Case of Short Times.*** Equation (3-15) is very useful for short times because it can be reduced to:

$$\frac{M_t}{M_\infty} = \frac{6}{R} \left(\frac{Dt}{\pi}\right)^{0.5} \tag{3-16}$$

***Remark 2. Case of Long Times.*** In the case of long times, when $M_t/M_\infty$ is sufficiently large, equation (3-13) is reduced to the first term of the series with $n = 1$.

$$\frac{M_\infty - M_t}{M_\infty} = \frac{6}{\pi^2} \cdot \exp\left(-\frac{\pi^2}{R^2} Dt\right) \tag{3-17}$$

This equation is very well known. The diffusivity can easily be calculated from the slope of the straight line obtained by plotting Log ($M_\infty - M_t/M_\infty$) as a function of time.

### 3-2-3 Sphere in Contact with a Stirred Solution of Limited Volume (Volume V)

**The diffusing substance is in the solution at the beginning of the process, and the sphere is empty.** This problem is similar to that of the plane sheet in a stirred solution of limited volume. In the case of absorption of diffusing substance by the sphere, the conditions are:

| | | | | | |
|---|---|---|---|---|---|
| Initial | $t = 0$ | $r < R$ | $C = 0$ | sphere | (3-18) |
| | | | $C = C_0$ | solution | |

**Table 3-1**

Roots of $\tan q_n = \dfrac{3q_n}{3 + aq_n^2}$ (equation 3-21)

| a | q1 | q2 | q3 | q4 | q5 | q6 | q7 | q8 | q9 | q10 |
|---|---|---|---|---|---|---|---|---|---|---|
| 9.0000 | 3.2410 | 6.3350 | 9.4559 | 12.5928 | 15.7291 | 18.8672 | 22.0063 | 25.1460 | 28.2861 | 31.4265 |
| 4.0000 | 3.3485 | 6.3978 | 9.5029 | 12.6254 | 15.7554 | 18.8892 | 22.0251 | 25.1625 | 28.3008 | 31.4398 |
| 2.3333 | 3.4650 | 6.4736 | 9.5567 | 12.6667 | 15.7888 | 18.9172 | 22.0492 | 25.1836 | 28.3196 | 31.4567 |
| 1.5000 | 3.5909 | 6.5664 | 9.6254 | 12.7204 | 15.8326 | 18.9541 | 22.0811 | 25.2117 | 28.3446 | 31.4792 |
| 1.0000 | 3.7264 | 6.6814 | 9.7156 | 12.7927 | 15.8924 | 19.0048 | 22.1251 | 25.2504 | 28.3793 | 31.5106 |
| 0.6666 | 3.8711 | 6.8246 | 9.8369 | 12.8940 | 15.9780 | 19.0785 | 22.1895 | 25.3075 | 28.4305 | 31.5570 |
| 0.4286 | 4.0236 | 7.0019 | 10.0039 | 13.0423 | 16.1082 | 19.1932 | 22.2914 | 25.3989 | 28.5131 | 31.6322 |
| 0.2500 | 4.1811 | 7.2169 | 10.2355 | 13.2689 | 16.3211 | 19.3898 | 22.4719 | 25.5647 | 28.6656 | 31.7732 |
| 0.1111 | 4.3395 | 7.4645 | 10.5437 | 13.6134 | 16.6831 | 19.7565 | 22.8350 | 25.9189 | 29.0082 | 32.1025 |
| 0.0000 | 4.4934 | 7.7253 | 10.9041 | 14.0662 | 17.2208 | 20.3713 | 23.5195 | 26.6661 | 29.8116 | 32.9564 |

Boundary $\quad t > 0$

$$V \cdot \frac{\partial C}{\partial t} = -D \cdot \left(\frac{\partial C}{\partial r}\right)_{r=R} \tag{3-19}$$

The concentration of diffusing substance within the sphere is given by:

$$\frac{C_{r,t} - C_\infty}{C_\infty} = \sum_{n=1}^{\infty} \frac{6(1+\alpha)\cdot\exp\left(-\frac{q_n^2}{R^2}Dt\right)}{9 + 9\alpha + q_n^2\alpha^2} \cdot \frac{R}{r} \cdot \frac{\sin \frac{q_n r}{R}}{\sin q_n} \tag{3-20}$$

where the $q_n$s are the nonzero roots of

$$\tan q_n = \frac{3q_n}{3 + \alpha q_n^2} \quad \text{(given in Table 3–1)} \tag{3-21}$$

and $\alpha$ is the ratio of the volumes of solution ($V$) and sphere:

$$\alpha = \frac{3V}{4\pi R^3} \tag{3-22}$$

When there is a partition factor $K$ between the diffusing substance in equilibrium in the sphere and the solution, $\alpha$ is rewritten:

$$\alpha = \frac{3V}{4K\pi R^3} \tag{3-22'}$$

The total amount, $M_t$, of diffusing substance in the sphere after time $t$ is determined in terms of the corresponding quantity after infinite time, $M_\infty$:

$$\frac{M_\infty - M_t}{M_\infty} = \sum_{n=1}^{\infty} \frac{6\alpha(1 + \alpha)}{9 + 9\alpha + q_n^2\alpha^2} \cdot \exp\left(-\frac{q_n^2}{R^2} Dt\right) \tag{3-23}$$

The fractional uptake of the solute by the sphere is obviously $M_\infty/VC_0$. It is expressed in terms of the rate $\alpha$. By writing that the initial amount of diffusing substance in the liquid ($VC_0$) is equal to the amount of diffusing substance in the solution ($VC_\infty/K$) and in the sphere ($\frac{4}{3}\pi R^3 C_\infty$) at the end of the process, it is easy to obtain the fractional uptake.

$$\text{Fraction uptake} = \frac{1}{1 + \alpha} \tag{3-24}$$

**Case of desorption of diffusing substance into the well-stirred solution.** Initially conditions are different from the case of absorption.

| Initial | $t = 0$ | $r < R$ | $C = C_0$ | sphere |
|---|---|---|---|---|
| | | | $C = 0$ | solution |

Equation (3-23) can be used in the case of desorption, while equation (3-20) expressing the concentration of diffusing substance within the sphere becomes:

$$\frac{C_\infty - C_{r,t}}{C_0 - C_\infty} = \sum_{n=1}^{\infty} \frac{6(1 + \alpha) \cdot \exp\left(-\frac{q_n^2}{R^2} Dt\right)}{9 + 9\alpha + q_n^2\alpha^2} \cdot \frac{R}{r} \cdot \frac{\sin \frac{q_n R}{R}}{\sin q_n} \tag{3-20'}$$

The fractional uptake of diffusing substance by the solution is $VC_0/M_\infty$. The balance of diffusing substance, by considering the amount of this substance initially in the sphere ($\frac{4}{3}\pi R^3 C_0$) and finally in the sphere and in the solution, is:

$$\frac{4}{3}\pi R^3 C_0 = \frac{4}{3}\pi R^3 C_\infty + \frac{VC_\infty}{K} \tag{3-25}$$

with $M_\infty = VC_\infty/K$.
The fractional uptake of the substance by the solution is then:

$$\text{Fractional uptake} = \frac{\alpha}{1 + \alpha} \tag{3-26}$$

***Remark 1. Amount of Substance Transported After Long Time.*** The amount of diffusing substance which has entered (or left) the sphere after a long time can be expressed as a function of time by equation (3-23) in which the first term becomes preponderant.

$$\frac{M_\infty - M_t}{M_\infty} = \frac{6\alpha(1 + \alpha)}{9 + 9\alpha + q_1\alpha^2} \cdot \exp\left(-\frac{q_1^2}{R^2} Dt\right) \tag{3-27}$$

The diffusivity can be obtained from the slope of the straight line obtained by plotting Log($M_\infty - M_t/M_\infty$) as a function of time.

### 3-2-4 Transport with Diffusion and Surface Evaporation. Sphere Initially at Uniform Concentration.

The sphere is initially at a uniform concentration $C_i$. The transport of diffusing substance is obtained either by diffusion within the polymer or by evaporation on the sphere surface.

The rate of diffusing substance which evaporates (or is absorbed after condensation) from the surface of the sphere is equal to the rate of transport by diffusion near the surface (Figure 3–2):

$$-D \cdot \left(\frac{\partial C}{\partial r}\right)_{r=R} = F_0 \cdot (C_0 - C_{ext}) \tag{3-28}$$

where the rate of evaporation is proportional to the difference between the actual concentration on the surface, $C_0$, and the concentration $C_{ext}$ which corresponds with the vapor pressure in the atmosphere near the surface, and $F_0$ is the rate of evaporation of the pure diffusing substance ($C_0 = 1$).

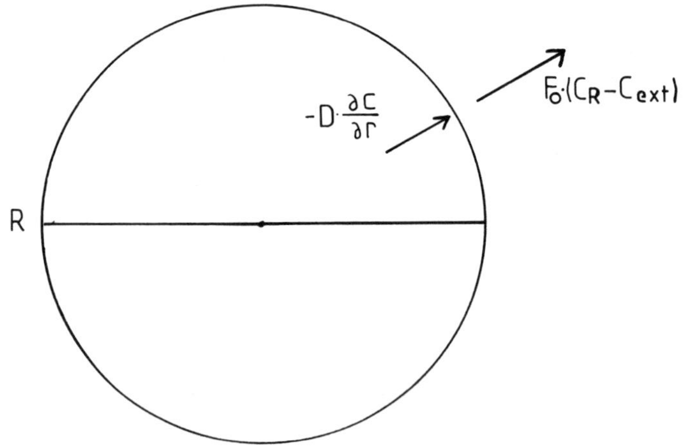


**Figure 3–2** Scheme of the circular cross section of a sphere, with diffusion in the solid and evaporation on the surface.

The initial conditions are for the sphere:

$$t = 0 \qquad 0 < r < R \qquad C = C_i \tag{3-29}$$

The analytical solution of the equation of diffusion

$$\frac{\partial U}{\partial t} = D \cdot \frac{\partial^2 U}{\partial r^2} \tag{3-4}$$

is

$$\frac{C_{r,t} - C_{ext}}{C_i - C_{ext}} = \frac{2MR}{r} \cdot \sum_{n=1}^{\infty} \frac{\exp\left(-\frac{\beta_n^2}{R^2} Dt\right)}{\beta_n^2 + M(M-1)} \cdot \frac{\sin \frac{\beta_n}{R} r}{\sin \beta_n} \tag{3-30}$$

where

$$M = \frac{R\alpha}{D} \tag{3-31}$$

and the $\beta_n$s are the roots of:

$$\beta_n \cdot \cot \beta_n + M - 1 = 0 \quad \text{(give in Table 3-2)} \tag{3-32}$$

**Table 3-2**

Roots of $\beta_n \cot \beta_n + L - 1 = 0$ (equation 3-32)

| L | $\beta_1$ | $\beta_2$ | $\beta_3$ | $\beta_4$ | $\beta_5$ | $\beta_6$ | $\beta_7$ | $\beta_8$ | $\beta_9$ | $\beta_{10}$ |
|---|---|---|---|---|---|---|---|---|---|---|
| 500.00 | 3.1353 | 6.2706 | 9.4059 | 12.5412 | 15.6766 | 18.8119 | 21.9472 | 25.0825 | 28.2178 | 31.3532 |
| 100.00 | 3.1102 | 6.2204 | 9.3308 | 12.4414 | 15.5521 | 18.6632 | 21.7746 | 24.8865 | 27.9987 | 31.1114 |
| 50.00 | 3.0788 | 6.1582 | 9.2384 | 12.3200 | 15.4034 | 18.4888 | 21.5764 | 24.6664 | 27.7589 | 30.8540 |
| 10.00 | 2.8363 | 5.7172 | 8.6587 | 11.6532 | 14.6869 | 17.7481 | 20.8282 | 23.9218 | 27.0250 | 30.1354 |
| 2.50 | 2.1746 | 5.0036 | 8.0385 | 11.1295 | 14.2421 | 17.3649 | 20.4934 | 23.6254 | 26.7595 | 29.8953 |
| 1.00 | 1.5708 | 4.7124 | 7.8540 | 10.9956 | 14.1372 | 17.2788 | 20.4203 | 23.5619 | 26.7035 | 29.8451 |
| 0.50 | 1.1656 | 4.6042 | 7.7899 | 10.9499 | 14.1017 | 17.2498 | 20.3958 | 23.5407 | 26.6848 | 29.8284 |
| 0.30 | 0.9208 | 4.5601 | 7.7641 | 10.9316 | 14.0875 | 17.2382 | 20.3860 | 23.5322 | 26.6773 | 29.8217 |
| 0.20 | 0.7593 | 4.5379 | 7.7511 | 10.9225 | 14.0804 | 17.2324 | 20.3811 | 23.5280 | 26.6736 | 29.8183 |
| 0.10 | 0.5423 | 4.5157 | 7.7382 | 10.9133 | 14.0733 | 17.2266 | 20.3762 | 23.5237 | 26.6698 | 29.8150 |
| 0.05 | 0.3854 | 4.5045 | 7.7317 | 10.9087 | 14.0697 | 17.2237 | 20.3738 | 23.5216 | 26.6679 | 29.8133 |
| 0.01 | 0.1730 | 4.4956 | 7.7265 | 10.9050 | 14.0669 | 17.2213 | 20.3718 | 23.5199 | 26.6664 | 29.8119 |
| 0.00 | 0.0000 | 4.4934 | 7.7253 | 10.9041 | 14.0662 | 17.2208 | 20.3713 | 23.5195 | 26.6661 | 29.8116 |

The total amount of diffusing substance entering (or leaving) the sphere is:

$$\frac{M_\infty - M_t}{M_\infty} = \sum_{n=1}^{\infty} \frac{6M^2 \cdot \exp\left(-\frac{\beta_n^2}{R^2} Dt\right)}{\beta_n^2 \cdot \{\beta_n^2 + M(M - 1)\}} \tag{3-33}$$

***Remark 1. Amount of Diffusing Substance Entering (or Leaving) the Sphere After a Long Time***

$$\frac{M_\infty - M_t}{M_\infty} = \frac{6M^2}{\beta_1^2} \cdot \frac{\exp\left(-\frac{\beta_1^2}{R^2} Dt\right)}{\beta_1^2 + M(M - 1)} \tag{3-34}$$

Equation (3-34) is obtained by keeping the first term of the series in equation (3-33). The diffusivity can be obtained from the slope of the straight line obtained by plotting $\mathrm{Log}(M_\infty - M_t/M_\infty)$ as a function of time.

***Remark 2***

Evaporation or condensation of vapor takes place, following the positive or negative sign of $(C_o - C_{ext})$ in equation (3-28), as well as that of $(C_i - C_{ext})$ in equation (3-30).

***Remark 3. Amount of Diffusing Substance Entering (or Leaving) the Sphere after Short Time***

Equation (3-16) can also be used for a short time, in order to determine the diffusivity $D$ when the amount of substance transported at equilibrium, $M_\infty$, is known.

$$\frac{M_t}{M_\infty} = \frac{6}{R}\left(\frac{Dt}{\pi}\right)^{0.5} \tag{3-16}$$

### 3-2-5 Hollow Sphere: Spherical Membrane

There are various cases of hollow spheres, located between the spherical surface of radii $R_i$ and $R_e$. However, the main application of hollow spheres is the spherical membrane separating the diffusing substance (located in the small sphere of radius $R_i$) from the external solution (in contact with the sphere of radius $R_e$).

The internal sphere is full of diffusing substance, and the concentration can be considered as constant on the internal spherical surface of radius $R_i$ (Figure 3–3). The external sphere of radius $R_e$ is kept at the constant concentration zero.

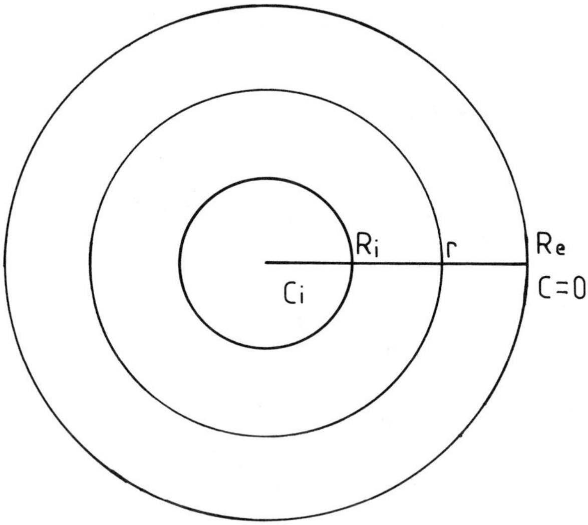


**Figure 3–3** Scheme of the circular cross section of an internal sphere of radii $R_i$ and $R_e$, with constant concentrations on the surfaces.

Initial and boundary conditions are:

$$\text{Initial} \quad t = 0 \quad r \leq R_i \quad C = C_i; \qquad r > R_i \quad C = 0 \tag{3-35}$$

$$\text{Boundary} \quad t > 0 \quad r = R_i \quad C = C_i; \qquad r = R_e \quad C = 0 \tag{3-36}$$

The concentration in the region of the membrane ($R_i \leq r \leq R_e$), is given by the relation:

$$C_{r,t} = \frac{C_i R_i}{r} \cdot \frac{R_e - r}{R_e - R_i} - \frac{2R_i \cdot C_i}{\pi r} \cdot \sum_{n=1}^{\infty} \frac{1}{n} \sin \frac{n\pi(r - R_i)}{R_e - R_i} \cdot \exp\left(-\frac{n^2\pi^2 Dt}{(R_e - R_i)^2}\right) \tag{3-37}$$

The amount of diffusing substance which escapes from the outer spherical surface of radius $R_e$ is given by:

$$\frac{M_t}{M_\infty} \cdot \frac{3R_i^2}{R_e(R_e - R_i)} = \frac{Dt}{(R_e - R_i)^2} - \frac{1}{6} - \frac{2}{\pi^2} \cdot \sum_{n=1}^{\infty} \frac{(-1)^2}{n^2} \cdot \exp\left(-\frac{n^2 \cdot \pi^2}{(R_e - R_i)^2} Dt\right) \tag{3-38}$$

## 3-3 DIFFUSION UNDER STEADY STATE: SPHERICAL MEMBRANE

The equation of radial diffusion

$$\frac{\partial C}{\partial t} = \frac{1}{r^2} \cdot \frac{\partial}{\partial r}\left[D \cdot r^2 \cdot \frac{\partial C}{\partial r}\right] \tag{3-1}$$

under steady state with $\partial C/\partial t = 0$, becomes:

$$\frac{d}{dr}\left[r^2 \cdot \frac{dc}{dr}\right] = 0 \tag{3-39}$$

The general solution is given by

$$C = A + \frac{B}{r} \tag{3-40}$$

where the constants $A$ and $B$ are determined from the boundary conditions.

Various cases of hollow spheres or spherical membranes are considered.

### 3-3-1 Spherical Membrane Limiting the Media with Constant Concentrations and Constant Diffusivity

The membrane is between $R_i$ and $R_e$, with the following initial and boundary conditions:

| | | | | | |
|---|---|---|---|---|---|
| Initial | $t = 0$ | $R_i < r < R_e$ | $C = 0$ | | (3-41) |
| Boundary | $t > 0$ | $r = R_i$ | $C = C_i$ | constant | (3-42) |
| | | $r = R_e$ | $C = C_e$ | constant | |

The concentration in the spherical membrane is:

$$C_{r,t} = \frac{R_i C_i (R_e - r) + R_e C_e (r - R_i)}{r(R_e - R_i)} \tag{3-43}$$

and the quantity of diffusing substance which passes through the spherical membrane in time $t$ is:

$$M_t = \frac{4\pi R_i R_e}{R_e - R_i}(C_i - C_e)Dt \tag{3-44}$$

### 3-3-2 Spherical Membrane Limiting the Internal Medium with Constant Concentration ($C_i$) and the External Medium with Concentration Zero, and Constant Diffusivity

This case is more simple than the previous one, but it is sometimes of higher interest. The membrane is between $R_i$ and $R_e$, with the following initial and boundary conditions:

$$\text{Initial} \quad t = 0 \quad R_i < r < R_e \quad C = 0 \tag{3-45}$$

$$\text{Boundary} \quad t > 0 \quad \begin{array}{ll} r = R_i & C = C_i \quad \text{constant} \\ r = R_e & C = 0 \end{array} \tag{3-46}$$

The concentration in the spherical membrane becomes:

$$C_{r,t} = \frac{(R_e - r)R_i}{(R_e - R_i)r} C_i \tag{3-47}$$

while the quantity of diffusing substance passing through the membrane in time $t$ is:

$$M_t = \frac{4\pi R_i \cdot R_e}{R_e - R_i} C_i D t \tag{3-48}$$

***Remark 1. Concentration-Dependent Diffusivity.*** As shown in the case of the plane sheet, two ways can be followed. One is by considering the mean value of diffusivity.

$$D = \frac{1}{C_i - C_e} \int_{C_e}^{C_i} D \cdot dC \tag{3-49}$$

The other is by using numerical methods. If there is a slight change in diffusivity from $C_i$ to $C_e$, equation (3-49) can be used. Otherwise numerical methods are of better use.

### 3-3-3 Spherical Membrane with Constant Concentration on the Internal Surface and Evaporation on the External Surface

This case is of interest because of some applications to control the evaporation. The membrane is between $R_i$ and $R_e$, with the following initial and boundary conditions.

$$\text{Initial} \quad t = 0 \quad \begin{array}{ll} r \le R_i & C = C_i \\ r > R_i & C = 0 \end{array} \tag{3-50}$$

Boundary $$-D\cdot\left(\frac{\partial C}{\partial r}\right)_{r=R_e} = F(C_2 - C_{ext}) \tag{3-51}$$

where $C_{ext}$ is the concentration of the vapor in the atmosphere, and $C_2$ is the concentration on the external surface.

The concentration of diffusing substance at the spherical surface of radius $r$ is:

$$C_r = \frac{R_i C_i \cdot \left\{\frac{F}{D} R_e^2 + r\left(1 - \frac{F}{D} R_e\right)\right\} + \frac{F}{D} R_e^2 \,(r - R_i) C_{ext}}{r \cdot \left\{\frac{F}{D} R_e^2 + R_i\left(1 - \frac{F}{D} R_e\right)\right\}} \tag{3-52}$$

and the amount of substance leaving the membrane in time $t$ is:

$$M_t = \frac{4\pi \frac{F}{D} R_i R_e^2 (C_i - C_{ext})}{\frac{F}{D} R_e^2 \,'\, R_i\left(1 - \frac{F}{D} R_e\right)} Dt \tag{3-53}$$

# Chapter Four

# Numerical Analysis in One-Dimensional Transient Transport: Plane Sheet

## 4-1 INTRODUCTION

In applying the numerical method, the continuous process of matter transport by diffusion is replaced by a stepwise one. The increment method is applied to one-dimensional transient diffusion through the thickness of the plane sheet.

Finite increments of space $\Delta x$ and time $\Delta t$ are considered in numerical analysis. The time at which an event occurs, as well as the place, are defined as follows:

$$\text{time } t = j \cdot \Delta t$$

$$\text{place} = n \cdot \Delta x$$

$n$ and $j$ being integers.

The two-dimension space-time diagram shown in Figure 4–1 is useful for understanding the process. At time 0, the initial conditions are shown for each place located in the solid, each concentration C being indiced with two integers characterizing space and time: $C_{n+1,0}$, $C_{n,0}$, $C_{n-1,0}$. At time $t$, defined by integer $j$, the concentrations for each place located in the solid are indiced as follows: $C_{n+1,j}$, $C_{n,j}$, $C_{n-1,j}$.

After the elapse of time $\Delta t$, the time is $t + \Delta t$, or $(j + 1)\Delta t$. The concentrations for each place in the solid are indiced in the same way:

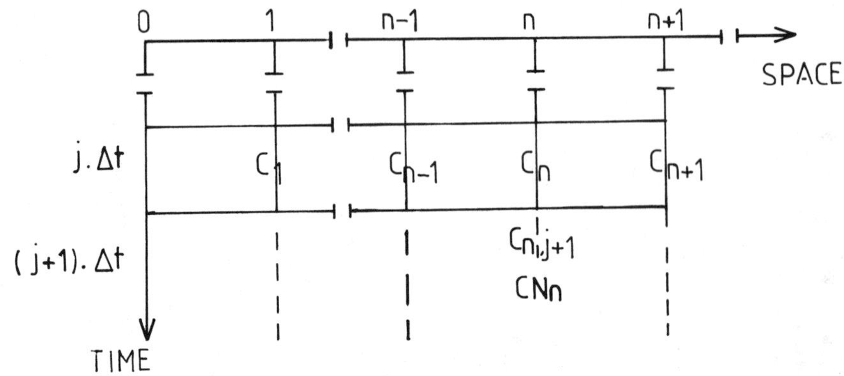


**Figure 4–1** Space-time diagram for numerical analysis. Concentration of diffusing substance for various places and times.

$C_{n+1,j+1}$, $C_{n,j+1}$, $C_{n-1,j+1}$. We have preferred, in a general way, to consider the concentration after elapse of time $\Delta t$ as the new concentration, and write it in a more simple way: $CN_{n+1}$, $CN_n$, $CN_{n-1}$.

## 4-2 DIFFUSION AND STORAGE IN THE INTERIOR OF THE SHEET

### 4-2-1 Diffusion within the Sheet with Constant Diffusivity

The cross section of a large slab of thickness $L$, having a uniform cross-sectional area $S$ is shown in Figure 4–2. The solid is divided into a number of equal finite slices of thickness $\Delta x$ by concentration-reference planes.

The matter balance is written on the crosshatched zone *abcd* of thickness $\Delta x$ during the increment of time $\Delta t$, by using Fick's first law on the planes *ab* and *cd*:

$$S\left[-D\cdot\left(\frac{\partial C}{\partial x}\right)_a + D\cdot\left(\frac{\partial C}{\partial x}\right)_b\right]\cdot\Delta t = (S\cdot\Delta x)\cdot(CN_1 - C_1) \qquad (4\text{-}1)$$

where $D$ is the value of diffusivity at positions $a$ and $b$, at time $t$; $(\partial C/\partial x)_a$ is the value of the gradient of concentration at position $a$ and time $t$; $C_1$ is the value of the concentration at position 1 at time $t$; and $CN_1$ is the value of the new concentration at position 1 after the elapse of the finite time increment $\Delta t$.

The gradient of concentration at plane *ac* (at position of abscissa 0.5)

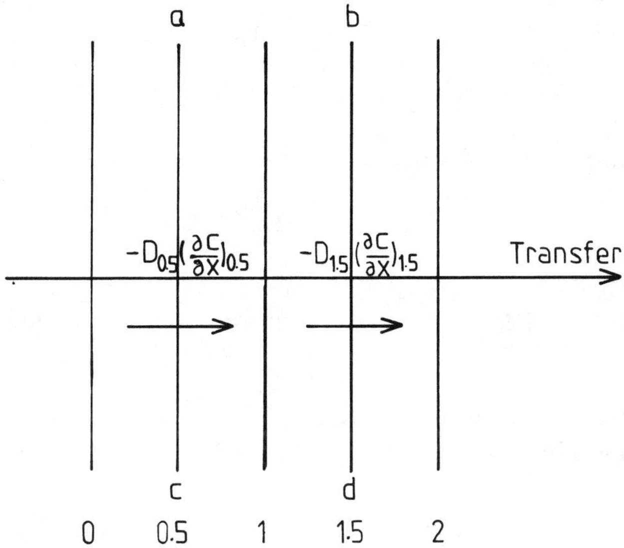


**Figure 4–2** Diffusion of the substance through the plane sheet.

is approximated by the chord slope between the positions 0 and 1 which are $\Delta x$ apart.

$$-\left(\frac{\partial C}{\partial x}\right)_a = \frac{C_0 - C_1}{\Delta x} \tag{4-2}$$

Similarly, the gradient of concentration at plane *bd* (at position of abscissa 1.5) is replaced by:

$$-\left(\frac{\partial C}{\partial x}\right)_b = \frac{C_1 - C_2}{\Delta x} \tag{4-2'}$$

The concentration at plane 1 approximates the average concentration of the crosshatched zone *abdc*. The resulting matter balance is:

$$\frac{D\cdot(C_0 - C_1)S\cdot\Delta t}{\Delta x} - \frac{D\cdot(C_1 - C_2)S\cdot\Delta t}{\Delta x} = (S\cdot\Delta x)(CN_1 - C_1) \tag{4-3}$$

Upon replacing the dimensionless ratio $(\Delta x)^2/D\cdot\Delta t$ by the modulus $M$,

$$M = \frac{(\Delta x)^2}{\Delta t}\cdot\frac{1}{D} \tag{4-4}$$

equation (4-3) becomes:

$$CN_1 = \frac{1}{M}[C_2 + (M - 2)\cdot C_1 + C_0] \tag{4-5}$$

Of course, the general equation for position $n$ is easily obtained:

$$CN_n = \frac{1}{M} \cdot [C_{n+1} + (M - 2) \cdot C_n + C_{n-1}] \tag{4-6}$$

or with the other way of indexing:

$$C_{n,j+1} = \frac{1}{M} \cdot [C_{n+1,j} + (M - 2) \cdot C_{n,j} + C_{n-1,j}] \tag{4-6'}$$

These equations express the new concentration at position $n$ after elapse of time $\Delta t$, $CN_n$, in terms of the concentrations obtained at previous time at position $n$, $C_n$, and at adjacent position $n + 1$, $C_{n+1}$, and $n - 1$, $C_{n-1}$.

While two ways of writing can be used, in the following pages, the first way of writing, with the new concentration $CN_n$, is used as much as possible.

### 4-2-2 Diffusion within the Sheet with Concentration-Dependent Diffusivity

In this important case, the diffusivity of the diffusing substance is concentration dependent. Then as the concentration of this substance varies with time and position, the diffusivity varies also with time and position.

The matter balance on the crosshatched zone *abcd* of thickness $\Delta x$ during the increment of time $\Delta t$, becomes for the cross-sectional area $S$ (Figure 4–2):

$$S \cdot \left[ -D_{0.5} \cdot \left(\frac{\partial C}{\partial x}\right)_{0.5} + D_{1.5} \cdot \left(\frac{\partial C}{\partial x}\right)_{1.5} \right] \cdot \Delta t = (S \cdot \Delta x) \cdot (CN_1 - C_1) \tag{4-7}$$

where $D_{0.5}$ and $D_{1.5}$ are the values of diffusivity at the same time $t$, and positions $a$ and $b$, respectively, and $C_1$ and $CN_1$ are the concentrations at position 1, at time $t$ and $t + \Delta t$, respectively.

By approximating the gradient of concentration at plane *ac* (or *bd*) by the chord slope between the positions 0 and 1 (or 1 and 2) which are $\Delta x$ apart, the matter balance is:

$$\left[ D_{0.5} \frac{C_0 - C_1}{\Delta x} - D_{1.5} \frac{C_1 - C_2}{\Delta x} \right] \Delta t = \Delta x \cdot (CN_1 - C_1) \tag{4-8}$$

By putting the number $J$ (which has the dimension of concentration),

$$J_{0.5} = \frac{\Delta t}{(\Delta x)^2} D_{0.5} \cdot (C_0 - C_1) \tag{4-9}$$

and of course

$$J_{1.5} = \frac{\Delta t}{(\Delta x)^2} D_{1.5} \cdot (C_1 - C_2) \qquad (4\text{-}9')$$

the basic equation is obtained:

$$CN_1 = C_1 + J_{0.5} - J_{1.5} \qquad (4\text{-}10)$$

or

$$CN_n = C_n + J_{n-0.5} - J_{n+0.5} \qquad (4\text{-}10')$$

***Remark 1. Diffusivity Varies Little with Concentration.*** Two cases of interest are expressed by a small change in the value of diffusivity.

- the concentration varies little within the sheet during the process
- the diffusivity varies little with concentration of diffusing substance

When diffusivity varies little, following one of the preceding two cases, the mean value of diffusivity can be used at any place:

$$D_{0.5} = \frac{1}{2}(D_0 + D_1) \qquad \text{at position 0.5, for instance.} \qquad (4\text{-}11)$$

The matter balance, calculated for the crosshatched zone (Figure 4–2) is:

$$\frac{1}{2}(D_0 + D_1)(C_0 - C_1) - \frac{1}{2}(D_1 + D_2)(C_1 - C_2) = \frac{(\Delta x)^2}{\Delta t}(CN_1 - C_1) \qquad (4\text{-}12)$$

By using the dimensionless number $M$,

$$M = \frac{(\Delta x)^2}{\Delta t} \cdot \frac{1}{D} \qquad (4\text{-}4)$$

calculated at time $t$ and at various places, the preceding equation becomes:

$$\frac{1}{2}\left(\frac{1}{M_0} + \frac{1}{M_1}\right)(C_0 - C_1) - \frac{1}{2}\left(\frac{1}{M_1} + \frac{1}{M_2}\right)(C_1 - C_2) = CN_1 - C_1 \qquad (4\text{-}13)$$

The final expression for the new concentration after the elapse of time, $CN_1$, is thus expressed in terms of the concentration obtained at the previous time at the same and adjacent places:

$$CN_1 = \frac{C_0}{2}\left[\frac{1}{M_0} + \frac{1}{M_1}\right] + \frac{C_1}{2}\left[2 - \frac{1}{M_0} - \frac{2}{M_1} - \frac{1}{M_2}\right] + \frac{C_2}{2}\left[\frac{1}{M_1} + \frac{1}{M_2}\right] \qquad (4\text{-}14)$$

More generally, for position $n$, it is obvious that:

$$CN_n = \frac{C_{n-1}}{2}\left[\frac{1}{M_{n-1}} + \frac{1}{M_n}\right] + \frac{C_n}{2}\left[2 - \frac{1}{M_{n-1}} - \frac{2}{M_n} - \frac{1}{M_{n+1}}\right] + \frac{C_{n+1}}{2}\left[\frac{1}{M_n} + \frac{1}{M_{n+1}}\right] \quad (4\text{-}14')$$

***Remark 2. Diffusivity Varies Largely with the Concentration of Diffusing Substance.*** Following one of these cases, the mean value of diffusivity cannot be used as in the previous case:

- When the concentration of diffusing substance varies largely in the sheet.
- When the diffusivity varies largely with the concentration of diffusing substance.

Very often, diffusivity is expressed in terms of the concentration of diffusing substance by one of the following relations:

$$D_{n,j} = D_0 \cdot \exp(cte \cdot C_{n,j}) \quad (4\text{-}15)$$

$$D_{n,j} = D_0 \cdot \exp(cte - C_{n,j}) \quad (4\text{-}15')$$

or

$$D_{n,j} = D_0 \cdot \exp\left(\frac{cte}{C_{n,j}}\right) \quad (4\text{-}15'')$$

In this case, it is necessary to determine the mean concentration at the intermediate plane, 0.5, 1.5, 2.5, and then to calculate the local value of diffusivity by using one of equations (4-15).

$$D_{0.5,j} = D_0 \cdot \exp\left[\frac{cte(C_{0,j} + C_{1,j})}{2}\right] \quad (4\text{-}16)$$

$$D_{0.5,j} = D_0 \cdot \exp\left[cte - \frac{C_{0,j} + C_{i,j}}{2}\right] \quad (4\text{-}16')$$

$$D_{0.5,j} = D_0 \cdot \exp\left[\frac{2 \cdot cte}{C_{0,j} + C_{1,j}}\right] \quad (4\text{-}16'')$$

The new concentration after the elapse of time $\Delta t$, at position $n$, can be obtained by using the preceding equations:

$$CN_n = C_n + J_{n-0.5} - J_{n+0.5} \quad (4\text{-}10)$$

with

$$J_{n-0.5} = \frac{\Delta t}{(\Delta x)^2} \cdot D_{n-0.5} \cdot [C_n - C_{n+1}] \quad (4\text{-}9'')$$

and the diffusivity, $D_{n-0.5}$ being obtained with one of the preceding equations (4-16), according to the general expression of diffusivity as a function of concentration.

***Remark 3. Finite Difference Solution with Explicit Method.*** As shown in Figure 4–1, the value of concentration at time $t = j \cdot \Delta t$ and space defined by $n$, is denoted $C_{n,j}$, $n$ and $j$ being integers.

By using Taylor's series with $t$ variable and constant position, it is possible to obtain the concentration at the following time and at position $n$: $C_{n,j+1}$.

$$C_{n,j+1} = C_{n,j} + \left(\frac{\partial C}{\partial t}\right)_{n,j} \cdot dt + \frac{1}{2}\left(\frac{\partial^2 C}{\partial t^2}\right)_{n,j} \cdot (dt)^2 + \dots \tag{4-17}$$

The following equation can be written

$$\left(\frac{\partial C}{\partial t}\right)_{n,j} = \frac{C_{n,j+1} - C_{n,j}}{\Delta t} + 0(\Delta t) \tag{4-18}$$

where the neglected term $0(\Delta t)$ is of the order of $\Delta t$.

Similarly, by applying Taylor's series with position variable and keeping time constant, the following is obtained:

$$C_{n+1,j} = C_{n,j} + \left(\frac{\partial C}{\partial x}\right)_{n,j} \cdot dx + \frac{1}{2}\left(\frac{\partial^2 C}{\partial x^2}\right)_{n,j} \cdot (dx)^2 + \dots \tag{4-19}$$

$$C_{n-1,j} = C_{n,j} - \left(\frac{\partial C}{\partial x}\right)_{n,j} \cdot dx + \frac{1}{2}\left(\frac{\partial^2 C}{\partial x^2}\right)_{n,j} \cdot (dx)^2 + \dots \tag{4-19'}$$

where $C_{n+1,j}$ and $C_{n-1,j}$ are concentrations at positions $n + 1$ and $n - 1$, both adjacent to the position $n$, and at the same time $j \cdot \Delta t$.

On adding the preceding equations, the following is obtained:

$$\left(\frac{\partial^2 C}{\partial x^2}\right)_{n,j} = \frac{C_{n+1,j} - 2C_{n,j} + C_{n-1,j}}{(\Delta x)^2} + 0(\Delta x)^2) \tag{4-20}$$

and by rearranging, and neglecting the error terms $0((\Delta x)^2)$, the basic equation is found:

$$C_{n,j+1} = \frac{1}{M} \cdot [C_{n+1,j} + (M - 2) \cdot C_{n,j} + C_{n-1,j}] \tag{4-6'}$$

or more simply

$$CN_n = \frac{1}{M} \cdot [C_{n+1} + (M - 2) \cdot C_n + C_{n-1}] \tag{4-6}$$

with the dimensionless number $M$

$$M = \frac{(\Delta x)^2}{\Delta t} \cdot \frac{1}{D} \tag{4-4}$$

and the new concentration at position $n$, $CN_n$ (or $C_{n,j+1}$).

## 4-3 DIFFUSION WITHIN THE SHEET AND EVAPORATION ON THE SURFACE

When a stream of dry air passes over the surface of a solid containing moisture, loss of moisture occurs by surface evaporation. In the same way, if the solid is initially dry and the atmosphere contains moisture (water vapor), the solid absorbs moisture.

In each case, the rate of exchange of moisture at any time depends on:

1. the moisture concentration on the surface of the solid and on the relative humidity of the atmosphere.
2. the property of the solid and especially the diffusivity of moisture in the solid.

On the surface, the boundary condition is related to the rate of transport of diffusing substance across the surface of the medium.

The rate of evaporation is constantly equal to the rate of diffusion of substance near the surface. The evaporation process is diffusion controlled in the sense that the rate of evaporation depends largely on the rate at which solvent is supplied to the evaporating surface by internal diffusion.

The best boundary condition is that the rate of evaporation of the vapor out of the solid surface is constantly equal to the rate of transport of substance by diffusion through the solid near the surface, per unit area.

$$-D \cdot \left(\frac{\partial C}{\partial x}\right)_{surface} = F_o \cdot (C_o - C_{ext}) \tag{4-21}$$

where $D$ and $(\partial C/\partial x)$ are the diffusivity and gradient of concentration for diffusing substance near the solid surface on which evaporation takes place; $F_o$ is the evaporation rate of the pure substance, when the concentration on the evaporating substance is 1; $C_o$ is the concentration of diffusing substance on the solid surface; and $C_{ext}$ is the concentration of substance in the external atmosphere near the solid surface. This concentration is related to the vapor pressure of the substance, or to the relative humidity in the case of water vapor. Various cases are considered, related to the way of calculating the gradient of concentration $\partial C/\partial x$ near the solid surface.

### 4-3-1 Diffusion-Evaporation on the Surface. The Material Balance is Done on the Last Slice. Constant Diffusivity.

As shown in Figure 4–3, the balance of diffusing substance is calculated during the increment of time $\Delta t$ in the slice next to the surface, this slice having a thickness of $\Delta x$ (crosshatched slice). The amount of substance entering the slice, during the time $\Delta t$ is, per unit area:

$$D\cdot \left(\frac{\partial C}{\partial x}\right)_1 \cdot \Delta t = D\cdot \frac{C_2 - C_0}{2\cdot \Delta x}\Delta t \tag{4-22}$$

by approximating the gradient of concentration at plane 1 by the slope between the planes at positions 2 and 0 (distant from each other by $2\cdot\Delta x$).

The amount of substance leaving the slice during the same time $\Delta t$ is evaporated, following the equation, per unit area:

$$F_0(C_0 - C_{ext})\cdot\Delta t \tag{4-22'}$$

Resulting from these two transports of substance, there is a change in the mean concentration of the crosshatched slice, noted: $CN_{0.5} - C_{0.5}$, 0.5 being the mean position of this slice.

The balance of diffusing substance during the time $\Delta t$ is for the crosshatched slice, per unit area:

$$\left[D\cdot \left(\frac{\partial C}{\partial x}\right)_1 - F_0\cdot(C_0 - C_{ext})\right]\cdot \Delta t = \Delta x\cdot[CN_{0.5} - C_{0.5}] \tag{4-23}$$

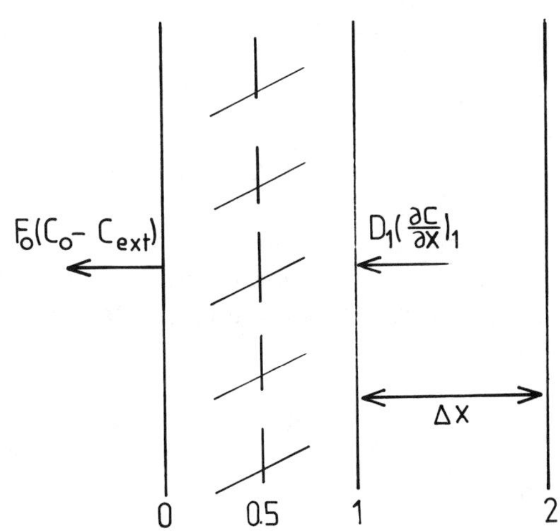


**Figure 4–3** Scheme of the slice next to the surface, with diffusion through the sheet and evaporation from the surface. Slice of thickness $\Delta x$.

The following approximation for the mean concentration in the cross-hatched slice is made. (In fact, another approximation can also be made. See Remark 1.)

$$CN_{0.5} - C_{0.5} = CN_0 - C_0 \tag{4-24}$$

By considering the following dimensionless numbers,

$$M = \frac{(\Delta x)^2}{\Delta t} \cdot \frac{1}{D} \tag{4-4}$$

$$N = \frac{F_0 \cdot \Delta x}{D} \tag{4-25}$$

and the assumption made in equation (4-24), equation (4-23) can be rewritten:

$$CN_0 = \frac{1}{2M} \cdot [C_2 + (2M - 1 - 2N) \cdot C_0 + 2N \cdot C_{ext}] \tag{4-26}$$

Equation (4-26) is of interest for calculating the new value of concentration of substance on the surface after the elapse of time $\Delta t$, $CN_0$, from the values of concentrations of substance obtained at the previous time: on the surface, ($C_0$), inside the solid at position 2, ($C_2$), and in the external atmosphere, ($C_{ext}$).

Of course, the new value of concentrations of substance after the elapse of time $\Delta t$ can also be calculated within the solid, by using equation (4-6):

$$CN_1 = \frac{1}{M} [C_0 + (M - 2)C_1 + C_2] \tag{4-5}$$

which is easily generalized:

$$CN_n = \frac{1}{M} [C_{n-1} + (M - 2) \cdot C_n + C_{n+1}] \tag{4-6}$$

the integer $n$ characterizing the $n$th slice.

The amount of substance evaporating the solid in time $t$, per unit area, on each face can be easily obtained by integrating equation (4-21) with respect to time:

$$M_t = \int_0^t F_0 \cdot (C_0 - C_{ext}) \cdot dt \tag{4-27}$$

which in fact becomes

$$M_t = \sum_{j=0}^{j-1} F_0 \cdot (C_{0,j} - C_{ext}) \cdot \Delta t \tag{4-28}$$

where $j$ is an integer characterizing the time:

$$t = j \cdot \Delta t$$

***Remark 1. Approximation for*** $\mathbf{CN_{0.5} - C_{0.5}}$. A better approximation can be made for calculating the change in concentration during the time $\Delta t$ in the crosshatched slice:

$$CN_{0.5} - C_{0.5} = \frac{1}{2}(CN_1 - C_1) + \frac{1}{2}(CN_0 - C_0) \tag{4-29}$$

$CN_0$ is then expressed in terms of the other concentrations, as shown in equation (4-26), and also in terms of the new concentration. $CN_1$ must thus be calculated before $CN_0$.

***Remark 2. Thinner Slice is Considered on the Surface.*** The thickness of the slice considered near the surface is $\Delta x/2$, and the mean abscissa of this slice is then 0.25 (Figure 4–4). The balance of diffusing substance is calculated during the increment of time $\Delta t$ in this slice, by considering the amount of substance which enters and leaves the slice per unit area.

$$\left[D \cdot \left(\frac{\partial C}{\partial x}\right)_{0.5} - F_0 \cdot (C_0 - C_{ext})\right] \cdot \Delta t = \frac{\Delta x}{2} \cdot [CN_{0.25} - C_{0.25}] \tag{4-30}$$

A simple approximation for the mean concentration in the thin slice at position 0.25 is:

$$CN_{0.25} - C_{0.25} = CN_0 - C_0 \tag{4-31}$$

The new value on the surface, after the elapse of time $\Delta t$, $CN_0$, can be obtained in the same way as in equation (4-26).

$$CN_0 = \frac{1}{M}[2C_1 + (M - 2 - 2N) \cdot C_0 + 2N \cdot C_{ext}] \tag{4-32}$$

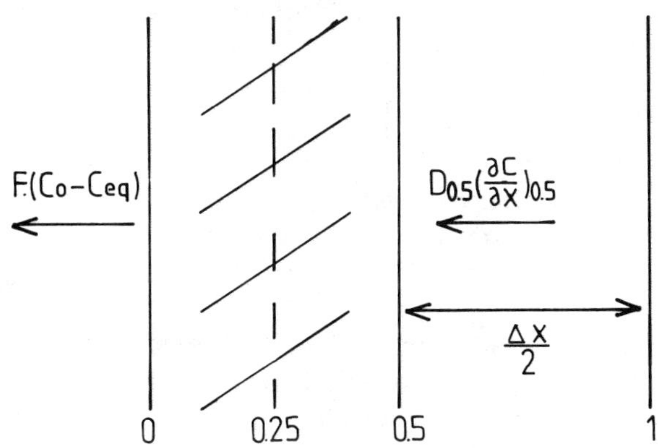


**Figure 4–4** Scheme of the slice next to the surface, with diffusion through the solid and evaporation from the surface. Slice of thickness of $\Delta x/2$.

***Remark 3. Thinner Slice and Approximation for*** $\mathbf{CN_{0.25} - C_{0.25}}$. Thinner slice of thickness $\Delta x/2$ near the surface is considered (Figure 4–4), and the following approximation for the mean concentration of this slice is made.

$$CN_{0.25} - C_{0.25} = \frac{1}{4}(CN_1 - C_1) + \frac{3}{4}(CN_0 - C_0) \tag{4-33}$$

The matter balance within this slice during the increment of time $\Delta t$, per unit area, becomes:

$$\left[D\cdot\frac{C_1 - C_0}{\Delta x} - F_0(C_0 - C_{ext})\right]\cdot\Delta t = \frac{3\Delta x}{8}(CN_0 - C_0) + \frac{\Delta x}{8}(CN_1 - C_1) \tag{4-34}$$

and then:

$$CN_0 + \frac{1}{3}(CN_1 - C_1) = \frac{8}{3M}\cdot\left[C_1 + \left(\frac{3M}{8} - 1 - N\right)\cdot C_0 + N\cdot C_{ext}\right] \tag{4-35}$$

with the dimensionless numbers $M$ and $N$ given in equations:

$$M = \frac{(\Delta x)^2}{\Delta t}\cdot\frac{1}{D} \tag{4-4}$$

$$N = \frac{F_0\cdot\Delta x}{D} \tag{4-25}$$

***Remark 4. Expression of the Rate of the Diffusing Substance.*** Equation (4-21) is written when the flux of the diffusing substance is in the same direction as the $x$-axis, and when the gradient of concentration is negative.

In the other case, when the flux of the diffusing substance is on the opposite direction of the $x$-axis, the gradient of concentration is $>0$, and equation (4-21) is then written:

$$+D\cdot\left(\frac{\partial C}{\partial x}\right) = F_o(C_o - C_{ext})$$

***Remark 5. The Concentration*** $\mathbf{C_{ext}}$. Generally the pressure of the vapor is used. The concentration $C_{ext}$ is the concentration of the diffusing substance on the surface of the solid which is at equilibrium with the vapor.

### 4-3-2 Diffusion-Evaporation on the Surface. Rate of Evaporation = Rate of Diffusion to the Surface. Constant Diffusivity.

The fundamental basis is that the rate of evaporation of diffusing substance is constantly equal to the rate of substance by diffusion through the solid near the surface, per unit area of the surface:

The following boundary condition

$$-D\cdot\left(\frac{\partial C}{\partial x}\right)_{\text{surface}} = F_o\cdot(C_o - C_{ext}) \tag{4-21}$$

is especially considered, by approximating the gradient of concentration on the surface by the chord slope between positions 0 and 1.

$$D\frac{C_1 - C_o}{\Delta x} = F_o(C_o - C_{ext}) \tag{4-36}$$

By putting the dimensionless number $N$,

$$N = \frac{F_0\cdot\Delta x}{D} \tag{4-25}$$

the boundary condition can be rewritten. The equation obtained expresses the concentration of substance on the surface, $C_0$, in terms of the concentration of this substance: at a distance $\Delta x$ from the surface in the solid, $C_1$, and in the external atmosphere, $C_{ext}$. All these concentrations are obtained at the same time.

$$C_0 = \frac{1}{N+1}C_1 + \frac{N}{N+1}C_{ext} \tag{4-37}$$

The concentration on the surface, $C_0$, is calculated by using the preceding equation, with the constant concentration $C_{ext}$, and concentration $C_1$, at the same time.

The new concentration at position 1, after the elapse of time $\Delta t$, $CN_1$, is given by:

$$CN_1 = \frac{1}{M}[C_0 + (M - 2)C_1 + C_2] \tag{4-5}$$

with the dimensionless number $M$:

$$M = \frac{(\Delta x)^2}{\Delta t}\cdot\frac{1}{D} \tag{4-4}$$

The scheme used for calculation is shown in Figure 4–4.

### 4-3-3 Diffusion-Evaporation on the Surface. Very High Rate of Evaporation. Constant Diffusivity.

Very often, and more often than permitted, the very simple assumption is made for the concentration of diffusing substance on the surface of the solid: The concentration of substance reaches the concentration at equilibrium as soon as the process of evaporation-diffusion starts.

The equation expressing the flux of substance evaporating out of the surface

$$-D\cdot\left(\frac{\partial C}{\partial x}\right)_{surface} = F_o\cdot(C_o - C_{ext}) \tag{4-21}$$

with the equality of concentration on the surface

$$C_o = C_{ext} \qquad t > 0 \tag{4-38}$$

shows evidently that the rate $F_0$ must be infinite.

The problem has thus the following initial and boundary conditions:

| | | | | | |
|---|---|---|---|---|---|
| Initial | $t = 0$ | $0 < x < L$ | $C = C_i$ | solid | (4-39) |
| | | | $C = C_{ext}$ | exterior | |
| Boundary | $t > 0$ | $x = 0$ | $C_o = C_{ext}$ | surface | (4-40) |
| | | $x = L$ | | | |

The solution of the equation of diffusion within the solid, with constant diffusivity

$$\frac{\partial C}{\partial t} = D\cdot\frac{\partial^2 C}{\partial x^2} \tag{4-41}$$

and the preceding initial and boundary conditions, is as follows. Within the solid

$$CN_n = \frac{1}{M}\left[C_{n+1} + (M - 2)C_n + C_{n-1}\right] \tag{4-6}$$

and on the plane next to the surface

$$CN_1 = \frac{1}{M}\left[C_2 + (M - 2)C_1 + C_{ext}\right] \tag{4-5}$$

### 4-3-4 Diffusion-Evaporation on the Surface. Concentration-Dependent Diffusivity. Matter Balance Done on the Last Slice of Thickness $\Delta x/2$.

In the case of concentration-dependent diffusivity, the matter balance done on the last slice of thickness $\Delta x/2$, is written as follows (per unit area):

$$\left[D_{0.5}\cdot\left(\frac{\partial C}{\partial x}\right)_{0.5} - F_0\cdot(C_0 - C_{ext})\right]\Delta t = [CN_{0.25} - C_{0.25}]\cdot\frac{\Delta x}{2} \tag{4-30}$$

By assuming that

$$CN_{0.25} - C_{0.25} = CN_0 - C_0 \tag{4-31}$$

and that the gradient of concentration at plane 0.5 is equal to the chord slope between planes 1 and 0, this equation becomes:

$$\left[D_{0.5}\frac{C_1 - C_0}{\Delta x} - F_0 \cdot (C_0 - C_{ext})\right]\Delta t = \frac{\Delta x}{2}[CN_0 - C_0] \qquad (4\text{-}42)$$

By putting the dimensionless numbers $M$ and $N$, at position 0.5 and at the previous time ($t = j \cdot \Delta t$),

$$M_{0.5} = \frac{(\Delta x)^2}{\Delta t} \cdot \frac{1}{D_{0.5}} \qquad (4\text{-}4)$$

$$N_{0.5} = \frac{F_0 \cdot \Delta x}{D_{0.5}} \qquad (4\text{-}25)$$

the equation expressing the new concentration (at time $t + \Delta t$, or $(j + 1)\Delta t$) on the surface, $CN_0$, in terms of the concentrations at the previous time, is obtained:

$$CN_0 = \frac{1}{M_{0.5}}[2C_1 + (M_{0.5} - 2N_{0.5} - 2) \cdot C_0 + 2N_{0.5} \cdot C_{ext}] \qquad (4\text{-}43)$$

Of course, the parameters $M_{0.5}$ and $N_{0.5}$ are concentration dependent. They can be calculated by using various ways, as shown previously (Chapter 2-2 Diffusion Within the Sheet with Concentration-Dependent Diffusivity).

Generally, high gradients of concentration are developed near the surface, and this is especially true at the beginning of the process. Thus, the simple assumption made in calculating the mean value of diffusivity

$$D_{0.5} = \frac{1}{2}(D_0 + D_1) \qquad (4\text{-}11)$$

cannot be used. It is better to calculate the diffusivity at position 0.5 by using the mean concentration at this plane.

The values of dimensionless numbers $M_{0.5}$ and $N_{0.5}$ are thus easily determined by using one of equations (1-16) expressing diffusivity in terms of concentration.

$$D_{0.5,j} = D_0 \cdot \exp\left[\frac{cte(C_{0,j} + C_{1,j})}{2}\right] \qquad (4\text{-}16)$$

$$D_{0.5,j} = D_0 \cdot \exp\left[cte - \frac{C_{0,j} + C_{1,j}}{2}\right] \qquad (4\text{-}16')$$

$$D_{0.5,j} = D_0 \cdot \exp\left[\frac{2 \cdot cte}{C_{0,j} + C_{1,j}}\right] \qquad (4\text{-}16'')$$

Of course, the new concentration $CN_n$ within the solid sheet is calculated by using the equations shown previously:

$$CN_n = C_n + J_{n-0.5} - J_{n+0.5} \tag{4-10'}$$

with

$$J_{n-0.5} = \frac{\Delta t}{(\Delta x)^2} \cdot D_{n-0.5} \cdot (C_{n-1} - C_n) \tag{4-9''}$$

and the equation expressing diffusivity in terms of concentration.

***Remark 1. Other Assumptions.*** If other assumptions are made for the thickness of the slice near the surface, and for the equality of $CN_{0.5} - C_{0.5}$, calculation is followed in the same way.

***Remark 2. Rate of Evaporation = Rate of Diffusion to the Surface. Concentration-Dependent Diffusivity.*** The case shown in Section 4-3-2, where the rate of evaporation of diffusing substance is constantly equal to the rate of substance through the solid near the surface, is considered with concentration-dependent diffusivity.

The equation expressing the boundary condition with $C_0$ being the concentration on the surface, becomes:

$$-D_0 \cdot \left(\frac{\partial C}{\partial x}\right)_{surface} = F_0 \cdot (C_0 - C_{ext}) \tag{4-21}$$

By assuming that the gradient of concentration on the surface is equal to the chord slope between positions 0 and 1 (Remark 3),

$$D_0 \cdot \frac{C_1 - C_0}{\Delta x} = F_0 \cdot (C_0 - C_{ext}) \tag{4-44}$$

With the dimensionless number,

$$N_0 = \frac{F_0 \cdot \Delta x}{D_0} \tag{4-25}$$

the equation expressing the concentration on the surface, $C_0$, in terms of the concentration at plane 1, $C_1$, and outside the sheet, $C_{ext}$, is obtained,

$$C_o = \frac{1}{N_o + 1} C_1 + \frac{N_o}{N_o + 1} C_{ext} \tag{4-45}$$

these concentrations being at the same time.

The new concentration at position 1, $CN_1$, is also given by:

$$CN_1 = \frac{1}{M} [C_0 + (M - 2) \cdot C_1 + C_2] \tag{4-5}$$

where the dimensionless number $M_1$ and diffusivity are constant.

$$M = \frac{(\Delta x)^2}{\Delta t} \cdot \frac{1}{D} \tag{4-4}$$

***Remark 3. Concentration Within the Sheet.*** Concentration within the sheet can be calculated by using either equation (4-10′):

$$CN_n = C_n + J_{n-0.5} - J_{n+0.5} \tag{4-10′}$$

with

$$J_{n-0.5} = \frac{\Delta t}{(\Delta x)^2} \cdot D_{n-0.5} \cdot (C_n - C_{n+1}) \tag{4-9″}$$

or equation (4-5),

$$CN_n = \frac{1}{M_n} [C_{n+1} + (M_n - 2) \cdot C_n + C_{n-1}] \tag{4-5}$$

with

$$M_n = \frac{(\Delta x)^2}{\Delta t} \cdot \frac{1}{D_n} \tag{4-4}$$

Equations (4-10′) and (4-9″) are obtained when the diffusivity is concentration dependent, while equations (4-5) and (4-4) are obtained with constant diffusivity or with diffusivity varying little with concentration.

# Chapter Five

# Numerical Analysis for Transient Transports in Cylinders

## 5-1 INTRODUCTION

The following cases are considered:

- solid cylinder of infinite length with radial transport.
- solid cylinder of finite length with radial and longitudinal transport.
- hollow cylinder (or tubing) of infinite length, with radial transport.
- hollow cylinder of finite length, with radial and longitudinal transport.

The equation of radial diffusion is:

$$\frac{\partial C}{\partial t} = \frac{1}{r} \cdot \frac{\partial}{\partial r}\left[r \cdot D \cdot \frac{\partial C}{\partial r}\right] \tag{5-1}$$

with concentration-dependent diffusivity.

## 5-2 SOLID CYLINDER OF INFINITE LENGTH. CONCENTRATION-DEPENDENT DIFFUSIVITY

### 5-2-1 Absorption of Liquid by the Cylinders

The circular cross section of the cylinder is considered, with radial diffusion. As shown in Figure 5–1, this circular cross section is divided into circular annuli with constant thickness $\Delta r$.

The matter balance is determined in the annulus centered on position $r$, during the increment of time $\Delta t$, by using Fick's first law.

Simple notation is used with integer $n$ such as:

$$r = n \cdot \Delta r \tag{5-2}$$

$$R = N \cdot \Delta r \tag{5-2'}$$

$R$ being the radius of cylinder.

$$\left[ -2\pi \left( r - \frac{\Delta r}{2} \right) \cdot D_{n-1/2} \cdot \left( \frac{\partial C}{\partial r} \right)_{n-1/2} + 2\pi \left( r + \frac{\Delta r}{2} \right) \cdot D_{n+1/2} \left( \frac{\partial C}{\partial r} \right)_{n+1/2} \right] \cdot \Delta t = 2\pi r \Delta r (CN_n - C_n) \tag{5-3}$$

The preceding equation can be rewritten:

$$\left[ \left( n - \frac{1}{2} \right) \Delta r \cdot D_{n-1/2} \frac{C_{n-1} - C_n}{\Delta r} - \left( n + \frac{1}{2} \right) \Delta r \cdot D_{n+1/2} \cdot \frac{C_n - C_{n+1}}{\Delta r} \right] \Delta t = n (\Delta r)^2 \cdot (CN_n - C_n) \tag{5-3'}$$

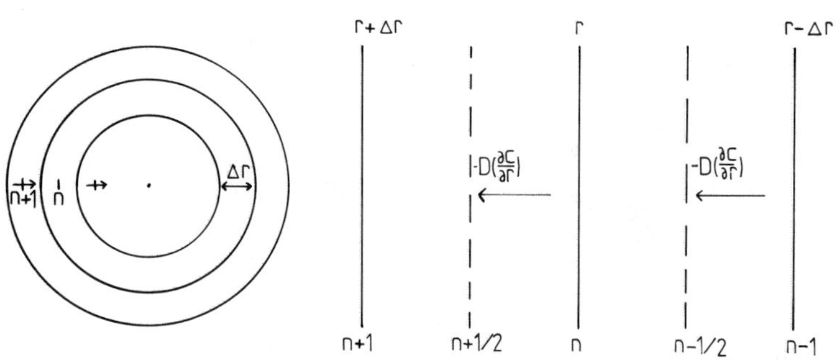


**Figure 5–1** Scheme of the circular cross section of the solid cylinder of infinite length, with radial diffusion within the solid.

By putting $I$ as the function of concentration and diffusivity,

$$I_{n-1/2} = \left(n - \frac{1}{2}\right) \cdot D_{n-1/2}(C_{n-1} - C_n) \tag{5-4}$$

the fundamental equation becomes:

$$CN_n = C_n + \frac{\Delta t}{n \cdot (\Delta r)^2} (I_{n-1/2} - I_{n+1/2}) \tag{5-5}$$

**Middle of cylinder: ($r = 0$ or $n = 0$).** As the preceding equation cannot be used for the center of cylinder ($r = 0$ or $n = 0$), the matter balance is calculated in the circle of radius $\Delta r/2$ (Figure 5–2).

$$-2\pi \cdot \frac{\Delta r}{2} \cdot D_{1/2} \cdot \left(\frac{\partial C}{\partial r}\right)_{1/2} \Delta t = \pi \left(\frac{\Delta r}{2}\right)^2 \cdot (CN_0 - C_0) \tag{5-6}$$

which becomes:

$$CN_0 = C_0 + \frac{8 \cdot \Delta t}{(\Delta r)^2} \cdot I_{1/2} \tag{5-7}$$

with the function $I$.

The value of diffusivity $D_{n-1/2}$ can be obtained from one of the following equations:

$$D_{n-1/2} = D_0 \cdot \exp\left[\frac{cte}{2} \cdot (C_{n-1} + C_n)\right] \tag{5-8}$$

$$D_{n-1/2} = D_0 \cdot \exp\left[cte - \frac{C_{n-1} + C_n}{2}\right] \tag{5-8'}$$

$$D_{n-1/2} = D_0 \cdot \exp\left(\frac{2cte}{C_{n-1} + C_n}\right) \tag{5-8''}$$

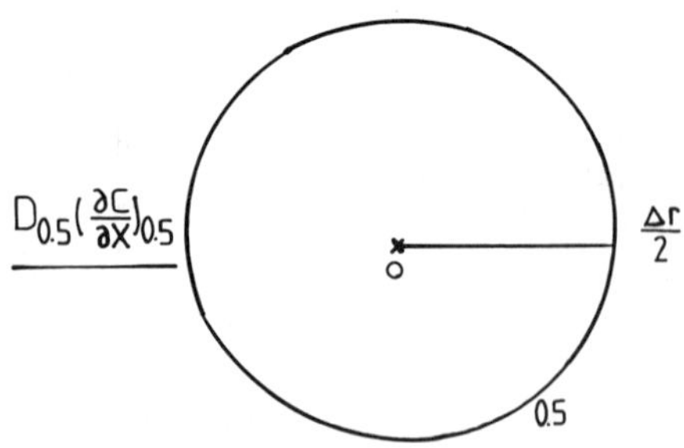


**Figure 5–2** Scheme of the circular cross section of the solid cylinder of infinite length, with radial diffusion next to the middle.

**Surface of cylinder.** Calculations are made in the case of constant concentration on the surface of the cylinder, this concentration being equal to the external atmosphere. This case occurs when the coefficient of matter transport on the surface is very high. Other cases are seen in Remarks.

The concentration at the surface is:

$$\text{Position } R \qquad C_N = C_R = C_{ext} \tag{5-9}$$

The fundamental equation (5-5) is rewritten as follows for the position $(R - \Delta r)$, or $N - 1$

$$CN_{N-1} = C_{N-1} + \frac{\Delta t}{(N-1)(\Delta r)^2} \cdot [I_{N-1.5} - I_{N-0.5}] \tag{5-10}$$

with

$$I_{N-0.5} = \left(N - \frac{1}{2}\right) \cdot D_{N-0.5}(C_{N-1} - C_{ext}) \tag{5-4}$$

### 5-2-2 Evaporation on the Surface. Concentration-Dependent Diffusivity

In the circular cross section of the cylinder drawn in Figure 5–3, the annulus near the surface is considered, with radial diffusion within the cylinder and evaporation on the surface. This annulus has a thickness of $\Delta r/2$ and is centered on position $N - 0.25$.

The matter balance is calculated in this annulus during the increment of time $\Delta t$, by using Fick's first law.

$$\left[-2\pi \cdot D_{N-0.5}\left(\frac{\partial C}{\partial r}\right)_{N-0.5} \cdot \left[\left(R - \frac{\Delta r}{2}\right) - F_0 \cdot C_R \cdot 2\pi R\right]\Delta t \right.$$
$$= 2\pi\left(R - \frac{\Delta r}{4}\right)\frac{\Delta r}{2}(CN_{N-0.25} - C_{N-0.25}) \tag{5-11}$$

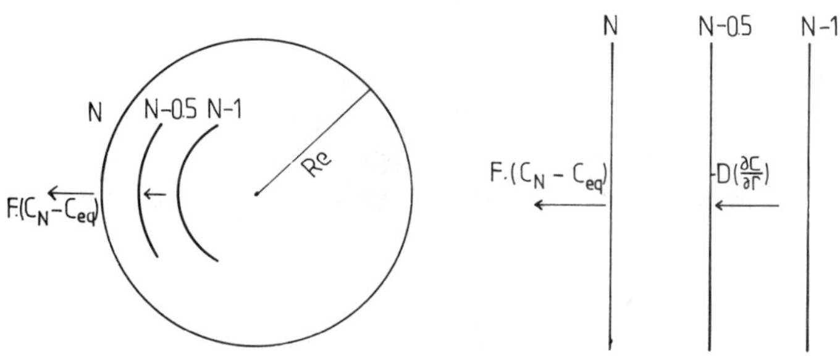


**Figure 5–3** Scheme of the circular cross section of the solid cylinder of infinite length, with radial diffusion within the solid and evaporation from the surface.

By assuming that

$$CN_{N-0.25} - C_{N-0.25} = CN_N - C_N \tag{5-12}$$

and that the gradient of concentration at position $N - 0.5$ is equal to the chord slope between positions $N$ and $N - 1$, it becomes:

$$CN_N = C_N + \left[2(C_{N-1} - C_N)\left(\frac{N - 0.5}{N - 0.25}\right) \cdot \frac{D_{N-0.5}}{(\Delta r)^2} - \frac{2F_0 \cdot N}{(N - 0.25)\Delta r} C_N\right]\Delta t \tag{5-13}$$

The new concentration on the surface after elapse of time $\Delta t$, $CN_N$, is obtained as a function of concentrations at the previous time and position $N$ (on the surface) and $N - 1$ and of the evaporation rate of pure diffusing substance $F_0$ (per unit area).

Within the cylinder, the concentration of diffusing substance is calculated as shown in the previous chapter. For instance, concentration at position $N - 1$ is determined by:

$$CN_{N-1} = C_{N-1} + \frac{\Delta t}{(N - 1)(\Delta r)^2} \cdot [I_{n-1.5} - I_{N-0.5}] \tag{5-10}$$

with

$$I_{N-0.5} = (N - 0.5) \cdot D_{N-0.5}(C_{N-1} - C_N) \tag{5-4}$$

the diffusivity $D_{N-0.5}$ being obtained by one of equations (5-8).

$$D_{N-0.5} = D_0 \cdot \exp\left[\frac{cte}{2}(C_N + C_{N-1})\right] \tag{5-8}$$

***Remark 1. Condensation of Vapor on the Surface.*** This case is similar to the previous one with evaporation, and the equations obtained for the evaporation-diffusion process can be used. In fact, the process of condensation or evaporation is defined by the values of the concentration at equilibrium and the concentration on the surface.

$$C_{eq} > C_{surface} \qquad \text{absorption, condensation} \tag{5-14}$$

$$C_{eq} < C_{surface} \qquad \text{evaporation} \tag{5-15}$$

## 5-3 SOLID CYLINDER OF FINITE LENGTH. RADIAL AND LONGITUDINAL TRANSPORT. CONCENTRATION-DEPENDENT DIFFUSIVITY.

The following two cases are considered:

1. When the concentration of diffusing substance on the surface is equal to the value at equilibrium as soon as the process starts. This is the case of a solid immersed into a liquid.
2. When the presaturated cylinder is exposed to air, enabling the liquid to evaporate. The evaporation rate on the surface is then considered, while the liquid diffuses through the solid.

### 5-3-1 SOLID CYLINDER OF FINITE LENGTH WITH CONCENTRATION-DEPENDENT DIFFUSIVITY.

The initial and boundary conditions are:

| | | | | | |
|---|---|---|---|---|---|
| Initial | $t = 0$ | $0 < r < R$ | $C = 0$ | The cylinder | (5-16) |
| | and | $0 < h < H$ | | is empty. | |
| Boundary | $t > 0$ | $r = R$ | $C_{eq}$ | radial surface | |
| | | $h = 0$ and | $C_{eq}$ | plane surface | (5-17) |
| | | $h = H$ | | | |

Various positions in the cylinder are considered successively by using the previous notation:

$$r = n \cdot \Delta r \tag{5-2}$$

$$R = N \cdot \Delta r \tag{5-2'}$$

**Within the solid.** The annulus of thicknesses $\Delta r$ and $\Delta h$, and of radius $r$ and position $h$ is drawn in Figure 5–4. The matter balance during the increment of time $\Delta t$ is obtained by considering the radial matter transport and the longitudinal transports.

$$\left[-D_{h-0.5,n} \cdot \left(\frac{\partial C}{\partial h}\right) + D_{h+0.5,n} \cdot \left(\frac{\partial C}{\partial h}\right)\right] S_{h-0.5,n} \cdot \Delta t + \left[-D_{h,n+0.5} \cdot \left(\frac{\partial C}{\partial r}\right) \cdot S_{h,n+0.5} + D_{h,n-0.5} \cdot \left(\frac{\partial C}{\partial r}\right) \cdot S_{h,n-0.5}\right] \cdot \Delta T = 2\pi n (\Delta r)^2 \cdot \Delta h \cdot [CN_{h,n} - C_{h,n}] \tag{5-18}$$

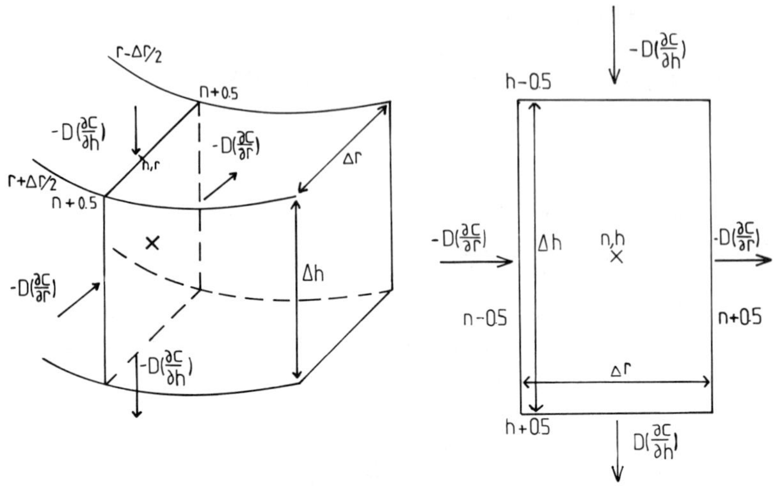


**Figure 5–4** Scheme of the annulus inside the solid cylinder of finite length, with radial and longitudinal diffusion.

By expressing the gradients of concentration in terms of the chord slope of concentration as previously done, and by expressing the values of the surfaces $S$ in various positions, the equation becomes:

$$\left[D_{h-0.5,n}\frac{C_{h-1,n}-C_{h,n}}{\Delta h}-D_{h+0.5,n}\frac{C_{h,n}-C_{h+1,n}}{\Delta h}\right]2\pi n(\Delta r)^2$$
$$+D_{h,n+0.5}\frac{C_{h,n+1}-C_{h,n}}{\Delta r}2\pi(n+0.5)\Delta r\cdot\Delta h$$
$$-D_{h,n-0.5}\frac{C_{h,n}-C_{h,n-1}}{\Delta r}2\pi(n-0.5)\Delta r\cdot\Delta h \tag{5-19}$$
$$=\frac{2\pi n(\Delta r)^2\cdot\Delta h}{\Delta t}[CN_{h,n}-C_{h,n}]$$

By putting the functions $I$ and $G$ as follows:

$$I_{h,n+0.5}=\left(n+\frac{1}{2}\right)\cdot D_{h,n+0.5}\cdot(C_{h,n+1}-C_{h,n}) \tag{5-4}$$

$$G_{h-0.5,n}=n\cdot D_{h-0.5,n}\cdot(C_{h-1,n}-C_{h,n}) \tag{5-4'}$$

the new concentration after elapse of time can be expressed in terms of the previous concentration, at the same place:

$$CN_{h,n} = C_{h,n} + \frac{\Delta t}{n(\Delta r)^2}[I_{h,n++0.5} - I_{h,n-0.5}] + \frac{\Delta t}{n(\Delta h)^2} \cdot [G_{h-0.5,n} - G_{h+0.5,n}] \tag{5-20}$$

**On the surface.** On the radial surface ($r = R$) and on the plane surface ($h = 0$, and $h = H$) of the cylinder, the concentrations of diffusing substance are, at any time:

$$C_{h,R} = C_{o,n} = C_{H,n} = C_{eq} \tag{5-21}$$

### 5-3-2 SOLID CYLINDER OF FINITE LENGTH WITH CONCENTRATION-DEPENDENT DIFFUSIVITY. DESORPTION OF LIQUID WITH EVAPORATION.

The initial and boundary conditions are:

| | | | | |
|---|---|---|---|---|
| Initial | $t = 0$ $0 < r < R$ $0 < h < H$ | $C = C_{in}$ | within the solid | (5-22) |
| Boundary | $t > 0$ $r = R$ | $F = F_o(C_s - C_{eq})$ | radial surface | |
| | $h = 0$ and $h = H$ | | plane surface | (5-23) |

Various places are considered in the cylinder of finite length: within the solid, the radial surface, and the plane surfaces.

**Within the solid.** The equation already found in the case of absorption can also be used in the case of desorption (Figure 5–4).

$$CN_{h,n} = C_{h,n} + \frac{\Delta t}{n\Delta r)^2}[I_{h,n+0.5} - I_{h,n-0.5}] + \frac{\Delta t}{n(\Delta h)^2} \cdot [G_{h-0.5,n} - G_{h+0.5,n}] \tag{5-20}$$

**Radial surface.** The annulus of thickness $\Delta r/2$ and $\Delta h$, on the radial surface, is considered (Figure 5–5). The matter balance is calculated during the increment of time $\Delta t$ in this annulus, by considering the radial transport

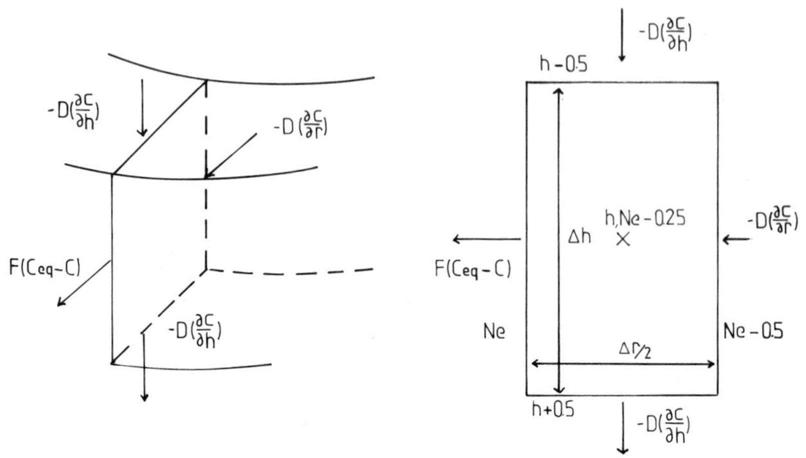


**Figure 5–5** Scheme of the annulus next to the external surface of the solid cylinder of finite length, with longitudinal and radial diffusion, and evaporation from the external surface.

(with the evaporation rate on the surface and diffusional transport) and the longitudinal transport by diffusion.

$$F\cdot(C_{eq} - C_{h,N})\cdot S_{h,N} - D_{h,N-0.5}\left(\frac{\partial C}{\partial r}\right)\cdot S_{h,N-0.5}$$

$$+ \left[-D_{h-0.5,N-0.25}\left(\frac{\partial C}{\partial h}\right) + D_{h+0.5,N-0.25}\left(\frac{\partial C}{\partial h}\right)\right]\cdot S \tag{5-24}$$

$$= 2\pi\,\frac{(N - 0.25)}{\Delta t}\,\frac{(\Delta r)^2}{2}\cdot\Delta h[CN_{h,N-0.25} - C_{h,N-0.25}]$$

By making the assumption

$$CN_{h,N-0.25} - C_{h,N-0.25} = CN_{h,N} - C_{h,N} \tag{5-12}$$

and replacing the gradient of concentration $\partial C/\partial r$ and $\partial C/\partial h$ by the chord slope of concentrations, the preceding equation becomes:

$$F(C_{\text{eq}} - C_{h,N})2\pi N\cdot\Delta r\cdot\Delta h - D_{h,N-0.5}\,\frac{C_{h,N} - C_{h,N-1}}{\Delta r}\,2\pi(N - 0.5)\Delta r\cdot\Delta h$$

$$+ \left[D_{h-0.5,N-0.25}\,\frac{C_{h-1,N-0.25} - C_{h,N-0.25}}{\Delta h} - D_{h+0.25,N-0.25}\,\frac{C_{h,N} - C_{h+1,N}}{\Delta h}\right] \tag{5-25}$$

$$2\pi(N - 0.25)\cdot\frac{(\Delta r)^2}{2} = \frac{2\pi(N - 0.25)}{\Delta t}\,\frac{(\Delta r)^2}{2}\,\Delta h[CN_{h,N} - C_{h,N}]$$

By using the functions $I$ and $G$

$$I_{h,N-0.5} = (N - 0.5)\cdot D_{h,N-0.5}(C_{h,N} - C_{h,N-1}) \tag{5-4}$$

$$G_{h-0.5,N-0.25} = (N - 0.25)\cdot D_{h-0.5,N-0.25}\cdot(C_{h-1,N-0.25} - C_{h,N-0.25}) \tag{5-4'}$$

the new concentration after the elapse of time $\Delta t$ on the radial surface is expressed in terms of the previous concentration at this place:

$$CN_{h,N} = C_{h,N} + \frac{N\cdot F_o}{N - 0.25}\cdot\frac{2\Delta t}{\Delta r}(C_{eq} - C_{h,N}) - \frac{N - 0.5}{N - 0.25} \cdot\frac{2\Delta t}{(\Delta r)^2} J_{h,N-0.5} + \frac{\Delta t}{(\Delta h)^2}\cdot[G_{h-0.5,N-0.25} - G_{h+0.5,N-0.25}] \tag{5-26}$$

**Plane surface.** The annulus of thickness $\Delta r$ and $\Delta h/2$, located on the plane surface, is shown in Figure 5–6. The matter balance is calculated during the increment of time $\Delta t$, by considering the radial transports (by diffusion) and the longitudinal transports (by diffusion and evaporation).

$$\left[F_o(C_{eq} - C_{o,n}) - D_{0.5,n}\left(\frac{\partial C}{\partial h}\right)\right] S + \left[-D_{0.25,n+0.5}\left(\frac{\partial C}{\partial r}\right) S_{n+0.5} + D_{0.25,n-0.5}\left(\frac{\partial C}{\partial r}\right) S_{0.25,n-0.5}\right] = \frac{2\pi n(\Delta r)^2\cdot\Delta h}{\Delta t\cdot 2}[CN_{0.25,n} - C_{0.25,n}] \tag{5-27}$$

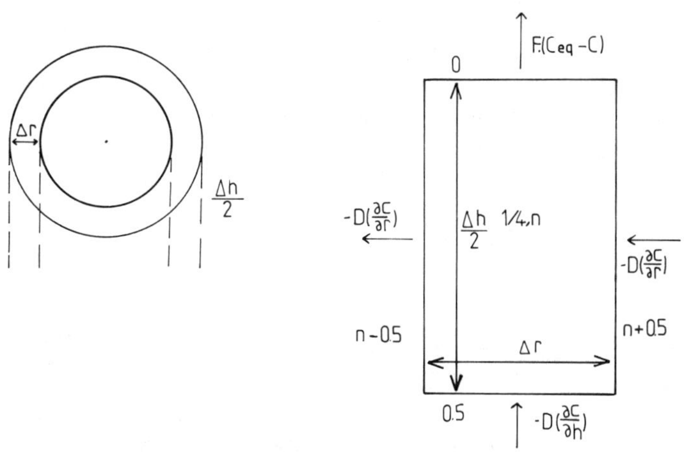


**Figure 5–6** Scheme of the annulus next to the plane surface of the cylinder of finite length, with radial and longitudinal diffusion, and evaporation from the plane surface.

By expressing the gradients of concentration, and the assumption of equation (5-13), this equation becomes:

$$\left[F_o(C_{eq} - C_{0,n}) - D_{0.5,n}\frac{C_{0,n} - C_{1,n}}{\Delta h}\right] 2\pi n(\Delta r)^2 + 2\pi\Delta r \cdot \frac{\Delta h}{2}\left[D_{0.25,n+0.5}\frac{C_{0.25,n+1} - C_{0.25,n}}{\Delta r}(n + 0.5) - D_{0.25,n-0.5}\frac{C_{0.25,n} - C_{0.25,n-1}}{\Delta r}(n - 0.5)\right] = \frac{2\pi n(\Delta r)^2\cdot\Delta h}{2\Delta t}[CN_{o,n} - C_{o,n}] \tag{5-28}$$

After introduction of the functions $I$ and $G$

$$I_{0.25,n+0.5} = (n + 0.5)\cdot D_{0.25,n+0.5}(C_{0.25,n+1} - C_{0.25,n}) \tag{5-4}$$

$$G_{0.5,n} = n\cdot D_{0.5,n}\cdot(C_{o,n} - C_{1,n}) \tag{5-4'}$$

the new concentration on the plane surface is thus expressed in terms of the previous concentration at the same place:

$$CN_{o,n} = C_{o,n} + F_o(C_{\text{eq}} - C_{o,n})\frac{2\Delta t}{\Delta h} - G_{0.5,n}\frac{2\Delta t}{n(\Delta h)^2} + \frac{\Delta t}{n(\Delta r)^2}(I_{0.25,n+0.5} - I_{0.25,n-0.5}) \tag{5-29}$$

***Remark 1. Absorption of Vapor on the Surface.*** The case of absorption of vapor can be studied by using the same equations as those obtained in the case of desorption of liquid with evaporation. In fact, the direction of the transport on the surface (absorption or evaporation) is determined by the value of the concentration of substance on the surface and of a concentration at equilibrium.

$$C_{eq} > C_{\text{surface}} \quad \text{absorption} \tag{5-14}$$

$$C_{eq} < C_{\text{surface}} \quad \text{evaporation} \tag{5-15}$$

## 5-4 HOLLOW CYLINDER OF INFINITE LENGTH (TUBING). CONCENTRATION-DEPENDENT DIFFUSIVITY.

This case is very important because long tubings are often used for the transport of liquids.

### 5-4-1 Hollow Cylinder of Infinite Length. Absorption of Liquid

As a long tubing is studied, only the radial transport of matter can be considered. The circular cross section of the tubing is drawn (Figure 5–7) with the internal and external radii $R_i$ and $R_e$.

The simple notation is used with the integer $n$ such as:

$$r = n \cdot \Delta r \tag{5-2}$$

$$R_i = N_i \cdot \Delta r \tag{5-2'}$$

$$R_e = N_e \cdot \Delta r \tag{5-2''}$$

$$N_i \leq n \leq N_e \quad \text{or} \quad R_i \leq r \leq R_e \tag{5-30}$$

The initial and boundary conditions are:

| | | | | | |
|---|---|---|---|---|---|
| Initial | $t = 0$ | $N_i \leq n \leq N_e$ | $C = 0$ | solid | (5.31) |
| Boundary | $t > 0$ | $n = N_i$ | $C_{Ni} = C_{eq}$ | internal surface | (5-32) |
| | | $n = N_e$ | $C_{Ne} = C_{eq}$ | external surface | |

**Within the solid.** The matter balance is calculated in the annulus of thicknesses $\Delta r$ and unit length, during the increment of time $\Delta t$:

$$\Delta t \left[ -D_{n-0.5} \cdot \left(\frac{\partial C}{\partial r}\right) \cdot S_{n-0.5} + D_{n+0.5} \cdot \left(\frac{\partial C}{\partial r}\right) \cdot S_{n+0.5} \right] = 2\pi n (\Delta r)^2 \cdot [CN_n - C_n] \tag{5-33}$$

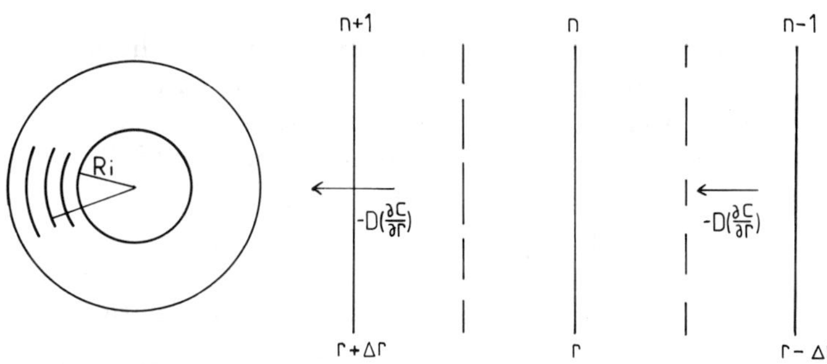


**Figure 5–7** Scheme of the circular cross section of the tubing of infinite length, with radial diffusion.

By expressing the gradients of concentration by the chord slope of concentration, as well as the surfaces, the equation becomes:

$$D_{n-0.5}\frac{C_{n-1} - C_n}{\Delta r}(n - 0.5)\Delta r - D_{n+0.5}\frac{C_n - C_{n+1}}{\Delta r}(n + 0.5)\Delta r = \frac{n\cdot(\Delta r)^2}{\Delta t}[CN_n - C_n] \tag{5-34}$$

By introducing the function $I$

$$I_{n-0.5} = \left(n - \frac{1}{2}\right)\cdot D_{n-0.5}(C_{n-1} - C_n) \tag{5-4}$$

the new concentration after the elapse of time is obtained as a function of the previous concentration, at the same place:

$$CN_n = C_n + \frac{\Delta t}{n\cdot(\Delta r)^2}[I_{n-0.5} - I_{n+0.5}] \tag{5-35}$$

This equation can be used throughout the tubing, with

$$N_i - 1 \leq n \leq N_e - 1 \tag{5-36}$$

**On the surfaces.** As the tubing is immersed in a liquid, with the liquid in contact with the internal and external surfaces, the concentration on both surfaces is:

$$C_{Ni} = C_{Ne} = C_{eq} \tag{5-32}$$

### 5-4-2 Hollow Cylinder of Infinite Length. Absorption of Liquid Through the Internal Surface. Evaporation of Substance from the External Surface.

The case of a tubing used for transportation of a liquid is of great concern. Generally, the internal surface is in contact with the liquid, while the external surface is exposed to the surrounding atmosphere (air). As the tubing is long, only the radial transport is considered.

The initial and boundary conditions are:

| | | | | | |
|---|---|---|---|---|---|
| Initial | $t = 0$ | $R_i < r < R_e$ | $C = 0$ | solid | (5-33) |
| Boundary | $t > 0$ | $r = R_i$ | $C = C^1_{eq}$ | internal surface | (5-34) |
| | | $r = R_e$ | $F_o(C_{R_e} - C^v_{eq})$ | external surface | |

The two values of concentration at equilibrium, $C^1_{eq}$ and $C^v_{eq}$ are quite different. $C^1_{eq}$ is very high, as the liquid enters the polymer. $C^v_{eq}$ is very low,

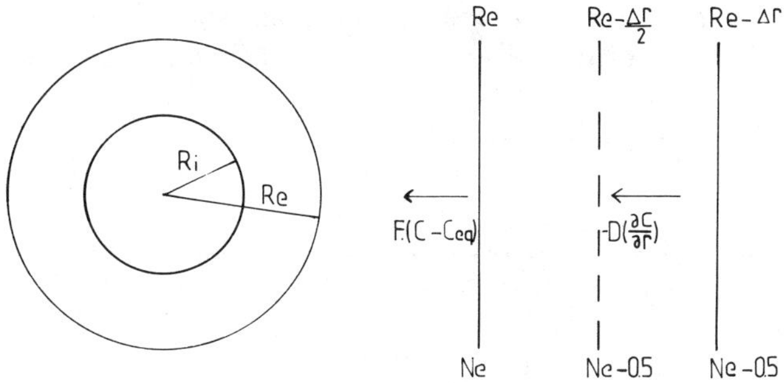


**Figure 5–8** Scheme of the circular cross section of the tubing of infinite length, with radial diffusion and evaporation from the external surface.

as this is the concentration of the liquid in the surrounding atmosphere. It can be zero, if air is in motion.

The circular cross section of the tubing is drawn (Figure 5–8). Three places are of interest: the external surface with the evaporation, inside the solid with diffusion, and the internal surface with concentration of liquid at equilibrium.

**External surface with evaporation-diffusion.** The matter balance is calculated in the annulus of thicknesses $\Delta r/2$ and unit length, located next to the external surface. The rate of evaporation and diffusion is considered.

$$\left[F(C_{eq}^{v} - C_{N_e})S_{N_e} - D_{N_e-0.5}\left(\frac{\partial C}{\partial r}\right) S_{N_e-0.5}\right]\Delta t = 2\pi\frac{(\Delta r)^2}{2}(N_e - 0.25)[CN_{N_e-0.25} - C_{N_e-0.25}] \tag{5-36}$$

By assuming that

$$CN_{N_e-0.25} - C_{N_e-0.25} = CN_{N_e} - C_{N_e}$$

and that the gradients of concentration are equal to the chord slope of concentration, the equation becomes:

$$\left[F(C_{eq}^{v} - C_{N_e})N_e(\Delta r) + (N_e - 0.5)\Delta r\cdot D_{N_e-0.5}\frac{C_{N_e-1} - C_{N_e}}{\Delta r}\right]\Delta t = (N_e - 0.25)\frac{(\Delta r)^2}{2}[CN_{N_e} - C_{N_e}] \tag{5-37}$$

By introducing the function $I$

$$I_{N_e-0.5} = (N_e - 0.5)\cdot D_{N_e-0.5}(C_{N_e-1} - C_{N_e}) \tag{5-4}$$

the new concentration on the external surface after the elapse of time $\Delta t$ is then expressed in terms of the previous concentrations in various places:

$$CN_{N_e} = C_{N_e} + \frac{2\Delta t}{(N_e - 0.25)\cdot(\Delta r)^2}[F(C^v_{eq} - C_{N_e})\cdot N_e\cdot\Delta r - I_{N_e-0.5}] \quad (5\text{-}38)$$

**Within the solid.** The substance is transported through the solid by radial diffusion, as in the case of absorption of liquid on both surfaces. The new concentration after the elapse of time $\Delta t$ at the position $n$ is then obtained by using the equation:

$$CN_n = C_n + \frac{\Delta t}{n(\Delta r)^2}[I_{n-0.5} - I_{n+0.5}]$$

**Internal surface in contact with the liquid.** The concentration of liquid on the internal surface is equal to the concentration at equilibrium, as soon as the process starts:

$$C_{N_i} = C^1_{eq}$$

***Remark 1. Transient and stationary conditions.*** The transport of liquid through the thickness of the hollow cylinder is obtained in transient

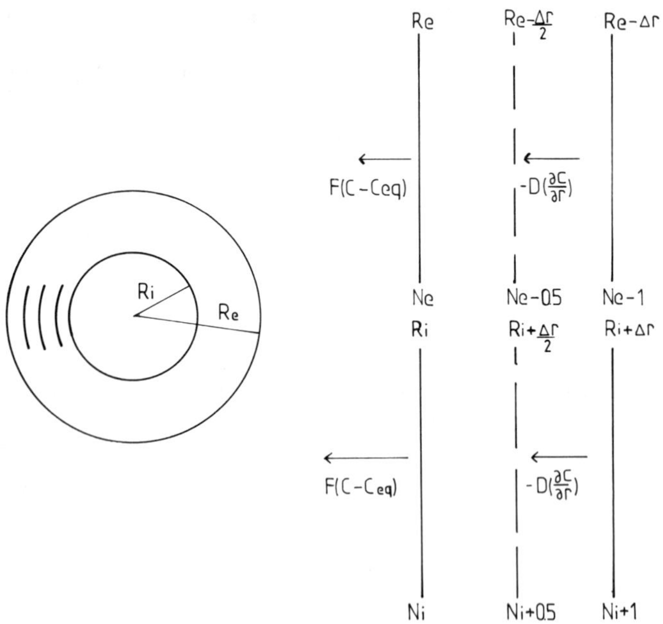


**Figure 5–9** Scheme of the circular cross section of the tubing of infinite length, with radial diffusion and evaporation from the internal and external surfaces.

conditions. After a definite time, the concentration of liquid at any place within the solid becomes constant, and the liquid diffuses under stationary conditions.

### 5-4-3 Evaporation of the Liquid Through the Internal and External Surfaces

This is the case of a tubing containing a definite amount of liquid, with a known profile of concentration of liquid, when the internal and external surfaces are exposed to atmosphere. The problem has already been studied in the previous chapters.

The circular cross section is considered, with the evaporation on the internal and external surfaces (Figure 5–9). The matter balance during the increment of time $\Delta t$ is calculated in various places.

**Within the solid.** The new concentration after the elapse of time $\Delta t$ is expressed in terms of the previous concentrations at various places: $n$, $n - 1$, and $n + 1$.

$$CN_n = C_n + \frac{\Delta t}{n(\Delta r)^2}[I_{n-0.5} - I_{n+0.5}] \tag{5-35}$$

**On the external surface.** The calculation is done in the annulus of thickness $\Delta r/2$ next to the external surface:

$$CN_{N_e} = C_{N_e} + \frac{2\Delta t}{(N_e - 0.25)(\Delta r)^2}[F(C_{eq} - C_{N_e})N_e \cdot \Delta r - I_{N_e - 0.5}] \tag{5-38}$$

**On the internal surface.** The calculation is done in the annulus of thickness $\Delta r/2$ next to the internal surface:

$$CN_{N_i} = C_{N_i} + \frac{2\,\Delta t}{(N_i + 0.25)(\Delta r)^2} \cdot [F(C_{eq} - C_{N_i})N_i \cdot \Delta r - I_{N_i + 0.5}] \tag{5-39}$$

***Remark 1. Amount of Matter Located in the Solid.*** The amount of matter located in the solid can easily be obtained at any time by integrating the concentration of matter with respect to space.

***Remark 2. Kinetics of the Release of Matter.*** The amount of matter releasing the solid can be calculated by using Fick's first law on the surface through which the matter is released. Another expression can be used, by integrating the rate of evaporation with respect to time.

$$M_{evaporated} = \sum_{t=0}^{t} F_o(C_{eq} - C_{surface}) \cdot \Delta t \cdot S \tag{5-40}$$

# Chapter Six

# Numerical Analysis in One-Dimensional Transient Transport in a Sphere Concentration-Dependent Diffusivity

## 6-1 INTRODUCTION

The diffusivity of diffusing substance is concentration dependent. The following cases are considered:

- Solid sphere with radial transport
- Hollow sphere with radial transport
- Concentric spheres of various materials with radial transport

The equation of radial diffusion is:

$$\frac{\partial C}{\partial r} = \frac{1}{r^2} \cdot \frac{\partial}{\partial r} \left[ D \cdot r^2 \frac{\partial C}{\partial r} \right] \tag{6-1}$$

where $C$ is concentration at time $t$ and at a distance $r$ from the center of the sphere.

## 6-2 CASE OF A SOLID SPHERE. CONCENTRATION-DEPENDENT DIFFUSIVITY

### 6-2-1 Radial Diffusion in Solid Sphere. Constant Concentration on the Surface.

**Within the sphere ($r,n$).** The circular cross section of the sphere is considered with radial diffusion. This cross section is divided into concentric annuli with constant thickness $\Delta r$ (corresponding with concentric spheres) (Figure 6–1).

The matter balance is calculated in the annulus of radius $r$ located between $r - \Delta r/2$ and $r + \Delta r/2$, during the increment of time $\Delta t$, by using Fick's first law:

$$\left[-4\pi\left(r - \frac{\Delta r}{2}\right)^2 \cdot D_{r-\Delta r/2}\left(\frac{\partial C}{\partial r}\right)_{r-\Delta r/2} + 4\pi\left(r + \frac{\Delta r}{2}\right)^2 \cdot D_{r+\Delta r/2}\left(\frac{\partial C}{\partial r}\right)_{r+\Delta r/2}\right]\Delta t = 4\pi\cdot r^2\,\Delta r(CN_r - C_r) \tag{6-2}$$

By using the following notation:

$$r = n\cdot\Delta r \tag{6-3}$$

$$R = N\cdot\Delta r \tag{6-3'}$$

with $n$ and $N$ being integers and $R$ the radius of the sphere, the equation of matter transport becomes:

$$\left[(\Delta r)^2\cdot\left(n - \frac{1}{2}\right)^2 \cdot D_{n-1/2}\cdot\frac{C_{n-1} - C_n}{\Delta r} - (\Delta r)^2\cdot\left(n + \frac{1}{2}\right)^2 \cdot D_{n+1/2}\frac{C_n - C_{n+1}}{\Delta r}\right]\cdot\Delta t = \pi^2\cdot(\Delta r)^3\cdot[CN_n - C_n]$$

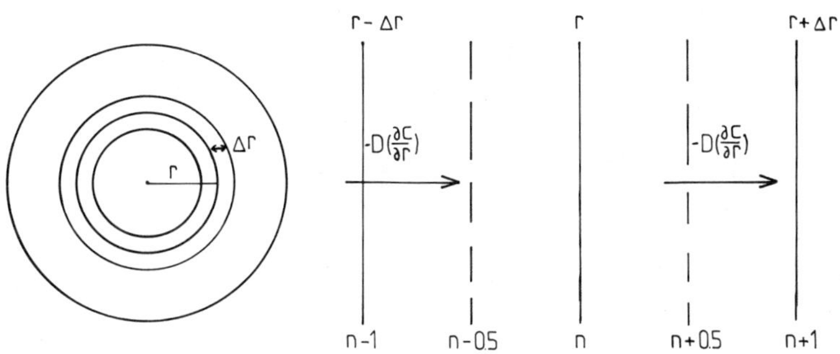


**Figure 6–1** Scheme of the circular cross section of the sphere, with diffusion of the substance through the solid.

By putting $J$ as a function of concentration and diffusivity,

$$J_{n-1/2} = \left(n - \frac{1}{2}\right)^2 \cdot D_{n-1/2} \cdot (C_{n-1} - C_n) \tag{6-5}$$

the new concentration $CN_n$ after elapse of time $\Delta t$ in position $n$ is obtained in terms of the concentration at the same position and previous time:

$$CN_n = C_n + \frac{\Delta t}{n^2 \cdot (\Delta r)^2} \cdot (J_{n-1/2} - J_{n+1/2}) \tag{6-6}$$

The diffusivity can be expressed in terms of concentration, for example, by one of the following equations:

$$D_n = D_0 \cdot \exp(cte\ C_n) \tag{6-7}$$

$$D_n = D_0 \cdot \exp \cdot cte(C - C_n) \tag{6-7'}$$

$$D_n = D_0 \cdot \exp \cdot \frac{cte}{C_n} \tag{6-7''}$$

**Center of the sphere ($r = 0$, or $n = 0$).** The preceding equation cannot be used for the center of the sphere. The matter balance is done for the sphere of radius $\Delta r/2$ as shown in Figure 6–2.

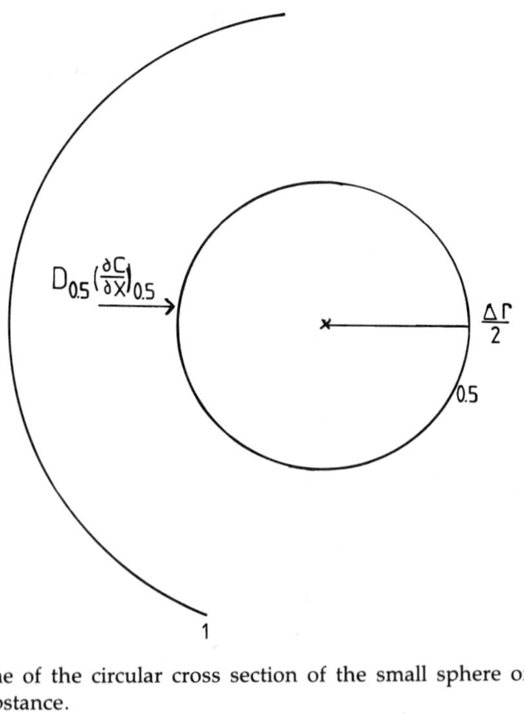


**Figure 6–2** Scheme of the circular cross section of the small sphere of radius $\Delta r/2$, with diffusion of the substance.

$$-4\pi\left(\frac{\Delta r}{2}\right)^2\cdot D_{\Delta r/2}\cdot\left(\frac{\partial C}{\partial r}\right)_{\Delta r/2}\cdot\Delta t = \frac{4}{3}\pi\left(\frac{\Delta r}{2}\right)^3\cdot(CN_0 - C_0) \qquad (6\text{-}8)$$

or

$$D_{1/2}\cdot\frac{C_0 - C_1}{\Delta r}\Delta t = \frac{1}{6}\cdot\Delta r(CN_0 - C_0) \qquad (6\text{-}9)$$

By putting the function $J$ as earlier

$$J_{1/2} = \left(\frac{1}{2}\right)^2\cdot D_{1/2}\cdot(C_0 - C_1)$$

the new concentration after elapse of time $\Delta t$ in the center of the sphere is:

$$CN_0 = C_0 + 24\cdot\frac{\Delta t}{(\Delta r)^2}\cdot J_{1/2} \qquad (6\text{-}10)$$

**Surface of the sphere (*R*, or *N*).** A simple condition is obtained for the concentration on the surface of the sphere:

$$CN_N = C_N = C_{eq} \qquad (6\text{-}11)$$

expressing that equilibrium is attained on the surface as soon as the process starts. This condition can be assumed in the following cases:

- The evaporation rate of diffusing substance is very high.
- The sphere is immersed into a well-stirred fluid in high quantity.

**Amount of liquid remaining in the sample.** The amount of diffusing substance remaining in the solid sphere at time $t$ is obtained by integrating the concentration of substance with respect to space within the sphere:

$$M_t = \int_o^R 4\pi r^2\cdot C(r,t)\cdot dr \qquad (6\text{-}12)$$

By using the finite differences, this expression is rewritten:

$$M_t = 4\pi(\Delta r)^3\cdot\left[\frac{C_0}{24} + \sum_{n=1}^{N-2} n^2\cdot C_n + \frac{9}{8}(N - 1)^2\cdot C_{N-1} + \frac{3}{8}N^2\cdot C_n\right] \qquad (6\text{-}13)$$

The contributions to the amount of substance remaining in the sphere are seen: the small central sphere of radius $\Delta r/2$, the part of the solid located between the spheres of radius $\Delta r/2$ and $R - \Delta r/2$, and the last part between the spheres of radius $R - \Delta r/2$ and $R$.

### 6-2-2 Evaporation-Diffusion in Solid Sphere. Concentration-Dependent Diffusivity.

**Concentration on the surface.** The diffusing substance is transported within the solid sphere by diffusion and out of the sphere by evaporation. The evaporation rate is proportional to the concentration of substance on the surface of the sphere. The matter balance is calculated in the spherical part of the sphere, of thickness $\Delta r/2$, located near the surface (Figure 6–3).

$$\left[-4\pi\left(R-\frac{\Delta r}{2}\right)^2\cdot D_{r-\Delta r/2}\left(\frac{\partial C}{\partial r}\right)_{R-\Delta r/2} - 4\pi R^2\cdot F_0\cdot(C_R - C_{ext})\right]\cdot\Delta t$$
$$= 4\pi\cdot\left(R-\frac{\Delta r}{4}\right)^2\left(\frac{\Delta r}{2}\right)\cdot[CN_{N-1/4} - C_{N-1/4}] \qquad (6\text{-}14)$$

By assuming that

$$CN_{N-1/4} - C_{N-1/4} = CN_N - C_N \qquad (6\text{-}15)$$

and using the preceding notations

$$R = N\cdot\Delta r \qquad (6\text{-}3')$$

$$r = n\cdot\Delta r \qquad (6\text{-}3)$$

the matter balance becomes:

$$\left[(\Delta r)^2\cdot\left(N-\frac{1}{2}\right)^2\cdot D_{N-1/2}\cdot\frac{C_{N-1}-C_N}{\Delta r} - (\Delta r)^2\cdot N^2\cdot F_o\cdot(C_N - C_{ex})\right.$$
$$\left[\cdot\Delta t = (\Delta r)^2\cdot\frac{\Delta r}{2}\cdot\left(N-\frac{1}{4}\right)^2\cdot[CN_N - C_N] \qquad (6\text{-}16)\right.$$

The new concentration after the elapse of time $\Delta t$ on the surface, $CN_N$, is then obtained in terms of the previous concentrations on the surface, $C_{Ni}$ and at position $C_{N-1}$:

$$CN_N = C_N + \frac{2\cdot\Delta t}{\left(N-\frac{1}{4}\right)^2(\Delta r)^2}J_{N-1/2} - \frac{2N^2\cdot\Delta t}{\left(N-\frac{1}{4}\right)^2\cdot\Delta r}F_0C_N \qquad (6\text{-}17)$$

with the function $J$:

$$J_{N-1/2} = \left(N-\frac{1}{2}\right)^2\cdot D_{N-1/2}\cdot(C_{N-1} - C_N) \qquad (6\text{-}5'')$$

**Concentration in the sphere.** The equations in Section 6-2-1 can be used for calculating the concentration of diffusing substance within the sphere.

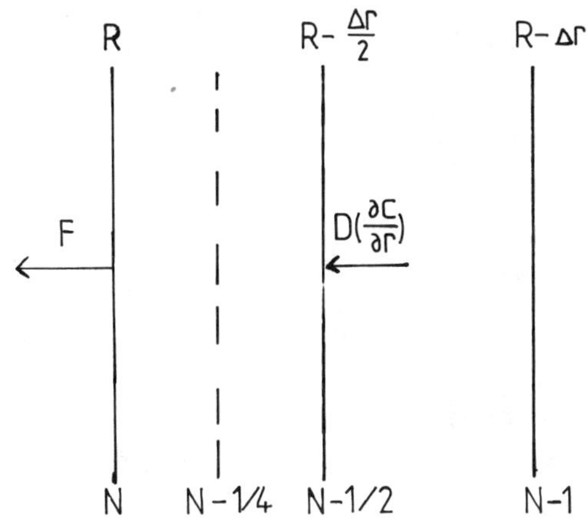


**Figure 6–3** Scheme of the circular cross section of the sphere next to the surface, with diffusion through the solid and evaporation from the surface.

**Amount of substance evaporating.** The amount of substance evaporating in time $t$ can be obtained by integrating the rate of evaporation with respect to time:

$$M_t = 4\pi R^2 \cdot \sum_{j=0}^{j} F_0 \cdot (C_N - C_{ext}) \cdot \Delta t \tag{6-18}$$

with the integer $j$:

$$t = j \cdot \Delta t \tag{6-19}$$

### 6-2-3 Radial Diffusion in Solid Sphere with Coefficient of Matter Transport on the Surface

**Concentration of substance on the surface.** The concentric spheres of radii, $R$, $R - \Delta r/2$ and $R - \Delta r$ are considered in Figure 6–3. The matter balance within the concentric spheres of radii $R$ and $R - \Delta r/2$ is calculated, during the increment of time $\Delta t$, by considering diffusion through the second surface $(R - \Delta r/2)$ and the coefficient of matter transport on the external surface $(R)$:

$$\left[-4\pi\left(R - \frac{\Delta r}{2}\right)^2 \cdot D_{R-\Delta r/2} \cdot \left(\frac{\partial C}{\partial r}\right)_{R-\Delta r/2} - 4\pi R^2 \cdot h \cdot (C_R - C_{ext})\right]\Delta t$$
$$= 4\pi\left(R - \frac{\Delta r}{4}\right)^2 \cdot \frac{\Delta r}{2} \cdot [CN_{R-\Delta r/4} - C_{R-\Delta r/4}] \tag{6-20}$$

By making the assumptions

$$CN_{R-\Delta r/4} - C_{R-\Delta r/4} = CN_R - C_R \tag{6-15}$$

and the previous indexation, the matter balance becomes:

$$\left[D_{N-1/2}\cdot \frac{C_{N-1} - C_N}{\Delta r}\left(N - \frac{1}{2}\right)^2 - N^2\cdot h(C_N - C_{ext})\right]\cdot\Delta t = \left(N - \frac{1}{4}\right)^2 \cdot \frac{\Delta r}{2}\,[CN_N - C_N] \tag{6-21}$$

The new concentration on the surface after the elapse of time $\Delta t$, $CN_N$, is thus obtained as a function of concentrations at the previous time on the surface and position $N - 1$:

$$CN_N = C_N + \frac{2\cdot\Delta t}{\left(N - \frac{1}{4}\right)^2\cdot(\Delta r)^2} J_{N-1/2} - \frac{N^2\cdot\Delta t\cdot}{\left(N - \frac{1}{4}\right)^2}\frac{2h}{\Delta r}(C_N - C_{ext}) \tag{6-22}$$

with the previous coefficient $J$.

$$J_{N-1/2} = \left(N - \frac{1}{2}\right)^2 \cdot D_{N-1/2}\cdot(C_{N-1} - C_N) \tag{6-5''}$$

**Concentration within the sphere.** Within the sphere, the equations obtained in Section 6-2-1 can be used for calculating the concentration.

**Amount of substance transported.** The amount of substance transported in time $t$ can be calculated by integrating the rate of matter transported with respect to time:

$$M_t = 4\pi R^2\cdot \sum_{j=0}^{j} h(C_N - C_{ext})\cdot\Delta t \tag{6-23}$$

***Remark 1. Amount of Substance Remaining in the Sphere.*** The total amount of diffusing substance initially located in the sphere ($\frac{4}{3}\pi R^3 C_i$) is equal to the sum of the amount of substance remaining in the sphere and the amount of substance transported. It is possible to obtain one of these two values:

- by integrating concentration with respect to space, for the amount of substance remaining in the solid.
- by integrating the rate of substance transported on the surface with respect to time, for the amount of substance transported.

***Remark 2. Frame of Reference.*** It often occurs that the solid swells and the sphere radius increases as the diffusing substance enters the solid, or shrinks as the radius decreases when the substance leaves the solid. A frame of reference is generally taken fixed with respect to the solid. This is the initial sphere in the absence of diffusing substance. The radius of the sphere is then constant and equal to the unswollen sphere.

***Remark 3. Absorption and Desorption of Substance.*** The equations shown in the previous chapters can be used not only for desorption of diffusing substance out of the sphere, but also for absorption of this substance. According to the relative values of concentrations on the surface and outside the sphere,

$$\text{absorption when} \qquad C_R < C_{ext} \tag{6-24}$$

$$\text{desorption} \qquad C_R > C_{ext} \tag{6-24'}$$

## 6-3 CASE OF A HOLLOW SPHERE. CONCENTRATION-DEPENDENT DIFFUSIVITY

The hollow sphere is located between the two spheres, the internal sphere of radius $R_i$ and the external of radius $R_e$. The following cases are considered:

- The concentration is kept constant on the two surfaces.
- The concentration is constant on the internal surface, and the substance evaporates from the external surface.

### 6-3-1 Concentration Constant on the Two Surfaces

The initial and boundary conditions are as follows:

| | | | | | |
|---|---|---|---|---|---|
| Initial | $t = 0$ | $R_i < r < R_e$ | $C$ | hollow sphere | (6-25) |
| Boundary | $t > 0$ | $r = R_i$ | $C = C_i$ | internal surface | (6-26) |
| | | $r = R_e$ | $C = C_e$ | external surface | |

The problem has already been solved previously with the solid sphere. The equation obtained can be used, with the condition for radius:

$$R_i < r < R_e \tag{6-27}$$

or

$$N_i < n < N_e \tag{6-27'}$$

with

$$r = n \cdot \Delta r \tag{6-3}$$

The amount of substance which enters (for $r = R_i$) and leaves (for $r = R_e$) the hollow sphere, in time $t$, can be obtained by integrating Fick's first law with respect to time. The gradient of concentration is taken near the internal or the external surface of the solid.

$$M_t = 4\pi R_i^2 \cdot \sum_{j=0}^{j} D_{N_i} \cdot \frac{C_{N_i} - C_{N_i+1}}{\Delta r} \Delta t \qquad \text{enters} \tag{6-28}$$

$$M_t = 4\pi R_e^2 \cdot \sum_{j=0}^{j} D_{N_e} \frac{C_{N_e-1} - C_e}{\Delta r} \Delta t \qquad \text{leaves} \tag{6-29}$$

### 6-3-2 Concentration Constant on Internal Surface. Evaporation on External Surface.

The initial and boundary conditions are:

| | | | | | |
|---|---|---|---|---|---|
| Initial | $t = 0$ | $R_i < r < R_e$ | $C$ | hollow sphere | (6-30) |
| Boundary | $t > 0$ | $r = R_i$ | $C = C_i$ | internal surface | (6-31) |
| | | $r = R_e$ | $F_0(C_e - C_{ext})$ | external surface | |

The evaporation rate on the external surface is assumed to be equal to the rate of substance diffusing to this surface.

$$-D_e \cdot \left(\frac{\partial C}{\partial r}\right)_e = F_0 \cdot (C_e - C_{ext}) \tag{6-32}$$

The equations obtained in the case of a solid sphere can be used for calculating the concentration of diffusing substance in the hollow sphere and on the external surface with the conditions:

$$R_i < r < R_e \tag{6-27}$$

$$N_i < n < N_e \tag{6-27$'$}$$

The amount of substance which evaporates the sphere in time $t$ is obtained by integrating the rate of evaporation with respect to time.

$$M_t = 4\pi R_e^2 \cdot \sum_{j=0}^{j} F_0(C_{Ne} - C_{ext}) \cdot \Delta t \tag{6-18}$$

The amount of substance which enters the hollow sphere in time $t$ is obtained by integrating Fick's first law with respect to time, near the internal surface.

$$M_t = 4\pi R_i^2 \cdot \sum_{j=0}^{j} D_{Ni} \cdot \frac{C_{Ni} - C_{Ni+1}}{\Delta r} \Delta t \tag{6-28$'$}$$

## Chapter Seven

# Numerical Analysis in Three-Dimensional Transient Transport: Cube, Parallelepipede, Isotropic, Anisotropic Media

### 7-1 INTRODUCTION

**Isotropic medium.** The element of volume in the form of a rectangular parallelepipede is considered in Figure 7–1. The sides are parallel to the axes of coordinates, with lengths of $dx$, $dy$, and $dz$, and the center of the volume is 0.

As the medium is isotropic, the direction of flow of diffusing substance is normal to the surface of constant concentration.

The rate at which diffusing substance enters the element of volume through the face *abcd* in the plane of abscissa $x$ is:

$$F_x = -D \cdot \frac{\partial C}{\partial x} \cdot dy \cdot dz \tag{7-1}$$

and the rate of transport through the face $a'b'c'd'$ of abscissa $x + dx$ is:

$$F_{x+dx} \tag{7-1'}$$

The contribution to the rate of increase of diffusing substance in the element of volume from these two faces is thus given by:

$$\frac{\partial C}{\partial t} \cdot dx \cdot dy \cdot dz = [F_x - F_{x+dx}]\, dy \cdot dz \tag{7-2}$$

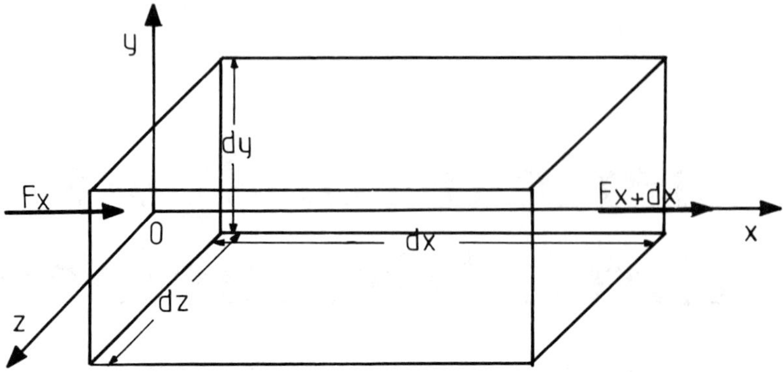


**Figure 7–1** Scheme of diffusion through a parallelepipede, with diffusion along the *x*-axis.

or

$$\frac{\partial C}{\partial t} = -\frac{\partial F}{\partial x} \tag{7-3}$$

By using Fick's first law, this relation becomes, with concentration-dependent diffusivity:

$$\frac{\partial C}{\partial t} = \frac{\partial}{\partial x}\left(D\cdot\frac{\partial C}{\partial x}\right) \tag{7-4}$$

By considering the rate of transfer through the other faces, the rate at which the amount of diffusing substance in the element increases is given by Fick's second law.

$$\frac{\partial C}{\partial t} = \frac{\partial}{\partial x}\left(D\cdot\frac{\partial C}{\partial x}\right) + \frac{\partial}{\partial y}\left(D\cdot\frac{\partial C}{\partial y}\right) + \frac{\partial}{\partial z}\left(D\cdot\frac{\partial C}{\partial z}\right) \tag{7-5}$$

In the case of constant diffusivity, this equation reduces to (Fick's second law):

$$\frac{\partial C}{\partial t} = D\cdot\left[\frac{\partial^2 C}{\partial x^2} + \frac{\partial^2 C}{\partial y^2} + \frac{\partial^2 C}{\partial z^2}\right] \tag{7-5'}$$

**Anisotropic media.** In anisotropic media, diffusion properties are not the same in different directions. The direction of flow of diffusing substance at any point is not normal to the surface of constant concentration through the point.

The rate at which the substance enters the volume through the face *abcd* becomes:

$$-F_x = \left[D_{11}\cdot\frac{\partial C}{\partial x} + D_{12}\cdot\frac{\partial C}{\partial y} + D_{13}\cdot\frac{\partial C}{\partial z}\right] dy\cdot dz \tag{7-6}$$

$D_{12}\partial C/\partial y$ being the contribution to the rate of transport in the $x$-direction due the gradient of concentration of substance in the $y$-direction.

Fick's second law is then:

$$\frac{\partial C}{\partial t} = D_{11}\cdot\frac{\partial^2 C}{\partial x^2} + D_{22}\cdot\frac{\partial^2 C}{\partial y^2} + D_{33}\cdot\frac{\partial^2 C}{\partial z^2} + (D_{12} + D_{21})\cdot\frac{\partial^2 C}{\partial x\cdot\partial y} + (D_{23} + D_{32})\cdot\frac{\partial^2 C}{\partial y\cdot\partial z} + (D_{31} + D_{13})\cdot\frac{\partial^2 C}{\partial x\cdot\partial y} \tag{7-7}$$

when the diffusivities are constant.

A transformation of this equation is obtained by considering the principal axes of diffusion $x$, $y$, $z$, and $D_1$, $D_2$, and $D_3$ as the principal diffusivities.

$$\frac{\partial C}{\partial t} = D_1\cdot\frac{\partial^2 C}{\partial x^2} + D_2\cdot\frac{\partial^2 C}{\partial y^2} + D_3\cdot\frac{\partial^2 C}{\partial z^2} \tag{7-8}$$

or with concentration-dependent diffusivities:

$$\frac{\partial C}{\partial t} = \frac{\partial}{\partial x}\left(D_1\cdot\frac{\partial C}{\partial x}\right) + \frac{\partial}{\partial y}\left(D_2\cdot\frac{\partial C}{\partial y}\right) + \frac{\partial}{\partial z}\left(D_3\cdot\frac{\partial C}{\partial z}\right) \tag{7-8$'$}$$

In the case of wood furniture and wood pieces (beams, boards), three main directions exist: longitudinal, radial, and tangential.

These main directions are easy to see on the tree which is cylindrical in shape: the longitudinal direction which is vertical, the radial direction which passes through the center of the circular cross section, and the tangential direction which is perpendicular to the radial direction in the circular cross section (Figure 7–2). The beams and boards are cut from the tree by keeping their longer direction along the longitudinal direction.

The preceding equation can be written as follows:

$$\frac{\partial C}{\partial t} = \frac{\partial}{\partial T}\left(D_T\cdot\frac{\partial C}{\partial T}\right) + \frac{\partial}{\partial R}\left(D_R\cdot\frac{\partial C}{\partial R}\right) + \frac{\partial}{\partial L}\left(D_L\cdot\frac{\partial C}{\partial L}\right) \tag{7-9}$$

The main directions are:

- longitudinal along $L$-axis, and longitudinal diffusivity $D_L$.
- radial along $R$-axis, and radial diffusivity $D_R$.
- tangential along $T$-axis, and tangential diffusivity $D_T$.

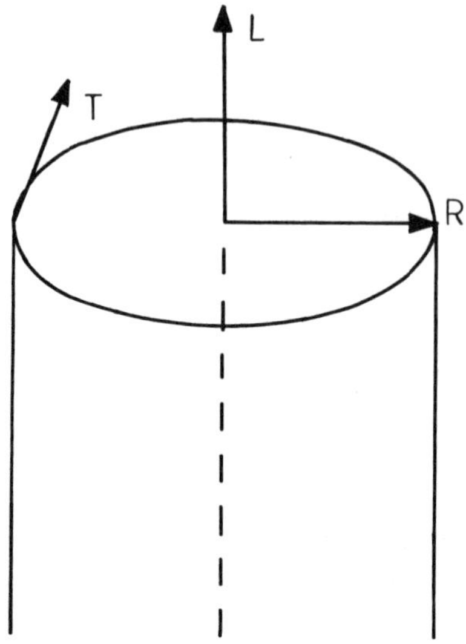


**Figure 7–2** Scheme of a tree with the three main directions of transport: longitudinal *L*, radial *R*, and tangential *T*.

These main diffusivities are quite different from each other, and they vary with the moisture concentration in a different way (as shown in Chapter 13). Two cases are of interest:

- a plane sheet, cut in such a way that the moisture transfer follows a main direction.
- a parallelepipede cut in such a way that each plane is perpendicular to a main direction.

## 7-2 MODELING OF MOISTURE TRANSPORT IN ONE DIRECTION. THIN PLANE SHEETS

This simple case is considered in the chapter devoted to moisture transport through wood pieces. It is of interest essentially in the following two cases.

- when a large board with a thin thickness is studied, the effect of the other two main diffusions being negligible.
- when thin plane sheets perpendicular to a main direction are used, for determining the main diffusivity and its concentration dependence.

Two types of transport are of concern:

1. The absorption of moisture (or of water).
2. The desorption of moisture.

### 7-2-1 Modeling of Absorption of Moisture in One Direction

The absorption of moisture is controlled by diffusion through the solid and by a transport through the air-solid interface. The diffusion of moisture in one dimension in the solid is governed by Fick's law with concentration-dependent diffusivity.

$$\frac{\partial C}{\partial t} = \frac{\partial}{\partial x}\left(D \cdot \frac{\partial C}{\partial x}\right) \tag{7-4}$$

where $D$ is a main diffusivity.

The flux of moisture through the air-solid interface is expressed by:

$$F = F_o(C_{eq} - C_o) \tag{7-10}$$

where $F_o$ is the coefficient of moisture transport, $C_{t,o}$ (or $C_o$) is the concentration of moisture on the surface of the solid at time $t$, and $C_{eq}$ is the concentration on the solid surface which is at equilibrium with the external atmosphere. (This concentration $C_{eq}$ is obtained on the surface after infinite time of contact.)

The rate of diffusion through the solid next to the surface is constantly equal to the rate of absorption of moisture onto the surface. The boundary condition is then:

$$-D\left(\frac{\partial C}{\partial x}\right)_{surface} = F_o(C_o - C_{eq}) \tag{7-11}$$

The thickness of the solid is divided into $N$ planes of finite thickness $\Delta x$, and the slice next to the surface has a thickness of $\Delta x/2$.

The matter balance is calculated on this last slice of thickness $\Delta x/2$, during the increment of time $\Delta t$, by considering the diffusion and absorption of moisture (Figure 7–3):

$$\left[+D_{0.5}\left(\frac{\partial C}{\partial x}\right)_{0.5} + F_o(C_{eq} - C_o)\right]\Delta t = \frac{\Delta x}{2}\,[CN_{0.25} - C_{0.25}] \tag{7-12}$$

where $CN_{0.25}$ and $C_{0.25}$ are the concentrations of moisture at position 0.25 after the elapse $\Delta t$ and before this elapse of time, respectively.

By assuming that

$$CN_{0.25} - C_{0.25} = CN_o - C_o \tag{7-13}$$

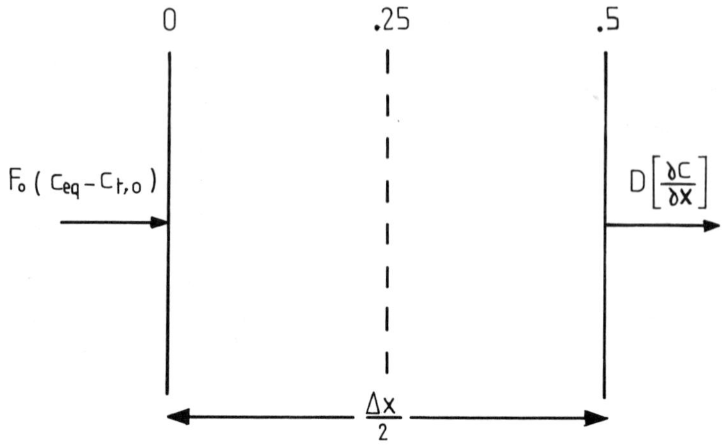


**Figure 7–3** Scheme of one-dimensional transport through a sheet of thickness of $\Delta x/2$, with absorption of moisture on the surface and diffusion within the solid.

and that the gradient of concentration at position 0.5 is equal to the chord slope between the planes 1 and 0, this equation becomes:

$$D_{0.5}\,\frac{C_1 - C_o}{\Delta x} + F_o(C_{eq} - C_o) = \frac{\Delta x}{2\,\Delta t}\,[CN_o - C_o] \tag{7-14}$$

By introducing the dimensionless numbers $M$ and $N$ at position 0.5 and at the previous time:

$$M_{0.5} = \frac{(\Delta x)^2}{\Delta t}\cdot\frac{1}{D_{0.5}} \tag{7-15}$$

$$N_{0.5} = \frac{F_o\cdot\Delta x}{D_{0.5}} \tag{7-16}$$

the new concentration (at time $t + \Delta t$) on the surface, $CN_o$, is then expressed in terms of the concentrations obtained at the previous time ($t$) at positions 0, 1 and of the concentration at equilibrium.

$$CN_o = \frac{1}{M_{0.5}}\,[2C_1 + (M_{0.5} - 2N_{0.5} - 2)\cdot C_o + 2N_{0.5}\cdot C_{ext}] \tag{7-17}$$

The parameters $M_{0.5}$ and $N_{0.5}$ are concentration dependent in the same way as the diffusivity. They can be obtained when the diffusivity at position 0.5, $D_{0.5}$, is calculated by using one of the following equations:

$$D_{0.5} = D_o\cdot\exp cte\left(\frac{C_o + C_1}{2}\right) \tag{7-18}$$

$$D_{0.5} = D_o \cdot \exp\left(Cte - \frac{C_o + C_1}{2}\right) \tag{7-18$'$}$$

$$D_{0.5} = D_o \cdot \exp\left(\frac{-2 \cdot Cte}{C_o + C_1}\right) \tag{7-18$''$}$$

or even by

$$D_{0.5} = \frac{1}{2}(D_o + D_1) \tag{7-19}$$

when the profiles of concentration are flat next to the surface of the solid.

Within the solid sheet, the new concentration is calculated by using the equation (already shown in Chapter 1).

$$CN_n = C_n + J_{n-0.5} - J_{n+0.5} \tag{7-20}$$

with

$$J_{n-0.5} = \frac{\Delta t}{(\Delta x)^2} D_{n-0.5} \cdot (C_n - C_{n+1}) \tag{7-21}$$

### 7-2-2 Modeling of Desorption of Moisture

In fact, the problem is the same as the problem of absorption of moisture. The process of matter transport is controlled by diffusion within the solid and evaporation on the surface.

The flux of matter through the air-solid interface expressed by:

$$F = F_o(C_{eq} - C_o) \tag{7-10}$$

is generally called the flux of evaporation, except in special cases.

The equations obtained in the previous case (modeling of absorption of moisture) can be used in the case of desorption of moisture, the only difference between these two processes being the relative value of the concentrations: $C_{eq}$ and $C_o$.

Absorption: $C_{eq} > C_o$ on the surface (7-22)

Desorption: $C_{eq} < C_o$ on the surface (7-23)

## 7-3 MODELING OF MOISTURE IN THREE DIRECTIONS WITH VARIOUS DIFFUSIVITIES

The process of matter transport in this case is very general. There is a three-dimensional transport with three main directions of diffusivities.

The solid, parallelepipedic in shape, is cut in such a way that each face of the solid is perpendicular to a main direction of diffusivity. This case corresponds to the problem of the pieces of wood, as shown in the introduction.

Of course, in this difficult case, no analytical solution is obtained for the matter transport because of the concentration dependence of the main diffusivities and because of the anisotropic properties of the solid.

Three main directions which are perpendicular to each other exist for the matter transport with three main diffusivities:

- longitudinal, with longitudinal diffusivity $D_L$.
- radial, with radial diffusivity $D_R$.
- tangential, with tangential diffusivity, $D_T$.

The following types of transports are considered:

1. Absorption of moisture, with condensation and diffusion.
2. Desorption of moisture, with evaporation and diffusion.
3. Absorption of moisture with diffusion, the equilibrium of concentration on the surface being attained as soon as the process starts.
4. Desorption of moisture with diffusion, the equilibrium of concentration on the surface being attained as soon as the process starts.

### 7-3-1 Modeling of Absorption of Moisture in Three Directions with Condensation and Diffusion

The flux of moisture through the air-solid interface is given by the relation:

$$F = F_o(C_{eq} - C_o) \tag{7-10}$$

where $F_o$ is the coefficient of matter transport, and $C_o$ and $C_{eq}$ are the concentration of moisture on the surface of the solid at the time of experiment, and at equilibrium (after infinite time), respectively.

The diffusion of moisture is described by Fick's transient law with the three main diffusivities and axes of diffusion:

$$\frac{\partial C}{\partial t} = \frac{\partial}{\partial T}\left(D_T \cdot \frac{\partial C}{\partial T}\right) + \frac{\partial}{\partial R}\left(D_R \cdot \frac{\partial C}{\partial R}\right) + \frac{\partial}{\partial L}\left(D_L \cdot \frac{\partial C}{dL}\right) \tag{7-9}$$

where $L$ is the longitudinal axis, with longitudinal diffusivity $D_L$, $T$ is the tangential axis, with tangential diffusivity $D_T$, and $R$ is the radial axis, with radial diffusivity $D_R$.

No analytical solution can be found for equations (7-9) and (7-10), and the problem is solved by using a numerical method with finite differences.

The solid, parallelepipedic in shape, is divided into small parallelepipedes of dimensions $\Delta T$, $\Delta R$, and $\Delta L$ (Figure 7–4). The position of each plane is defined by the integers $i$, $j$, $k$, as follows:

$$T = i \cdot \Delta T \qquad L = j \cdot \Delta L \qquad R = k \cdot \Delta R \tag{7-24}$$

**Within the sample.** The matter balance is calculated in the parallelepipede centered on $i$, $j$, $k$, during the increment of time $\Delta t$, by using Fick's first law (Figure 7–5). Only the two axes $L$ and $T$ are drawn in Figure 7–5 in order to simplify the drawing.

The change in the moisture content in the parallelepipede of volume $\Delta T \cdot \Delta L \cdot \Delta R$ is responsible for a change in the concentration of moisture, during the increment of time $\Delta t$:

$$(\Delta T \cdot \Delta L \cdot \Delta R) \cdot [CN(i, j, k) - C(i, j, k)] \tag{7-25}$$

where $CN(i, j, k)$ is the new concentration in the parallelepipede after the elapse of time $\Delta t$, and $C(i, j, k)$ is the concentration at the same place at the previous time.

This change in the moisture content in the parallelepipede results from the three moisture transports along each main axis.

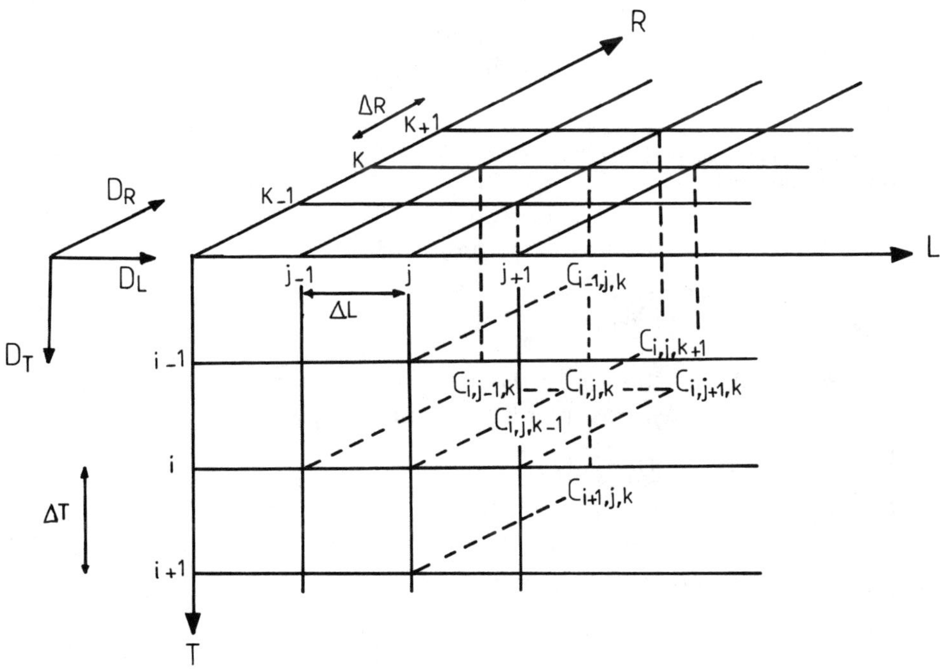


**Figure 7–4** Scheme of the three-dimensional transport through the solid, with the three main directions longitudinal $L(j)$, radial $R(k)$, and tangential $T(i)$.

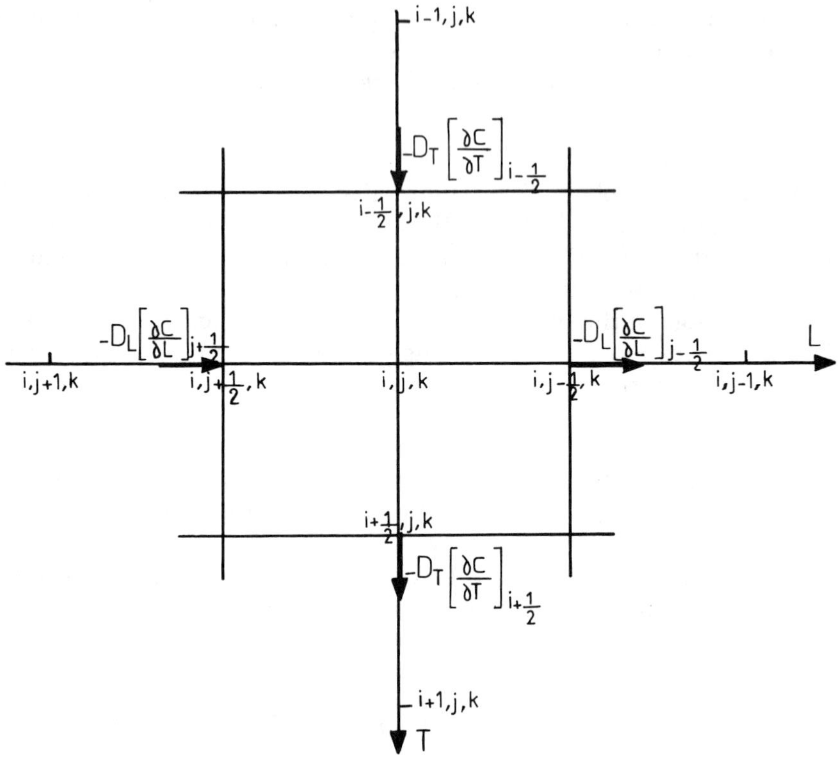


**Figure 7–5** Scheme of the transport of moisture through the solid, by considering the main directions: longitudinal $L$ (with the integer $j$), tangential $T$ (with the integer $i$), and radial $R$ (with the integer $k$).

The contribution of the longitudinal diffusion is:

$$\left[-D_L\left(\frac{\partial C}{\partial L}\right) \text{ at } (j+\tfrac{1}{2}) + D_L\left(\frac{\partial C}{\partial L}\right) \text{ at } (j-\tfrac{1}{2})\right]\Delta R\cdot\Delta T\cdot\Delta t \tag{7-26}$$

By calculating the gradients of concentration at the positions $(i, j + \frac{1}{2}, k)$ and $(i, j - \frac{1}{2}, k)$, this quantity becomes

$$\left[D_L(j+\tfrac{1}{2})\cdot\frac{C(j+1)-C(j)}{\Delta L} - D_L(j-\tfrac{1}{2})\cdot\frac{C(j)-C(j-1)}{\Delta L}\right]\Delta R\cdot\Delta T\cdot\Delta t \tag{7-27}$$

This term can be rewritten

$$[G_L(j+\tfrac{1}{2}) - G_L(j-\tfrac{1}{2})]\,\frac{\Delta R\cdot\Delta T\cdot\Delta t}{\Delta L} \tag{7-28}$$

by introducing the function $G(j)$.

$$G_L(j+\tfrac{1}{2}) = D_L(j+\tfrac{1}{2})\cdot[C(j+1) - C(j)] \tag{7-29}$$

The integers $i$ and $k$ are not written, in order to simplify: $D_L(j + \frac{1}{2})$ instead of $D_L(i, j + \frac{1}{2} k\cdot)$ and $G_L(j + \frac{1}{2})$ instead of $G_L(i, j + \frac{1}{2}, k)$.

By considering the three contributions, the matter balance enables one to express the new concentration after the elapse of time $\Delta t$ in terms of the previous concentration:

$$CN(i, j, k) = C(i, j, k) + [G_L(i, j + \tfrac{1}{2}, k) - G_L(i, j - \tfrac{1}{2}, k)]\frac{\Delta t}{(\Delta L)^2} + [G_T(i + \tfrac{1}{2}, j, k) - G_T(i - \tfrac{1}{2}, j, k)]\cdot\frac{\Delta t}{(\Delta T)^2} + [G_R(i, j, k + \tfrac{1}{2}) - G_R(i, j, k - \tfrac{1}{2})]\cdot\frac{\Delta t}{(\Delta R)^2} \tag{7-30}$$

Each diffusivity can be calculated by using one of these expressions:

$$D_L(i, j + \tfrac{1}{2}, k) = D_{Lo}\cdot\exp\, Cte\left(\frac{C(i, j + 1, k) + C(i, j, k)}{2}\right) \tag{7-31}$$

$$D_L(i, j + \tfrac{1}{2}, k) = D_{Lo}\cdot\exp\left(Cte - \frac{C(i, j + 1, k) + C(i, j, k)}{2}\right) \tag{7-31'}$$

$$D_L(i, j + \tfrac{1}{2}, k) = D_{Lo}\cdot\exp\left(\frac{-2Cte}{C(i, j + 1, k) + C(i, j, k)}\right) \tag{7-31''}$$

**Face of the sample (Figure 7–6).** Every face is in contact with the surrounding atmosphere, but the rate of transport is different on each face

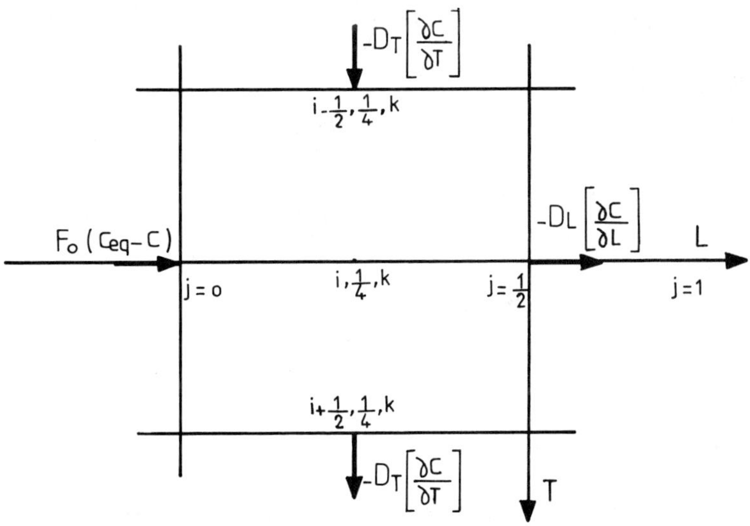


**Figure 7–6** Scheme of the matter transport through the slice next to the surface, with diffusion within the solid and evaporation (or condensation) of vapor on the surface. The longitudinal and tangential diffusion are drawn, the radial being perpendicular to the sheet.

because of the various diffusivities. When the solid is in contact with a liquid (for instance, a sample of wood immersed into water), the coefficient of matter transport onto the surface of the solid is very high. It can be assumed that the concentration of moisture on the surface reaches the value at equilibrium as soon as the process starts.

$$C(o, j, k) = C(i, o, k) = C(i, j, o) = C_{eq} \tag{7-32}$$

When the solid is in contact with a vapor (for instance, a piece of wood in contact with air), a coefficient of moisture transport is introduced, and the flux of moisture transport through the air-solid interface is:

$$F = F_o(C_{eq} - C_o) \tag{7-10}$$

The matter balance is calculated in the parallelepipede of dimensions $\Delta R$, $\Delta T$, and $\Delta L/2$ next to the surface with $j = 0$. The three contributions are then obtained. For the longitudinal diffusion, we have:

$$\left[D_L\left(\frac{\partial C}{\partial x}\right) \text{ at } (j = \tfrac{1}{2}) + F_o(C_{eq} - C(i, o, k))\right] \Delta R \cdot \Delta T \cdot \Delta t \tag{7-33}$$

which becomes with the finite differences:

$$[D_L(i, \tfrac{1}{2}, k) \cdot \frac{C(i, 1, k) - C(i, o, k)}{\Delta L} + F_o(C_{eq} - C(i, o, k))]\, \Delta R \cdot \Delta T \cdot \Delta t \tag{7-34}$$

For the tangential diffusion, we have:

$$\left[D_T\left(\frac{\partial C}{\partial T}\right) \text{ at } (i - \tfrac{1}{2}) + D_T\left(\frac{\partial C}{\partial T}\right) \text{ at } (i + \tfrac{1}{2})\right] \frac{\Delta L \cdot \Delta R \cdot \Delta t}{2} \tag{7-35}$$

which becomes:

$$\frac{\Delta L \cdot \Delta R \cdot \Delta t}{2}\left[D_T(i - \tfrac{1}{2}, \tfrac{1}{4}, k) \cdot \frac{C(i-1, \tfrac{1}{4}, k) - C(i, \tfrac{1}{4}, k)}{\Delta T} - D_T(i + \tfrac{1}{2}, \tfrac{1}{4}, k)\frac{C(i, \tfrac{1}{4}, k) - C(i+1, \tfrac{1}{4}, k)}{\Delta T}\right] \tag{7-36}$$

By putting the functions $G_T(i - \frac{1}{2}, \frac{1}{4}, k)$ and $G_L(i, \frac{1}{2}, k)$ as follows:

$$G_L(i, \tfrac{1}{2}, k) = D_L(i, \tfrac{1}{2}, k) \cdot [C(i, 1, k) - C(i, o, k)] \tag{7-37}$$

$$G_T(i - \tfrac{1}{2}, \tfrac{1}{4}, k) = D_T(i - \tfrac{1}{2}, \tfrac{1}{4}, k) \cdot [C(i-1, \tfrac{1}{4}, k) - C(i, \tfrac{1}{4}, k)] \tag{7-38}$$

the equations are rewritten:

$$[G_L(i, \tfrac{1}{2}, k) - F_o \cdot (C_{eq} - C(i, o, k))\Delta L]\, \frac{\Delta R \cdot \Delta T}{\Delta L}\, \Delta t \tag{7-39}$$

$$[G_T(i - \tfrac{1}{2}, \tfrac{1}{4}, k) - G_T(i + \tfrac{1}{2}, \tfrac{1}{4}, k)]\, \frac{\Delta L \cdot \Delta R}{2\Delta T}\, \Delta t \tag{7-40}$$

The transport of moisture by radial diffusion is also obtained by:

$$[G_R(i, \tfrac{1}{4}, k - \tfrac{1}{2}) - G_R(i, \tfrac{1}{4}, k + \tfrac{1}{2})] \frac{\Delta L \cdot \Delta T}{2\Delta R} \Delta t \tag{7-41}$$

with the function $G_R(i, \frac{1}{4}, k - \frac{1}{2})$.

$$G_R(i, \tfrac{1}{4}, k - \tfrac{1}{2}) = D_R(i, \tfrac{1}{4}, k - \tfrac{1}{2}) \cdot [C(i, \tfrac{1}{4}, k - 1) - C(i, \tfrac{1}{4}, k)] \tag{7-42}$$

From the matter balance in the parallelepipede of volume $\Delta L \cdot \Delta R \cdot \Delta T/2$, the new concentration on the surface after the elapse of time $\Delta t$ is expressed in terms of the concentrations at the previous time and various places: $(i, o\ k)$, $(i, 1, k)$, $(i - 1, 1, k)$, $(i + 1, 1, k)$, $(i, 1, k - 1)$ and $(i, 1, k + 1)$.

$$\begin{aligned} CN(i, o, k) = C(i, o, k) &+ [G_L(i, \tfrac{1}{2}, k) - F_o \cdot (C_{eq} - C(i, o, k))\Delta L] \\ &\cdot \frac{\Delta t}{2(\Delta L)^2} + [G_T(i - \tfrac{1}{2}, \tfrac{1}{4}, k) - G_T(i + \tfrac{1}{2}, \tfrac{1}{4}, k)] \frac{\Delta t}{(\Delta T)^2} \\ &+ [G_R(i, \tfrac{1}{4}, k - \tfrac{1}{2}) - G_R(i, \tfrac{1}{4}, k + \tfrac{1}{2})] \cdot \frac{\Delta t}{(\Delta R)^2} \end{aligned} \tag{7-43}$$

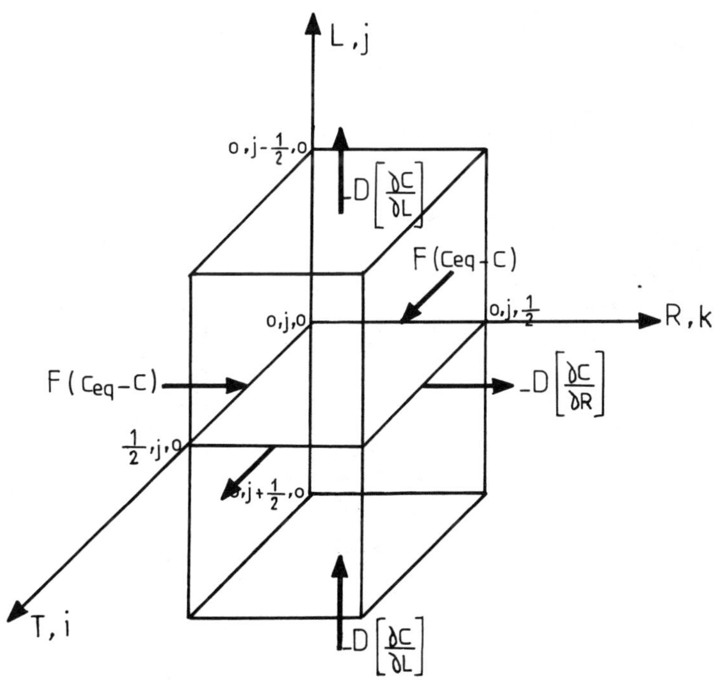


**Figure 7–7** Scheme of the slice at the intersection of two surfaces (radial and tangential) with diffusion within the solid and evaporation (condensation) on the surface.

by assuming that:

$$CN(i, o, k) - C(i, o, k) = CN(i, \tfrac{1}{4}, k) - C(i, \tfrac{1}{4}, k) \tag{7-44}$$

The expressions for the other faces are obtained by circular permutation of the integers.

This equation can also be written more simply:

$$\begin{aligned} CN(i, o, k) = C(i, o, k) &+ [G_L(i, \tfrac{1}{2}, k) - F_o[C_{eq} - C(i, o, k)\Delta L] \frac{\Delta t}{2(\Delta L)^2} \\ &+ [G_T(i - \tfrac{1}{2}, o, k) - G_T(i + \tfrac{1}{2}, o, k)] \frac{\Delta t}{(\Delta T)^2} \\ &+ [G_R(i, o, k - \tfrac{1}{2}) - G_R(i, o, k + \tfrac{1}{2})] \frac{\Delta t}{(\Delta R)^2} \end{aligned} \tag{7-45}$$

**Line at intersection of two faces (Figs. 7–7 and 7–7′).** The parallelepipede of dimensions $\Delta R/2$, $\Delta T/2$, and $\Delta L$ at position $(o, j, o)$ is considered. It is centered on the position $(\tfrac{1}{4}, j, \tfrac{1}{4})$.

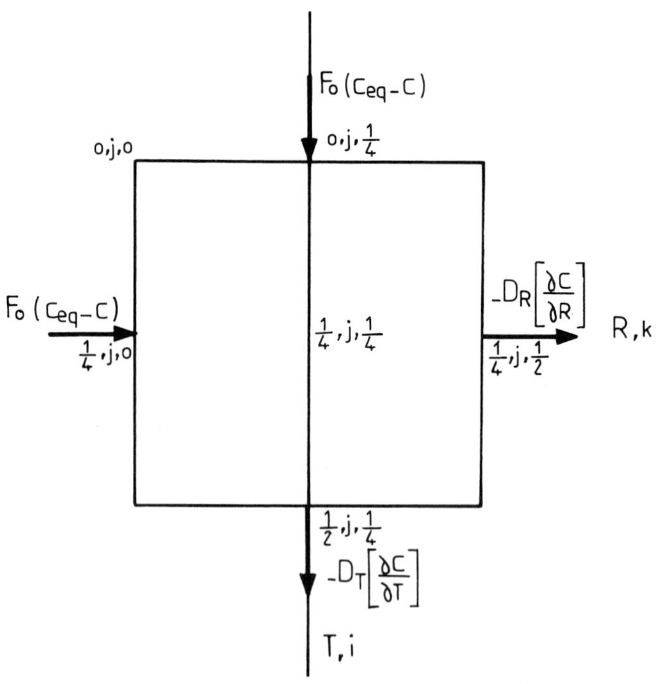


**Figure 7–7′** Scheme of the slice at the intersection of two surfaces (radial and tangential) with diffusion and evaporation (or condensation) on the surface.

The matter balance is calculated as follows:

$$\left[F_o[C_{eq} - C(\tfrac{1}{4}, j, o)] + D_R(\tfrac{1}{4}, j, \tfrac{1}{2})\cdot\left(\frac{\partial C}{\partial R}\right)\right]\frac{\Delta L\cdot\Delta T}{2}\,\Delta t$$
$$+\left[F_o[C_{eq} - C(o, j, \tfrac{1}{4})] + D_T(\tfrac{1}{2}, j, \tfrac{1}{4})\cdot\left(\frac{\partial C}{\partial T}\right)\right]\cdot\frac{\Delta R\cdot\Delta L}{2}\,\Delta t \tag{7-46}$$
$$+\left[-D_L(\tfrac{1}{4}, j + \tfrac{1}{2}, \tfrac{1}{4})\cdot\left(\frac{\partial C}{\partial L}\right) + D_L(\tfrac{1}{4}, j - \tfrac{1}{2}, \tfrac{1}{4})\cdot\left(\frac{\partial C}{\partial L}\right)\right]\frac{\Delta R\cdot\Delta T}{4}\,\Delta t$$
$$= [CN(\tfrac{1}{4}, j, \tfrac{1}{4}) - C(\tfrac{1}{4}, j, \tfrac{1}{4})]\cdot\frac{\Delta T\cdot\Delta L\cdot\Delta R}{4}$$

This equation is rewritten by expressing the gradients of concentration in various positions by the chord slope between the following planes:

$$\left[F_o[C_{eq} - C(\tfrac{1}{4}, j, o)]\right.$$
$$\left.+ D_R(\tfrac{1}{4}, j, \tfrac{1}{2})\,\frac{C(\tfrac{1}{4}, j, 1) - C(\tfrac{1}{4}, j, o)}{\Delta R}\right]\frac{\Delta L\cdot\Delta T}{2}\,\Delta t$$
$$+\left[F_o[C_{eq} - C(o, j, \tfrac{1}{4})] + D_T(\tfrac{1}{4}, j, \tfrac{1}{4})\frac{C(1, j, \tfrac{1}{4}) - C(o, j, \tfrac{1}{4})}{\Delta T}\right]\frac{\Delta R\cdot\Delta L}{2}\,\Delta t \tag{7-47}$$
$$+\frac{\Delta R\cdot\Delta T}{4}\left[-D_L(\tfrac{1}{4}, j + \tfrac{1}{2}, \tfrac{1}{4})\,\frac{C(\tfrac{1}{4}, j + 1, \tfrac{1}{4}) - C(\tfrac{1}{4}, j, \tfrac{1}{4})}{\Delta L}\right.$$
$$\left.+ D_L(\tfrac{1}{4}, j - \tfrac{1}{2}, \tfrac{1}{4})\,\frac{C(\tfrac{1}{4}, j, \tfrac{1}{4}) - C(\tfrac{1}{4}, j - 1, \tfrac{1}{4})}{\Delta L}\right]$$
$$= [CN(\tfrac{1}{4}, j, \tfrac{1}{4}) - C(\tfrac{1}{4}, j, \tfrac{1}{4})]\cdot\frac{\Delta T\cdot\Delta L\cdot\Delta R}{4}$$

By putting the function $G$

$$G_R(\tfrac{1}{4}, j, \tfrac{1}{2}) = D_R(\tfrac{1}{4}, j, \tfrac{1}{2})\cdot[C(\tfrac{1}{4}, j, 1) - C(\tfrac{1}{4}, j, o)] \tag{7-42}$$

$$G_T(\tfrac{1}{2}, j, \tfrac{1}{4}) = D_T(\tfrac{1}{2}, j, \tfrac{1}{4})\cdot[C(1, j, \tfrac{1}{4}) - C(o,j, \tfrac{1}{4})] \tag{7-42$'$}$$

$$G_L(\tfrac{1}{4}, j + \tfrac{1}{2}, \tfrac{1}{4}) = D_L(\tfrac{1}{4}, j + \tfrac{1}{2}, \tfrac{1}{4})\cdot[C(\tfrac{1}{4}, j + 1, \tfrac{1}{4}) - C(\tfrac{1}{4}, j, \tfrac{1}{4})] \tag{7-42$''$}$$

the equation is rewritten as follows:

$$CN(\tfrac{1}{4}, j, \tfrac{1}{4}) = C(\tfrac{1}{4}, j, \tfrac{1}{4}) + F_o[C_{eq} - C(\tfrac{1}{4}, j, o)]\cdot\frac{2\cdot\Delta t}{\Delta R}$$
$$+ G_R(\tfrac{1}{4}, j, \tfrac{1}{2})\cdot\frac{2\Delta t}{(\Delta R)^2} + F_o[C_{eq} - C(o,j, \tfrac{1}{4})]\cdot\frac{2\cdot\Delta t}{\Delta T} \tag{7-48}$$
$$+ G_T(\tfrac{1}{2}, j, \tfrac{1}{4})\cdot\frac{2\Delta t}{(\Delta T)^2} + [G_L(\tfrac{1}{4}, j - \tfrac{1}{2}, \tfrac{1}{4}) - G_L(\tfrac{1}{4}, j + \tfrac{1}{2}, \tfrac{1}{4})]\,\frac{\Delta t}{(\Delta L)^2}$$

This preceding equation is very complex. Moreover, the equation has to be rewritten in order to express $i$, $j$, $k$ in terms of integers by using the following rules:

$$C(\tfrac{1}{4}, j, \tfrac{1}{2}) = \tfrac{1}{2}[C(\tfrac{1}{4}, j, 1) + C(\tfrac{1}{4}, j, o)] \tag{7-49}$$

$$C(\tfrac{1}{4}, j, 1) = \tfrac{3}{4}\, C(o, j, 1) + \tfrac{1}{4}\, C(1, j, 1) \tag{7-50}$$

Another assumption can also be made:

$$CN(\tfrac{1}{4}, j, \tfrac{1}{4}) - C(\tfrac{1}{4}, j, \tfrac{1}{4}) = CN(o, j, o) - C(o, j, o) \tag{7-44'}$$

with

$$C(\tfrac{1}{4}, j, o) = C(o, j, \tfrac{1}{4}) = C(o, j, o) \tag{7-51}$$

And the following equation is thus obtained:

$$\begin{aligned} CN(o, j, o) = {} & C(o, j, o) + \Big[F_o\cdot[C_{eq} - C(o, j, o)]\Big]\left[\frac{2\cdot\Delta t}{\Delta R} + \frac{2\cdot\Delta t}{\Delta T}\right] \\ & + G_R(0, j, \tfrac{1}{2})\cdot\frac{2\Delta t}{(\Delta R)^2} + G_T(\tfrac{1}{2}, j, 0)\cdot\frac{2\Delta t}{(\Delta T)^2} \\ & + \Big[G_L(0, j - \tfrac{1}{2}, 0) - G_L(0, j + \tfrac{1}{2}\cdot 0)\Big]\frac{\Delta t}{(\Delta L)^2} \end{aligned} \tag{7-52}$$

**Corner, for instance (*o, o, o*).** The volume of dimensions $\Delta R/2$, $\Delta L/2$, and $\Delta T/2$ is considered, with the center of abscissa $(\tfrac{1}{4}, \tfrac{1}{4}, \tfrac{1}{4})$ (Figure 7–8).

The matter balance is obtained from the three contributions along each main axis of diffusion:

$$\begin{aligned} \Big[CN(\tfrac{1}{4}, \tfrac{1}{4}, \tfrac{1}{4}) - C(\tfrac{1}{4}, \tfrac{1}{4}, \tfrac{1}{4})\Big]\frac{\Delta R\cdot\Delta T\cdot\Delta L}{8} = {} & \Big[F_o\cdot[C_{eq} - C(o, \tfrac{1}{4}, \tfrac{1}{4})] \\ & - D_T(\tfrac{1}{2}, \tfrac{1}{4}, \tfrac{1}{4})\frac{C(o, \tfrac{1}{4}, \tfrac{1}{4}) - C(1, \tfrac{1}{4}, \tfrac{1}{4})}{\Delta T}\Big]\frac{\Delta R\cdot\Delta L}{4}\Delta t + \Big[F_o\cdot[C_{eq} \\ & - C(\tfrac{1}{4}, o, \tfrac{1}{4})] - D_L(\tfrac{1}{4}, \tfrac{1}{2}, \tfrac{1}{4})\frac{C(\tfrac{1}{4}, o, \tfrac{1}{4}) - C(\tfrac{1}{4}, 1, \tfrac{1}{4})}{\Delta L}\Big]\frac{\Delta R\cdot\Delta T}{4}\Delta t \\ & + \Big[F_o\cdot[C_{eq} - C(\tfrac{1}{4}, \tfrac{1}{4}, o)] \\ & - D_R(\tfrac{1}{4}, \tfrac{1}{4}, \tfrac{1}{2})\frac{C(\tfrac{1}{4}, \tfrac{1}{4}, o) - C(\tfrac{1}{4}, \tfrac{1}{4}, 1)}{\Delta R}\Big]\frac{\Delta T\cdot\Delta L}{4}\Delta t \end{aligned} \tag{7-53}$$

By assuming that

$$CN(\tfrac{1}{4}, \tfrac{1}{4}, \tfrac{1}{4}) - C(\tfrac{1}{4}, \tfrac{1}{4}, \tfrac{1}{4}) = CN(o, o, o) - C(o, o, o) \tag{7-44''}$$

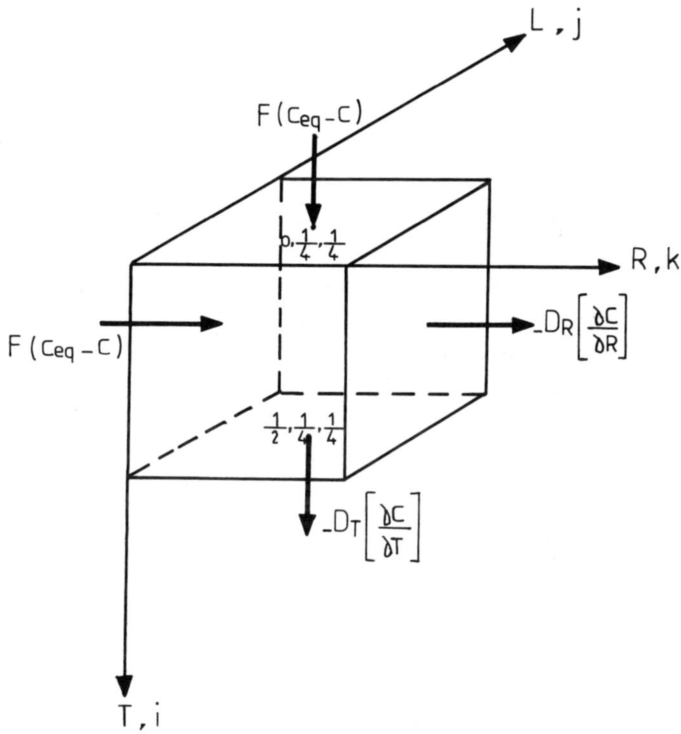


**Figure 7–8** Scheme of the solid next to the corner (at the intersection of the three surfaces) with diffusion in the solid and condensation (or evaporation) on the surfaces. Only the transports along the radial and tangential axes are drawn.

and

$$C(o, \tfrac{1}{4}, \tfrac{1}{4}) = C(\tfrac{1}{4}, o, \tfrac{1}{4}) = C(\tfrac{1}{4}, \tfrac{1}{4}, o) = C(o, o, o) \qquad (7\text{-}51')$$

the equation becomes more simply:

$$CN(o, o, o) = C(o, o, o) + F_o \cdot [C_{eq} - C(o, o, o)] \cdot \left[\frac{2}{\Delta T} + \frac{2}{\Delta L} + \frac{2}{\Delta R}\right] \Delta t$$

$$+ G_T(\tfrac{1}{2}, o, o) \cdot \frac{2\Delta t}{(\Delta T)^2} + G_L(o, \tfrac{1}{2}, o) \frac{2\Delta t}{(\Delta L)^2} + G_R(o, o, \tfrac{1}{2}) \cdot \frac{2\Delta t}{(\Delta R)^2} \qquad (7\text{-}54)$$

### 7-3-2 Modeling of Desorption of Moisture in Three Dimensions by Evaporation and Diffusion

The desorption and absorption of moisture are similar, and they can be studied by using the equation written in the previous chapter. The main difference appears in the value of the flux of matter absorbed or desorbed, which depends on the relative values of the concentration on the surface at time $t$ and at equilibrium.

$$\text{Absorption} \qquad C_{eq} > C \text{ surface} \tag{7-55}$$

$$\text{Desorption} \qquad C_{eq} < C \text{ surface} \tag{7-56}$$

### 7-3-3 Modeling of Absorption of Liquid

The main assumption in this case is that the concentration on the surface reaches the value at equilibrium as soon as the process starts. The problem can be solved in a very simple way, by using the numerical treatment made in the previous chapter.

**On the face**

$$C(i, j, o) = C(i, o, k) = C(o, j, k) = C_{eq} \tag{7-57}$$

**Within the solid.** The equation obtained for the solid in the previous case, can be used for:

$$1 \le i \le N_i - 1 \tag{7-58}$$

$$1 \le j \le N_j - 1 \tag{7-58'}$$

$$1 \le k \le N_k - 1 \tag{7-58''}$$

$$\begin{aligned} CN(i, j, k) = C(i, j, k) &+ [G_L(i, j + \tfrac{1}{2}, k) - G_L(i, j - \tfrac{1}{2}, k)] \frac{\Delta t}{(\Delta L)^2} \\ &+ [G_T(i + \tfrac{1}{2}, j, k) - G_T(i - \tfrac{1}{2}, j, k)] \cdot \frac{\Delta t}{(\Delta T)^2} \\ &+ [G_R(i, j, k + \tfrac{1}{2}) - G_R(i, j, k - \tfrac{1}{2})] \cdot \frac{\Delta t}{(\Delta R)^2} \end{aligned} \tag{7-30}$$

## 7-4 CONCLUSIONS

Two places are considered: within the solid, and on the surface of the solid. The equations of transport of matter within the solid are relatively simple. The transport to the surface (or from the surface) is more complex. In fact, two cases are of interest when the wood is studied:

1. When the wood is in contact with liquid water, the concentration on the surface reaches the value at equilibrium as soon as the process starts.
2. When the wood is in contact with vapor water, a coefficient of matter transport is considered.

# Chapter Eight

# Matter Transports Between Plasticized PVC and Liquids

## 8-1 PREVIOUS STUDIES ON PLASTICIZED PVC IN CONTACT WITH LIQUIDS

### 8-1-1 Introduction

In the United States and Japan, PVC is second to polyethylene (PE), but in other plastics-producing countries it is the leading plastic material. However, PVC use in packaging is growing fast in the United States [1], from 300 million pounds in 1987 for film and sheet forms in packaging to 360 million pounds in 1988. Use in bottles for a wide variety of household products is growing even faster, but from a smaller base, from 160 million pounds in 1987 to 225 million pounds in 1988.

The glass transition temperature of PVC is readily dropped from around 80°C down to −50°C by incorporating the suitable concentration of plasticizer. Then the main role of the plasticizer is to improve the flexibility of PVC compounds. But the presence of the plasticizer is responsible for various matter transports. The liquid components of food penetrate into plastics and the additives as plasticizer, UV-stabilizers, and antioxidants diffuse from the polymer into the liquid, with the following results:

1. Because of loss of plasticizer and additives, the polymer shows a considerable decrease in mechanical properties.
2. The liquid in contact with PVC (food, blood) is contaminated.

**Packaging.** A few decades ago, food was sold mainly from bulk containers, the packaging being done by the retailer after selling. Prepackaging was used in the early 1900s with the production of sealable glass containers and airtight metal cans. Later, flexible materials such as greaseproof paper, metal oils, and cellophane were developed. Today, since they exhibit outstanding usage properties, plastics are used preferentially for packaging foodstuffs.

Plastics packaging is capable of retarding and sometimes preventing detrimental changes in the packed material due to external influences (oxygen, light, micro-organisms). Plastics are also able to greatly reduce the loss of components in the packed material (water, flavor). Resulting from this protection, plastic packaging enables the consumer to use foodstuffs in perfectly hygienic conditions, and to store them without loss in quality for an extended period of time, especially when the material has been previously sterilized by pasteurization.

But plastics contain low-molecular compounds as monomers and oligomers, and additives as plasticizers, lubricants, stabilizers, and antioxidants, which are absolutely necessary for the processing and the stability of the final materials. The drawback appears with the migration of the additives from the packaging into the packed material.

**Legislation.** As the problems of interaction of liquid food, cosmetics, and blood with plastic packagings have become of increasing importance, legislation and control procedures have been developed in the United States with the Food and Drug Administration (1977) and in the European Economic Community with an EEC Commission in 1978, and elsewhere. Recent reports regarding the possible carcinogenicity of plasticizers used in PVC, essentially phthalate esters, have motivated a wider adoption of alternative products or improved plasticizer PVC for blood storage and heating [2].

**Plasticizers in PVC.** The residual monomer is of great importance in some plastics, such as PVC and polystyrene (PS), and it has been of concern for some time. Because of the toxicity of vinyl chloride, migration studies have been undertaken by considering the total extractables from PVC into various food simulating solvents (3–9).

Di-2 - ethylhexyl phthalate (DEHP) is the most used plasticizer among around 20 di-esters of phthalic acid. Bottles used for liquid food are usually made of PVC with a very low percentage of plasticizer, and PVC films for packaging contain a higher percentage. In some countries DEHP is considered as an alimentary plasticizer, meaning that it can be in contact with the food, while in other countries it is not considered as biologically inert.

The toxicity of DEHP is partly known. Its metabolism follows four steps in the human body:

1. Absorption, especially by lipoproteins after hydrolysis of DEHP [10].
2. Biodistribution, as the DEHP transported through transfusion or hemodialysis has been observed in liver, spleen, or grease [11].
3. Transformation, and up to 14 metabolites have been found after action of the enzyme lipoprotein lipase [12, 13] on DEHP.
4. Excretion of DEHP and its metabolites by way of glycuronides through the kidneys [14].

The dissolution temperature of PVC in the plasticizer is increased by increasing the alcohol chain length, which reduces the plasticizing capability and the polarity. By increasing alcohol radical chain length, some properties of PVC compounds are changed:

- the volatility and migration of plasticizer are reduced.
- lower extraction by polar liquids such as water, and higher extraction by nonpolar ones such as oils, are observed.

Phthalates containing higher isoalcohols are recommended for special applications, when a reduced volatility and resistance to elevated temperatures are desired.

Another group of phthalates are the esters of the ortho-phthalic acid with alcohols characterized by entirely linear chains. Production of linear phthalates is realized in connection with production of mixed alcohols by means of oxysynthesis. These plasticizers exhibit lower viscosity, higher efficiency, and reduced migration as compared with the branched phthalates (15).

**Nature of the liquid.** The nature of the liquid plays an important role for plasticizer transport, every liquid having a particular behavior toward plastics and plasticizers [16]. The migration of phthalate plasticizer from PVC into various liquids is widely reported in recent literature [17–19], and a few examples are given:

1. The migration is strongly affected by the nature of petroleum oils, and an increase in the aromaticity increases the transport [20].
2. Various types of alcohols [methanol, ethanol, n-propanol, isopropanol, n-butanol, isobutanol, 2-ethyl-hexanol] have been tested at different temperatures (10°C; 30°C; 60°C) with various concentrations

of the plasticizer: dibutyl phthalate, diocytl phthalate [21]. The kinetics of transport have been studied and diffusivities have been measured with various alcohols. In the case of methanol [22], it is done by determining the kinetics of transport and the profiles of concentration of plasticizer developed through PVC sheets. Two different transports have been observed and studied for the liquid and plasticizer, when the liquid is benzyl alcohol [23]. The effect of the water percent in ethanolic solutions on the plasticizer transports has been found to be decisive, the presence of water provoking a reduction in the transport [24].

3. Transports of liquid and plasticizer have been studied with vinegar and other acetic acid solutions, an increase in the water percent decreasing these transports [25, 26].
4. Various oils have been tested with plasticized PVC. The test with $^{14}C$ labeled DEHP and fat oil HB 307 is worth noting [17]. Other edible oils have been considered [6, 27, 28]. Transports between plasticized PVC and arachis oils have been studied, in emphasizing the different behavior of this oil and n-heptane used as fat oil simulant [16, 29, 30].
5. Blood components interact strongly with plasticized PVC, the plasticizer being transported into the blood with various inconveniences for health [2–6, 31–37].

**Food and food simulants.** The problems of identifying the migrating species in the presence of chemically complex structures are very difficult in case of real foodstuffs: edible oils proposed as test fats such as olive oil [38, 39], sunflower oil [39], coconut oil [17], and others present great analytical difficulties because of their accompanying substances.

Food simulants with a simple structure having a resemblance to the real food in its behavior and interaction with PVC have been tested [3]. Two main requirements are for simulants to be uniform and readily available. Some of the simulants used for various types of foods are shown in Table 8–1.

**TABLE 8–1** Food and food simulants

| Group | Food | Simulants |
|---|---|---|
| I | Watery, nonacidic food (pH >5) | Distilled water |
| II | Watery, acidic food (pH ≤5) | 3 wt % acetic acid |
| III | Watery, acidic or nonacidic fatty food | Group I, II, IV |
| IV | Fatty foods | Heptane, diethyl, ether, edible oil. |
| V | Liquids with 5% ethanol | 20–50 vol % ethanol |

It was proposed in 1963 to use distilled water, 3 percent by weight acetic acid, and 10 percent by volume ethanol as simulants for various types of foods [18]. Alcoholic beverages are simulated by 15 percent and 96 percent ethanol by a few users, while other users prefer using 20 percent and 50 percent aqueous ethanol [18]. The effect of the water percent in ethanol solutions on the transports is found to be of importance in the case of packaging made of plasticized PVC [24].

From the studies carried out on a simulation of fatty foods [17, 18], it emerges that of the simulants in common use, that is, heptane, diethyl ether and paraffin oil, none is in agreement with the migration figure obtained with real foodstuffs [9]. Other studies have found that coconut oil gives migration figures in close agreement to actual fatty oils [18]. The behavior of n-heptane, the simulant of oil, has been found to be quite different from that of arachis oil when contacted with plasticized PVC [29, 30, 40], with a higher rate of transport of plasticizer in the case of heptane. Compared with the triglycerides of edible oils, these preceding simulants have a completely different chemical structure and thus different interaction properties toward plastics packagings [6, 9, 41, 42].

A large study has been aimed at developing a synthetic fat simulant, with the description of the simulant HB 307 [43, 44]. This simulant was built up of a mixture of triglycerides with fully saturated fatty acids obtained from coconut oil. This material has the advantage of being free from interfering impurities, and having high optical transmission allowing UV measurements.

As shown by the use of milk as a food simulant [18], it is difficult to imitate a real foodstuff because milk is not simply an oil but an emulsion in water of many other things.

Generally, simulants give at best only an overall picture of what is actually happening during the migration process. Various parameters must be considered.

- The effect of the stabilizer to plasticizer ratio on the rate of migration has been found to be significant [45].
- The thickness of the film is of importance, and the use of the same thickness of film for experiments as for packaging has been suggested [43].
- The nature of diffusing material as well as the nature of the liquid in contact with the packaging are also of importance, as shown in alcoholic solutions [24] and acetic solutions [25, 26] where the effect of water percent on the rate of transport is of concern.

Finally although simulants may not simulate real foods very well, no better idea has been yet suggested.

**Simulation of storage conditions.** Following the statement that "the test should imitate the migration in the most unfavorable conditions of use" [3], it would be reasonable to study the matter transports over a period of time equivalent to the longest shelf life at the maximum temperature. But, of course, experiments over six months are highly time consuming, and many attempts have been made in order to reduce the time of experiments, essentially by increasing the temperature. A ten-day storage test at 45°C has produced the same amount of matter transported as the storage for six months at 25°C in case of edible oil systems [3, 45–48]. A simple equation has been proposed where the diffusivity is proportional to the temperature [3].

However, results obtained from simulated tests did not always correlate with real tests [18]. From studies done on the migration of n-butyl stearate from polystyrene film into fat simulant HB 307 [27], more than one difference has been observed between the transports at 20°C and 40°C. In the whole, there is a danger that unrealistically strong migration occurs at too high test temperatures.

In fact, the process of matter transport is generally very complex, and short tests performed at a high temperature cannot always represent the long real test conducted at a lower temperature. The following reasons can be advanced:

1. The diffusivity increases with temperature.
2. The amount of matter transported may increase with temperature, in the same way as the solubility.
3. The amount of matter transported increases with the amount of liquid, and a partition coefficient is sometimes introduced.
4. The properties of the polymer may change with temperature, especially when the test temperature is not far from the transition temperature $Tg$.
5. The properties of the polymer may be altered by a decrease in the plasticizer content in PVC.

### 8-1-2 Experimental

The experiments performed on plasticized PVC with various liquids are described.

**Analysis in liquid and PVC.** Measurements are done in liquid for determining the plasticizer concentration and in the PVC itself to determine the profiles of concentration.

At intervals, the PVC sample is weighed and a small volume of liquid is taken [1 ml] for analysis. The concentration of plasticizer in the liquid is generally determined by spectrometry UV or by gas chromatography.

Gas chromatography is an interesting method for separation and identification (by using retention indexes with n-paraffins) and for quantitative determination. Various stationary phases, based on silicone elastomers have been tested: Silicone SE 30, Silicone OV 17, Silicone QF 1, and a porous polymer Tenax [49]. Liquid chromatography can also be used [50].

Radioactively labeled samples are often of interest, as the amount of the component transported is directly proportional to the radioactivity [17]. $^{14}C$ labeling techniques are useful to follow the compound or metabolite containing the $^{14}C$ labeled carbon.

**Analysis of plasticizer and liquid in PVC.** Plasticized PVC samples are prepared by mixing plasticizer and PVC in powder form, and pressing the compound into sheets by using a mold operated by a press heated to about 150°C. Discs are cut from the sheets and immersed in the liquid. At intervals, a disc is extracted and peeled into 17 layers parallel to the flat face. The concentrations of plasticizer and liquid are determined in every layer after dissolution, extraction, and addition of an internal standard by using gas chromatography. The whole mixture may also be injected in the chromatography without selective precipitation of PVC [16, 24, 49].

**Apparatus for diffusion measurements.** Experiments [3, 16, 22–24] are carried out by using 20 PVC discs immersed in 200 ml of liquid in a closed flask (500 ml). Experiments are repeated twice, once with controlled stirring and the other with motionless liquid. Samples are taken at intervals: 1 PVC disc, and 10 ml of liquid, so as to keep the weight ratio between the PVC and liquid constant.

### 8-1-3 Theoretical

The following points have been studied in the case of plasticized PVC compounds:

1. Processes of transport, which are different following the amount of plasticizer.
2. Kinetics of transport of plasticizer into a large volume of liquid.
3. Kinetics of transport of plasticizer into a finite volume of liquid.
4. Problem of simultanteous transport of the liquid into and plasticizer out of the PVC.
5. Conclusions drawn from these previous studies.

**Process of transport in PVC and plasticized PVC.** Diffusion of liquid in polymer can be visualized as a series of jumps in which Brownian motion of chain segments of the polymer produces transient voids in the vicinity of the liquid penetrant, enabling it to move within the polymer. Of course, the size of the molecule of liquid is related to the size of the void required to enable a jump. The temperature dependence of diffusivity is described by the Arrhenius law, and the activation energy is related to the amount of energy required to form the transient voids necessary for the process of diffusion. If the molecule of liquid is greater than the mobile segment of the polymer [51–55], the activation energy may be independent from the size of the liquid molecule, and this is especially true for a series of homologous organic compounds. The activation energy of diffusion generally decreases with increasing temperature above the glass transition temperature of the polymer $Tg$. Three classes of diffusion are distinguished as follow.

***Case 1 with Fickian Diffusion.*** The rate of diffusion is much less than that of relaxation. The amount of the material transported at time $t$ in a sheet is described by:

$$\frac{M_t}{M_\infty} = \frac{4}{L}\left(\frac{Dt}{\pi}\right)^n \tag{8-1}$$

with $n = \frac{1}{2}$.

***Non-Fickian Diffusion.*** The rates of diffusion and relaxation are comparable. This behavior needs two or more parameters to describe the interacting diffusion and relaxation effects inherent in it. This system lies between case 1 and case 2, and the power of $t$ takes an intermediate value between 0.5 and 1 for equation (8-1).

***Case 2 Diffusion.*** The rate of diffusion is faster than that of the relaxation process. The parameter of interest is the constant velocity of the advancing front of liquid located at the boundary between swollen gel and glassy core [56]. This is a limiting case characterized by equation (8-1) with $n = 1$.

**Transport of plasticizer into a large volume of liquid.** When sheets of PVC samples are immersed in a volume of liquid so large that the concentration of plasticizer in the solution conditions are:

| | | | | | |
|---|---|---|---|---|---|
| Initial | $t = 0$ | $0 < x < L$ | $C = C_{in}$ | PVC | (8-2) |
| Boundary | $t > 0$ | $x = 0$ and $x = L$ | $C = 0$ | PVC faces | (8-3) |

The process of plasticizer transport is described by Fick's law, expressed in one dimension:

$$\frac{\partial C}{\partial t} = \frac{\partial}{\partial x}\left(D \cdot \frac{\partial C}{\partial x}\right) \tag{8-4}$$

The solution of this equation is given in the form of a trigonometric series for the plasticizer concentration through the thickness of the PVC sheet, when the diffusivity $D$ is constant.

$$\frac{C_{t,x}}{C_{in}} = \frac{4}{\pi} \cdot \sum_{n=0}^{\infty} \frac{1}{2n+1} \cdot \sin\left(\frac{(2n+1)\pi x}{L}\right) \exp\left[-\frac{(2n+1)^2\pi^2}{L^2} Dt\right] \tag{8-5}$$

The total amount of plasticizer which leaves the PVC sheet during time $t$ is expressed in the form of a series, where $M_\infty$ is the corresponding amount at infinite time.

$$\frac{M_\infty - M_t}{M_\infty} = \frac{8}{\pi^2} \cdot \sum_{n=0}^{\infty} \frac{1}{(2n+1)^2} \cdot \exp\left[-\frac{(2n+1)^2\pi^2}{L^2} Dt\right] \tag{8-6}$$

Another expression for the kinetics of the transport of plasticizer is given in terms of error function:

$$\frac{M_t}{M_\infty} = \frac{4}{L} (Dt)^{0.5} \cdot \left\{\pi^{0.5} + 2 \sum_{n=1}^{\infty} (-1)^n \cdot ierfc \frac{nL}{2(Dt)^{0.5}}\right\} \tag{8-7}$$

For low values of time, the preceding equation can be reduced to:

$$\frac{M_t}{M_\infty} = \frac{4}{L}\left(\frac{Dt}{\pi}\right)^{0.5} \tag{8-8}$$

For high values of time, the series become very convergent and the first term is retained.

$$\frac{M_\infty - M_t}{M_\infty} = \frac{8}{\pi^2} \cdot \exp\left[-\frac{\pi^2}{L^2} Dt\right] \tag{8-9}$$

**Transport of plasticizer into a limited volume of liquid.** Generally, experiments are performed with finite volumes of liquid, and the concentration of plasticizer in the liquid increases as the transport proceeds. When the liquid is stirred well, the liquid is homogeneous and the plasticizer concentration in the liquid depends only on time.

The problem is controlled by transient diffusion through the PVC sheet of thickness $2L$. The initial and boundary conditions are:

| | | | | | |
|---|---|---|---|---|---|
| Initial | $t = 0$ | $-L < x < L$ | $C = C_{in}$ | PVC | |
| | | | $C = 0$ | liquid | (8-10) |
| Boundary | $t > 0$ | $x = \pm L$ | | | |

$$a \cdot \frac{\partial C}{\partial t} = -D \cdot \frac{\partial C}{\partial x} \tag{8-11}$$

A solution of the problem is obtained [57] in a form of a series, expressing the amount of plasticizer transported in the liquid at time, $M_t$, as a fraction of the corresponding quantity after infinite time $M_\infty$, when the diffusivity is constant.

$$\frac{M_\infty - M_t}{M_\infty} = \sum_{n=1}^{\infty} \frac{2\alpha(1 + \alpha)}{1 + \alpha + \alpha^2 q^2} \cdot \exp \frac{q_n^2}{L^2} Dt \tag{8-12}$$

where the $q_n$s are the nonzero positive roots of

$$\tan q_n = - \alpha \cdot q_n \tag{8-13}$$

and $\alpha = a/L$ is the ratio of the volumes of liquid and PVC.

**Simultaneous diffusion of liquid and plasticizer.** A mathematical study deals with the simultaneous diffusion of the liquid into and plasticizer out of the polymer, when the polymer is immersed into a well-stirred liquid of infinite volume [58]. As the general problem is very complex, the following assumptions are made.

1. The concentration of plasticizer is small.
2. The degree of swelling of the polymer due to the liquid is small.
3. The concentration of liquid and plasticizer on the PVC surface reaches the equilibrium as soon as the process starts.
4. The process of diffusion is described by two diffusivities which depend only on the liquid concentration in PVC.

$$D_1 = D_{lo} \cdot \exp \alpha(C - C_1) \tag{8-14}$$

$$D_p = D_{lo} \cdot \beta \cdot \exp \alpha(C - C_1) \tag{8-14'}$$

with constant values for $D$, $\alpha$, $\beta$ and $C$.
The equations are coupled

$$\frac{\partial C_1}{\partial t} = \frac{\partial}{\partial x}\left(D_1 \cdot \frac{\partial C_1}{\partial x}\right) \tag{8-4}$$

$$\frac{\partial C_P}{\partial x} = \frac{\partial}{\partial x}\left(D_p \cdot \frac{\partial C_p}{\partial x}\right) \tag{8-4'}$$

Various results are obtained according to the assumptions. One of them is concerned with the ratio of the fluxes of plasticizer and liquid which is independent of time.

**Conclusions about the previous studies.** Some conclusions can be drawn from these previous theoretical studies.

1. The ratio of the volumes of liquid and polymer is of interest. When this ratio is less than 20, the equation with finite volume must be used [22, 23].

2. For very short times, equation (8-8) can be used, whatever the ratio.
3. Very often, a partition coefficient exists, and at equilibrium the concentrations of plasticizer in the liquid and PVC are not the same.
4. The problem may be complex, with simultaneous transport of liquid and plasticizer, even if the whole process has not been considered.
5. Diffusivity for each of these two transports may depend largely on the plasticizer concentration in PVC, as well as on the concentration of liquid located in the polymer.
6. It is difficult to extract all the plasticizer from PVC compound in a single run of experiments, and generally the concentration of plasticizer at equilibrium on the PVC surface is far from being zero [57, 59].
7. The whole process can be described only by using mathematical models based on numerical methods with finite differences, and by taking into account all the facts.

### 8-1-4 Results

In the following results obtained from previous studies, the effect of parameters such as the stirring of the liquid, the temperature, the initial concentration of plasticizer in PVC, and the nature of the liquid is of concern.

**Effect of the stirring of the liquid.** The kinetics of transport of plasticizer in methanol are determined either in motionless liquid or in well-stirred liquid. Experiments are done with PVC discs of 35.5 wt % plasticizer (DEHP), at 30°C (22).

The significant effect of stirring is appreciated especially for short times (less than 50 h) when the transport of plasticizer is controlled by boundary layer phenomenon in the liquid phase in contact with the PVC surface.

**Effect of temperature on matter transports.** Experiments are carried out at various temperatures, 30°C, 40°C and 55°C, with well-stirred liquid: benzyl alcohol (BA) and PVC discs with 35.5 wt % DEHP (23).

The amount of matter transported is expressed in g/l for the DEHP

**TABLE 8–2** Effect of stirring on DEHP transport (g/l. in methanol at 30°C. DEHP = 35.5 wt %)

| Time (h) | 2 | 12 | 24 | 48 |
|---|---|---|---|---|
| Motionless liquid | 0.56 | 1.7 | 3.25 | 5.34 |
| Stirred liquid | 0.87 | 2.6 | 3.75 | 5.35 |

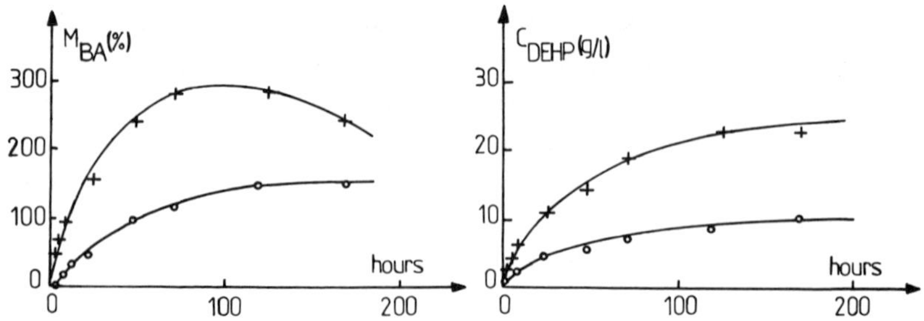


**Figure 8–1** Experimental kinetics of transports at two temperatures: 30°C (○), 55°C (+) with plasticized PVC (25 wt % DEHP). Left: benzyl alcohol (mg/g). Right: DEHP (g/l). (Reprinted from *Polym-Plast. Technol. Eng.*, 20, (1983), 1–22, J. M. Vergnaud, ref. 19, by courtesy of Marcel Dekker Inc.)

in benzyl alcohol and in mg/g of benzyl alcohol in PVC discs (Figure 8–1). Diffusivities are calculated with the help of experiments and equation (8-12), and their temperature dependence is described by Arrhenius law:

$$D = D_0 \cdot \exp - \left(\frac{E}{RT}\right)$$

with the activation energy $E$.

Some conclusions can be drawn:

1. Two matter transports are observed: the one with the liquid which enters the PVC, the other with the plasticizer which leaves the polymer.
2. As the rate of transport is higher for the liquid than for the plasticizer, a swelling is shown at the beginning of the process ($t < 150$ h).
3. The amount of liquid transported in PVC increases at the beginning of the process and passes through a maximum.
4. The effect of temperature on diffusivity is described by the Arrhenius law.

**Effect of concentration of plasticizer.** The kinetics of matter transports are determined [23] by immersing PVC discs with various DEHP concentrations (25, 38, and 50 wt %) in benzyl alcohol at two temperatures:

**TABLE 8–3** Diffusivity ($D.10^8$ cm²/s)

| | 30°C | 40°C | 55°C | $E$(cal/mol) | $D_0$ |
|---|---|---|---|---|---|
| DEHP | 0.8 | 2 | 6 | 16 | $2.5 \times 10^{11}$ |
| BA | 5 | | 10 | 7 | $3.6 \times 10^{14}$ |

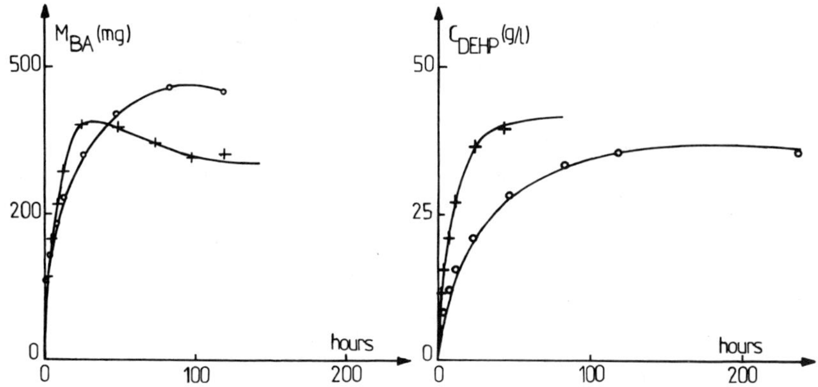


**Figure 8–2** Experimental kinetics of transports at two temperatures: 30°C (○), 55°C (+) with plasticized PVC (38 wt % DEHP). Left: benzyl alcohol (mg/g). Right: DEHP (g/l). (Reprinted from *Polym-Plast. Technol. Eng.*, 20, (1983), 1–22, J. M. Vergnaud, ref. 19, by courtesy of Marcel Dekker Inc.)

30°C and 55°C (Figure 8–1 with 25 wt %, Figure 8–2 with 38 wt %, Figure 8–3 with 50 wt %).

The following conclusions can be drawn:

1. The transport of liquid is coupled with the loss in plasticizer of PVC.
2. The amount of liquid in PVC passes through a maximum value. This value depends on the plasticizer concentration in PVC: the higher the initial plasticizer concentration, the shorter the time at which the maximum is obtained.

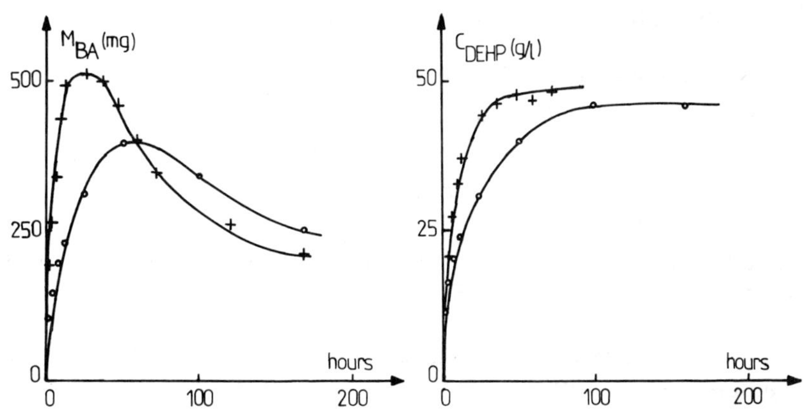


**Figure 8–3** Experimental kinetics of transports at two temperatures: 30°C (○), 55°C (+) with plasticized PVC (50 wt % DEHP). Left: benzyl alcohol (mg/g). Right: DEHP (g/l). (Reprinted from *Polym-Plast. Technol. Eng.*, 20, (1983), 1–22, J. M. Vergnaud, ref. 19, by courtesy of Marcel Dekker Inc.)

**TABLE 8–4** Diffusivity ($D.10^8$ cm²/s) for DEHP and BA.

| wt % | DEHP | 30°C | 55°C | $E$(kcal/ml) | $D_0$ |
|---|---|---|---|---|---|
| DEHP | 25 | 0.8 | 6 | 16 | $2.5 \times 10^{11}$ |
| DEHP | 38 | 8 | 23 | 8.3 | $1.3 \times 10^7$ |
| DEHP | 50 | 12 | 30 | 6.7 | $9 \times 10^5$ |
| BA | 25 | 5 | 10 | 6.7 | $3.6 \times 10^4$ |
| BA | 38 | 12 | 30 | 9 | $3.4 \times 10^7$ |
| BA | 50 | 16 | 68 | 14 | $1.7 \times 10^{11}$ |

3. The effect of the temperature on diffusivities of liquid and plasticizer can be described by Arrhenius law.
4. The increase in plasticizer concentration in PVC is accompanied by a decrease in the activation energy for the plasticizer, as well as in the value of $D_0$.

**Influence of the nature of the liquid.** The nature of the liquid in contact with the plasticized PVC plays an important role in the transport.

The behavior of peanut oil is quite different from that of n-heptane, in spite of the fact that n-heptane is considered as an oil simulant [16, 30]. Thus, the activation energy is nearly zero in the case of n-heptane and around 5 kcal/mole for peanut oil.

The effect of the percent water in alcoholic solutions illustrates also this difference in the behavior of liquids. The kinetics of DEHP transport are obtained at 45°C with PVC initially plasticized at 35.5 wt %, and with the following liquids:

- pure ethanol (0.2 wt % water)
- ethanol water (80/20 wt %)
- ethanol water (50/50 wt %)
- ethanol water (8/92 wt %) (Figure 8–4)

The following conclusions can be noticed from the effect of water in alcoholic solution.

1. There is no transport into liquids having 8 percent ethanol or less, at temperature of 45°C.
2. A significant transport takes place when the percent ethanol is more than 50 percent.
3. The rate of transport is decreased (as well as diffusivity) by increasing the water percent in solutions.

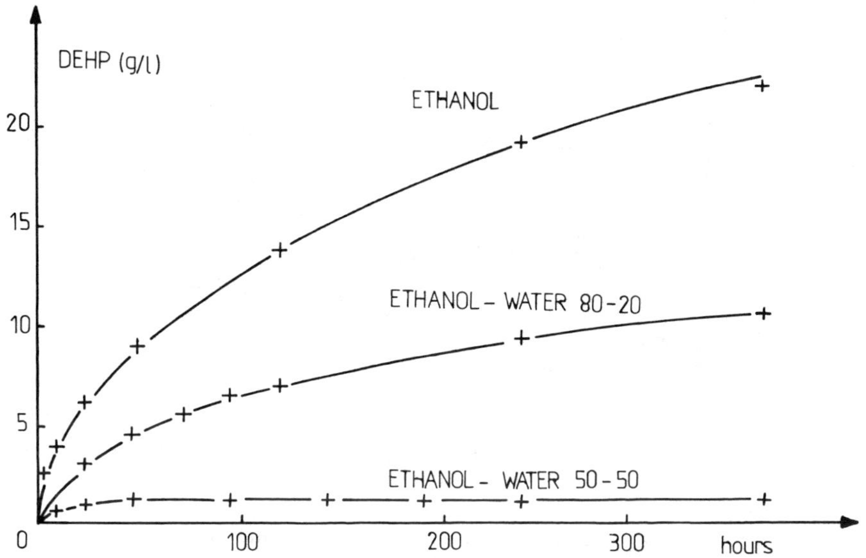


**Figure 8–4** Experimental kinetics of transport of plasticizer from plasticized PVC (35.5 wt % DEHP) into water-ethanol mixtures at 45°C. (Reprinted from *Polym-Plast. Technol. Eng.*, 20, (1983), 1–22, J. M. Vergnaud, ref. 19, by courtesy of Marcel Dekker Inc.)

**4.** The migration process of liquid in PVC is also affected by the water percent in solutions: the lower the water percent, the higher the transfer of alcohol into PVC.

**Profiles of concentration developed within the PVC.** The profiles of concentration have been measured for various liquids, as methanol [22] benzyl alcohol [23] alcoholic solutions [24].

Each PVC disc is peeled into 17 layers parallel to the plane faces. After weighing and dissolution in acetone-carbon disulfide mixture, analysis is performed by gas chromatography after addition of dioctyl adipate and tetradecane as internal standards.

Experiments carried out at various times for the initial plasticizer concentration of 38 clearly show the progression of DEHP and BA through the thickness of the PVC discs (Figure 8–5). The following interesting results can be drawn from these profiles of concentration.

**TABLE 8–5** Diffusivity of DEHP versus water in alcoholic solutions (45°C, 35.5 wt % DEHP)

| % water | 0.2 | 20 | 50 | 92 | 100 |
|---|---|---|---|---|---|
| $D.10^8$ ($cm^2/s$) | 2.7 | 0.2 | 0.01 | 0 | 0 |

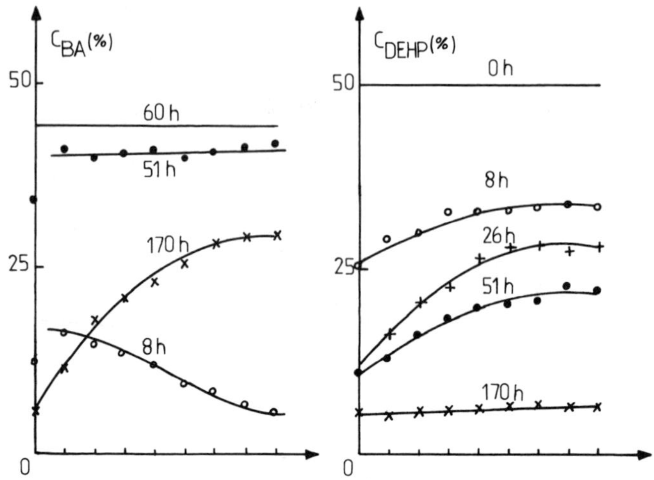


**Figure 8–5** Profiles of concentration measured through the thickness of plasticized PVC (38 wt % DEHP) at 30°C. Left: profiles of benzyl alcohol. Right: profiles of plasticizer. (Reprinted from *Polym-Plast. Technol. Eng.*, 20, (1983), 1–22, J. M. Vergnaud, ref. 19, by courtesy of Marcel Dekker Inc.)

1. Before the maximum value attained at around 60 h for BA, high concentration of BA is observed on the surface.
2. After this maximum value, gradients of concentration are grown with lower concentration of BA on the surface.

**Conclusions.** The whole process of matter transport in the case of a plasticized PVC is very complex, and it is difficult to compare the results produced by different laboratories. In order not to lose too many experimental results, a standardization of technique is necessary [19]. Various parameters are of concern:

- *Stirring*: Stirring the liquid is necessary to reduce the effect of diffusion of plasticizer in the liquid surrounding the PVC.
- *Temperature*: An increase in temperature provokes an increase in the rate of transport, and the effect of temperature on diffusivity can be expressed by Arrhenius law.
- *Plasticizer concentration*: The initial plasticizer concentration is the main parameter. For very low amounts of plasticizer the polymer is in a glassy state, while the PVC behaves like a rubbery polymer when it is plasticized.

Of course, many attempts have been made in order to reduce the transport of plasticizer from plasticized PVC, especially when the blood is contacted with the polymer [60]. One of them is by submitting the polymer to physical treatments [61–63]; another method has been used with complex systems [64–66] of polymers. But these studies have not been successful. Another method has been used, by creating steep gradients of concentrations of plasticizer in plasticized PVC next to the surface by immersing the polymer in alcohol and by drying it after its extraction from the liquid [67]. This method was first developed by using n-heptane as the liquid [68–70].

## 8-2 MATTER TRANSPORTS BETWEEN PVC AND LIQUIDS: MODELING AND SHORT TEST EXPERIMENTS

Studies of the matter transports taking place between plasticized PVC and liquids are difficult because of the problems of identifying the migrating species in the presence of chemically complex structures, and the inconvenience of experiments lasting six months or more. Thus, the simulation of real foodstuffs has become necessary, and shorter tests have been proposed to get the same results for migration: 10 d at 45°C giving the same transport as 6 m at 25°C [18]. But such studies undertaken on short time experiments at higher temperature have shown that this way of working (shorter time with higher temperature) can only give a rough idea of the real-life tests, and results do not always correlate [18, 45, 47, 71, 72].

In fact, the migration of liquids between plasticized PVC and liquids is a complex problem, and the process cannot be described by a single equation derived from the results obtained by two or three experiments conducted at various temperatures. For many liquids, double matter transports take place simultaneously with the plasticizer leaving the PVC while the liquid enters the polymer [16, 19, 22–26].

A new method has been elaborated to resolve this difficult problem, with the following three steps:

1. A good knowledge of the process of transports in real situations as well as of the effects of various parameters (temperature, percent plasticizer, type of liquid) is first needed to determine valuable conditions for short tests. The total amount of liquid and plasticizer transported after infinite time is also obtained.
2. Short tests are of interest for measuring the value of diffusivities either for the liquid or plasticizer, because the concentrations of plasticizer and liquid in PVC are about constant during these short times. Experiments conducted at various concentrations of plasticizer in PVC are of help for determining the dependence of diffusivity with the concentration.
3. Calculations are made by using numerical methods with finite differences. These models take into account all the facts, and for in-

stance, the coupled transport of plasticizer and liquid and concentration-dependent diffusivities. Of course, the validity of the model is tested by comparing the theoretical results with those of the experiments obtained with the help of long real tests.

The model is used in the present chapter in the rather simple case when the amount of liquid in PVC does not pass through a maximum. This case is generally observed with PVC having a high percentage of plasticizer [73]. The more complex case, when the kinetics of liquid entering the PVC pass through a maximum during the process, is considered in the following chapter [74].

### 8-3-1 Experimental

**Materials.** Plane sheets of plasticized PVC are prepared with diethylhexyl phthalate as plasticizer ranging from 20 to 50 wt %. The components are blended either by using a Plastograph, or by soaking the PVC in powder form into a solution of plasticizer and evaporating the solvent from the mixture. Sheets are pressed in a steel mold operated by a press at 150°C for 10 min under pressure of 50 bars. Discs of 18 mm in diameter and 3.0 mm thick are cut from the PVC sheets.

**Apparatus for experiments.** Experiments are carried out by immersing 20 PVC discs in 200 ml of liquid in a closed flask of 500 ml capacity, with a controlled stirring (19). Samples are taken at intervals for analysis, one PVC disc and 10 ml of liquid, so that the weight ratio between the liquid and PVC discs remains constant in the flask.

**Analysis of liquid and plasticizer.** Analysis of plasticizer in the liquid is performed by gas chromatography (Intersmat IGC 16) after addition of dioctyladipate as the internal standard.

The weight of the PVC disc is measured at the same time.

These experiments are conducted either as short tests or as long real ones. These results enable us to determine the kinetics of both transports: the liquid entering the PVC and the plasticizer leaving the polymer.

### 8-3-2 Theoretical

**Assumptions.** The following assumptions are made.

1. A thin sheet of plasticized PVC is considered, with a one-dimensional transport through the thickness.
2. A simultaneous transport of the liquid takes place into the PVC, and the previously dispersed plasticizer out of the PVC.

3. Both these transports are controlled by transient diffusion.
4. Diffusivities for the liquid and plasticizer depend on the concentration of the liquid and plasticizer.
5. No special assumption is made for the concentration of liquid and plasticizer on the PVC surface.

**Mathematical treatment.** The transient one-dimensional diffusion through the thickness of the sheet is expressed by Fick's second law, either for the liquid or plasticizer:

$$\frac{\partial C}{\partial t} = \frac{\partial}{\partial x}\left(D \cdot \frac{\partial C}{\partial x}\right) \tag{8-4}$$

For very short times, the small amount of substance transported $M_t$ at time $t$, as a function of $M_\infty$, the quantity after infinite time, is related to the diffusivity by the simple equation:

$$\frac{M_t}{M_\infty} = \frac{4}{L}\left(\frac{D \cdot t}{\pi}\right)^{0.5} \tag{8-8}$$

where $D$ is the diffusivity and $L$ is the sheet thickness.

For long times, when the diffusivity is constant, the concentration in PVC and the amount of matter transported are given in series form.

**Numerical method.** No analytical solution can be obtained, and the problem is solved by using a numerical method with finite differences.

The thickness of the PVC sheet is divided into equal finite slices of thickness $\Delta x$ by concentration-time reference planes $(n, i)$ (Figure 8–6).

The matter within the sheet balance calculated on the plane $n$ gives the new concentration $CN$ after the elapse of time $\Delta t$ expressed in terms of the concentrations at time $i \cdot \Delta t$ at the same plane $n$ and the adjacent two planes $n - 1$ and $n + 1$:

***Inside the Sheet***

For the liquid

$$CN_n^l = \frac{1}{M_n^l}\left[C_{n-1}^l + (M_n^l - 2)C_n^l + C_{n+1}^l\right] \tag{8-15}$$

with the dimensionless number $M_n^l$

$$M^l = \frac{(\Delta x)^2}{\Delta t} \cdot \frac{1}{D^l} \tag{8-16}$$

and

$$D^l = \exp\left(-\frac{A^l}{C^p + C^l \cdot \alpha} - B^l\right) \tag{8-17}$$

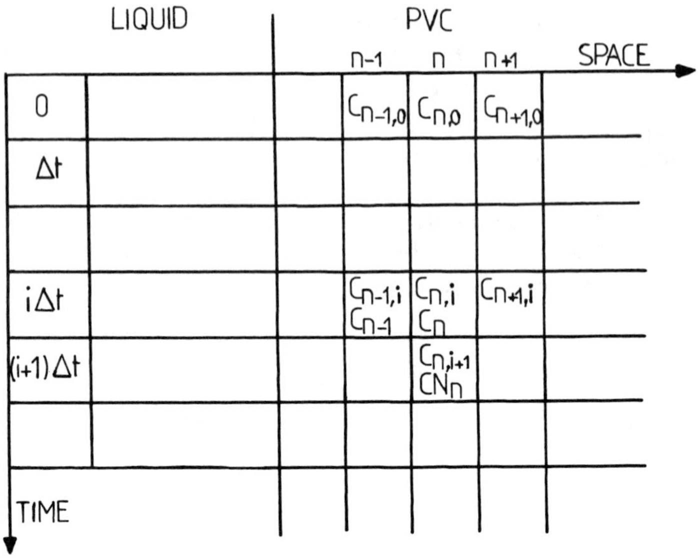


**Figure 8–6** Space-time diagram for calculation of matter concentrations with explicit method with finite differences.

For the plasticizer

$$CN_n^p = \frac{1}{M_n^p}\,[C_{n-1}^p + (M_n^p - 2)\,C_n^p + C_{n+1}^p] \tag{8-15'}$$

with

$$M^p = \frac{(\Delta x)^2}{\Delta t} \cdot \frac{1}{D^p} \tag{8-16'}$$

and

$$D^p = \exp\left(-\frac{A^p}{C^p + C^l \cdot \beta} - B^p\right) \tag{8-17'}$$

***PVC Faces.*** The matter balance is calculated on the slice of thickness $\Delta x$ on the PVC face. The new concentration on the PVC surface $CN_s$ is then obtained:

$$CN_s^l = \frac{1}{M^l}\,[C_{eq}^l + (M_s^l - 2)\,C_s^l + C_1^l] \tag{8-18}$$

$$CN_s^p = \frac{1}{M_s^p}\,[C_{eq}^p + (M_s^p - 2)\,C_s^p + C_1^p] \tag{8-18'}$$

where:

$C^l_{eq}$ and $C^p_{eq}$ are the concentration of liquid and plasticizer at equilibrium and $C^l_1$ and $C^p_1$ are the concentration of liquid and plasticizer at the adjacent plane.

***Amount of Matter Transported.*** The amount of matter (liquid, plasticizer) located in the PVC sheet is determined by integrating the concentration with respect to space. The values of coefficients $\alpha$ and $\beta$ introduced in diffusivities express the effect of the concentration of liquid in PVC on each diffusivity.

***Condition of Stability for Calculation.*** The following condition is necessary for the stability of calculation for the liquid and plasticizer.

$$M - 2 > 0 \tag{8-19}$$

### 8-2-3 Results

**Determination of data with short tests.** The amount of matter transported is determined at various times either for the liquid and plasticizer, by using short tests. During these experiments, the time is short enough so that the amount transported is very low, and the concentration of matter in PVC varies little from its initial value.

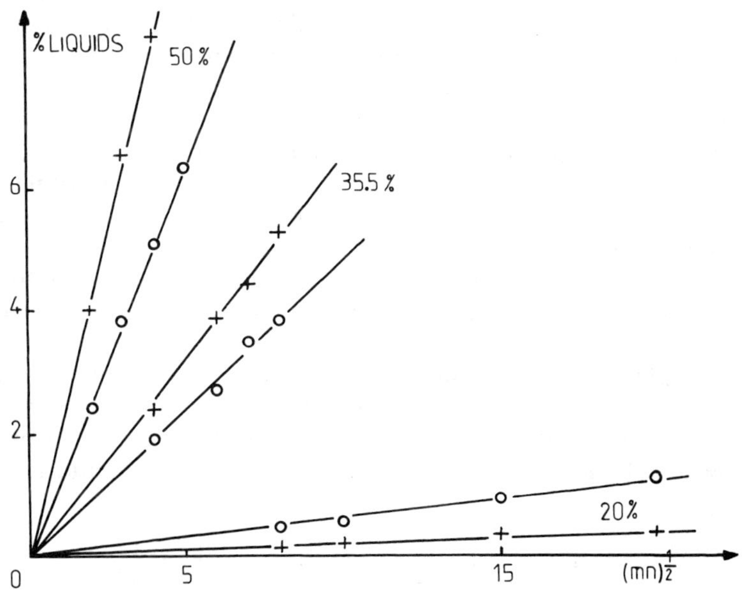


**Figure 8–7** Short-test experiments. Matter transported as a function of the square root of time, for various plasticizer concentrations in PVC sheet: plasticizer (+), n-heptane (O). (Reprinted from *J. Appl. Polym. Sci.*, 29, (1984), 3391–3400, J. L. Taverdet and J. M. Vergnaud, ref. 73, by courtesy of John Wiley and Sons, Inc.)

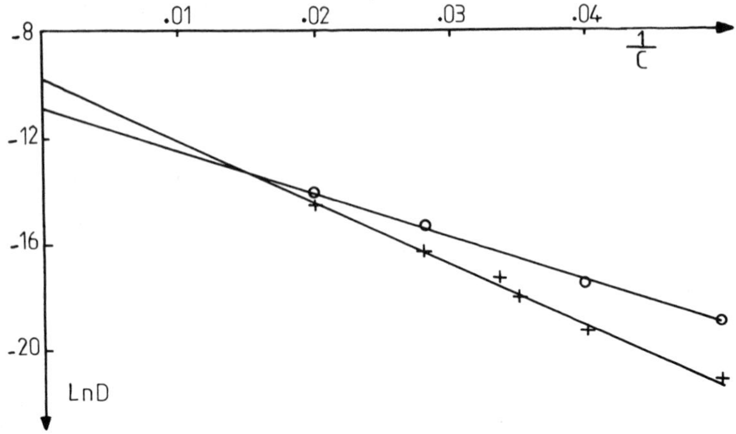


**Figure 8–8** Log *D* as a function of reciprocal of plasticizer concentration in PVC sheet: plasticizer (+), n-heptane (○). (Reprinted from *J. Appl. Polym. Sci.*, 29, (1984), 3391–3400, J. L. Taverdet and J. M. Vergnaud, ref. 73, by courtesy of John Wiley and Sons, Inc.)

For the kinetics treatment, the amount of matter transferred $M_t$ as a fraction of the total amount $M_\infty$, is plotted against the square root of time (Figure 8–7). Straight lines are obtained, proving Fickian diffusion. From the slope of these lines, the diffusivity is calculated by using equation (8-8) and the total amount of matter transported after infinite time which is determined from long real tests.

The values of diffusivity for the plasticizer and liquid (n-heptane) are expressed as a function of the concentration of plasticizer (Figure 8–8). The linear relationship between the log of diffusivity and the reciprocal of plasticizer concentration in PVC is thus written (Table 8–6) in the first line, without coefficient $\alpha$ and $\beta$. Values of $\alpha$ and $\beta$ are then tried and tested by comparing experimental kinetics with the kinetics calculated with them. The best values for these coefficients are shown in the following table.

**Validity of the model with long real experiments.** The profiles of concentration of plasticizer developed through the PVC thickness are calculated at different times (Figure 8–9). The concentration on PVC faces is not equal to zero, whatever the time, because the volume of liquid is not in-

**TABLE 8–6** Diffusivity of plasticizer and n-heptane

| | |
|---|---|
| $\log D^p = -\dfrac{234}{C^p} - 9.7$ | $\log D^1 = -\dfrac{160}{C^p} - 10.9$ |
| $\log D^p = \dfrac{-234}{C^p + 1.6\,C^1} - 9.7$ | $\log D^1 = \dfrac{-160}{C^p + 0.1\,C^1} - 10.9$ |

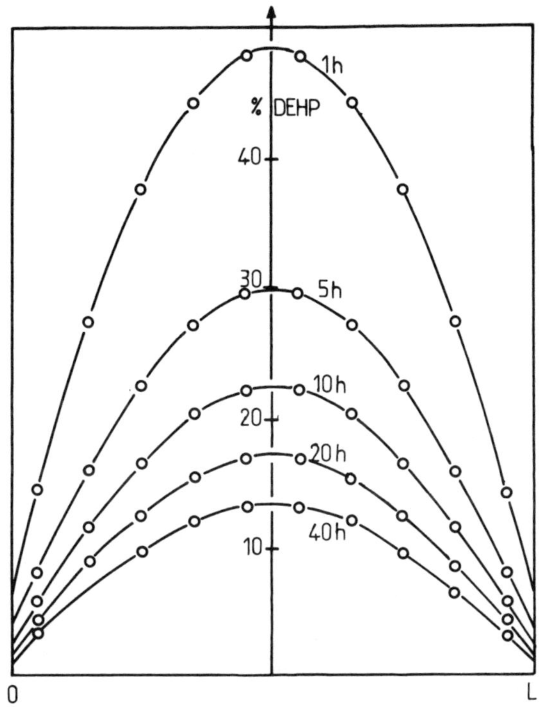


**Figure 8–9** Profiles of concentration of plasticizer developed through the thickness of PVC sheet calculated at various times, at 30°C. Initial percent plasticizer = 50. (Reprinted from *J. Appl. Polym. Sci.*, 29, (1984), 3391–3400, J. L. Taverdet and J. M. Vergnaud, ref. 73, by courtesy of John Wiley and Sons, Inc.)

finite. This fact has also been proved by using attenuated total reflectance measurements carried out on PVC faces [59].

The profiles of concentration through the thickness of PVC sheets are also calculated for the liquid (Figure 8–10). The concentration of n-heptane on PVC surfaces reaches the equilibrium value after only 1 hour of immersion of the PVC in the liquid.

These profiles of concentration have about the same slope as those obtained from experiments by using various liquids as methanol [22] or benzyl alcohol [23]. In the case at hand with n-heptane for the liquid, no experiments are carried out in order to test the validity of these profiles, as has been done previously [22–24]. However, a proof of the validity of these profiles exists with the kinetics of the matter transported, because the amount of matter transported is obtained by integration of the concentration with respect to space.

The validity of the model is tested by comparing the kinetics of matter transported (plasticizer, n-heptane) determined either by experiments or by calculation. A good superimposition of these curves can be appreciated

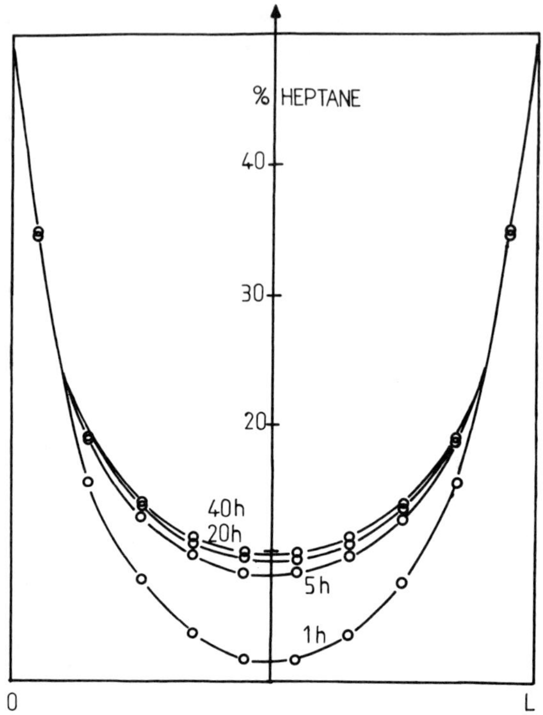


**Figure 8–10** Profiles of concentration of n-heptane developed through the thickness of PVC sheet calculated at various times, at 30°C. Initial percent plasticizer = 50. (Reprinted from *J. Appl. Polym. Sci.*, 29, (1984), 3391–3400, J. L. Taverdet and J. M. Vergnaud, ref. 73, by courtesy of John Wiley and Sons, Inc.)

(Figure 8–11) for long real tests, not only for the plasticizer but also for the liquid. The rate of transport is higher for the liquid than for the plasticizer, and a swelling of about 20 percent results from this fact.

**Conclusions.** The following conclusions can be drawn from these results.

1. The process is complex with a double transport of liquid and plasticizer, controlled by transient diffusion.
2. Short tests done with PVC sheets of various initial concentrations of plasticizer are very convenient to determine the diffusivity and the dependence of diffusivity with the concentration of plasticizer.
3. An extension of the concentration dependence is tried by considering the liquid located in the PVC sheet. In fact, the liquid has also a plasticizing effect in the same way as the plasticizer itself, but with a difference. The concentration of the liquid in PVC does not play

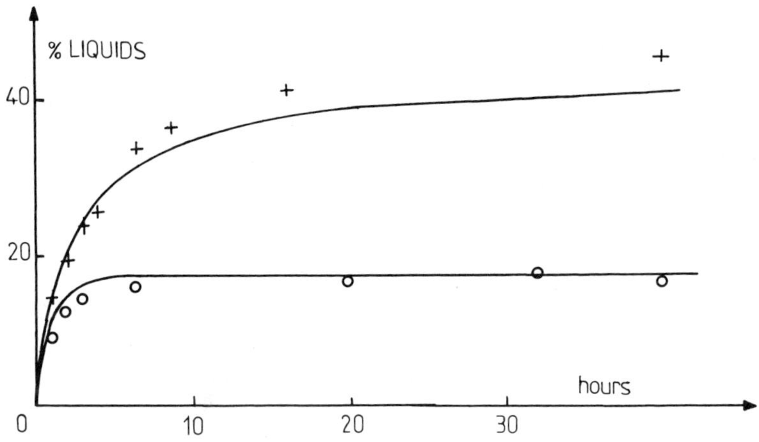


**Figure 8–11** Kinetics of transport of plasticizer out of, and n-heptane into, the PVC. Experiments: plasticizer (+), liquid (○). Calculated: full line −30°C −Initial % plasticizer = 50. (Reprinted from *J. Appl. Polym. Sci.*, 29, (1984), 3391–3400, J. L. Taverdet and J. M. Vergnaud, ref. 73, by courtesy of John Wiley and Sons, Inc.)

exactly the same role for the diffusivity of the liquid and of the plasticizer.

4. The model described previously, taking into account the simultaneous transport, is able to represent the process of long real tests in giving information on the kinetics of matter transport and on the profiles of concentration of matter expanded through the thickness of a PVC sheet.
5. The method, based on the short-test technique and on the simulation with the model, is then able to considerably reduce the time of experiments.

## 8-3 MATTER TRANSPORTS BETWEEN PVC AND LIQUIDS IN CASE OF A MAXIMUM FOR LIQUID-TIME HISTORY

In all applications of plasticized PVC for packaging, the polymer is in contact with the liquid: blood and blood components [2, 3, 10–14, 31–37] with extensively plasticized PVC, petroleum oils and aromatic hydrocarbons [20], alcohols [21–24], drugs and cosmetics or pharmaceutical products, and liquid foods [7–9, 17–19, 27–30]. Under these conditions, a migration of liquid and plasticizer may take place with the following results:

1. The surrounding liquid is contaminated by the plasticizer and additives.

2. Because of a loss in plasticizer, there is a decrease in mechanical properties of the PVC, essentially the flexibility.

In a variety of contexts, there is a simultaneous transport of the liquid into, and the plasticizer out of, the PVC [22–26]. Both these transports are controlled by transient diffusion, with diffusivities depending on the concentration of the liquid and plasticizer in PVC [30, 73].

A model based on a numerical method with finite differences takes into account the following facts:

1. The double transport controlled by transient diffusion.
2. The diffusivity which depends on the concentration of liquid and plasticizer.
3. The concentration of plasticizer and liquid reaches the values at equilibrium, after infinite time. These values of concentration at equilibrium are constant, and it describes the whole process when no maximum is observed in the kinetics of liquid transported.

But very often, the amount of liquid transported into PVC follows a more complex phenomenon [23, 74]. This amount increases rapidly, rises to a maximum value, and then decreases. In all cases, the maximum of liquid absorbed depends on the plasticizer content in the PVC. The following model is an extension of the previous model with finite differences [73], with two additional assumptions applied to the surface conditions:

1. The concentration at equilibrium of the liquid located in the PVC depends on the value of plasticizer content in polymer.
2. Moreover, a coefficient of matter transport is considered.

### 8-3-1 Experimental

**Materials.** PVC compounds are prepared with diethylhexylphthalate (DEHP) as plasticizer ranging from 20 percent to 30 percent by weight, and a commercial PVC resin, Lucovyl (Rhone-Poulenc), in white powder form. The two components are mixed by immersing the polymer in a solution of the plasticizer and evaporating the solvent. PVC compounds are pressed into sheets by using a steel mold operated by a press at 150°C for 10 min under a pressure of 50 bars. Discs of 18 mm in diameter and of thickness 3 mm and 1 mm are cut from these various sheets.

**Apparatus for experiments.** Experiments are carried out by soaking PVC discs in 200 ml of n-heptane as the liquid in a closed flask with a controlled rate of stirring. At intervals, the PVC disc is weighed, and a small amount (1 ml) of liquid is taken for analysis of plasticizer. The concentration of plasticizer in the liquid is measured by gas chromatography (IGC 16, Intersmat) after addition of dioctyladipate as internal standard. Experiments are conducted either as short tests or as long real tests.

### 8-3-2 Theoretical

**Assumptions.** The assumptions made in this chapter are the same as those in the previous chapter (without maximum for the amount of liquid transported in PVC), with an exception for the assumptions concerned with the PVC surface.

1. As a thin sheet of plasticized PVC is used, a one-dimensional transport through the thickness is considered.
2. Two matter transports take place simultaneously: The liquid enters the PVC and the plasticizer leaves the polymer.
3. Both these transports are controlled by transient diffusion.
4. Diffusivities of the liquid and plasticizer are expressed in terms of the concentration of liquid and plasticizer.
5. The concentration of liquid at equilibrium in the PVC is determined as a function of the concentration of plasticizer.
6. A coefficient of matter transport is introduced at the liquid-PVC interface.

**Mathematical treatment.** Transient diffusion through the thickness of a PVC sheet is described by Fick's second equation:

$$\frac{\partial C}{\partial t} = \frac{\partial}{\partial x}\left(D \cdot \frac{\partial C}{\partial x}\right) \tag{8-4}$$

with concentration-dependent diffusivity, either for the liquid or for plasticizer.

Under the preceding conditions, there is no analytical solution for the problem. However, for very short times, the small amount of substance transported $M_t$ at time $t$ is so low that the concentration of plasticizer and liquid may be considered constant during the test. The amount of matter transported at time $t$, as a fraction of the quantity transported after infinite time $M_\infty$, can be expressed in terms of the square root of time. The following equation is of interest for the determination of the diffusivity.

$$\frac{M_t}{M_\infty} = \frac{4}{L}\left(\frac{D \cdot t}{\pi}\right)^{0.5} \tag{8-8}$$

**Numerical analysis.** The problem is solved with the help of a numerical method with finite differences.

The thickness of the PVC sheet is divided into finite slices of thickness $\Delta x$ by concentration-time reference planes with the integers $n$ and $i$ characterizing the position and time (Figure 8–12).

The balance of matter (liquid, plasticizer) calculated on the plane $n$ enables one to express the new concentration $CN$ after the elapse of time $\Delta t$ in terms of the concentration at the previous time $i \cdot \Delta t$ at the same plane $n$ and the adjacent two planes $n - 1$ and $n + 1$:

***Inside the Sheet***

For the liquid

$$CN_n^l = \frac{1}{M_n^l} [C_{n-1}^l + (M_n^l - 2)C_n^l + C_{n+1}^l] \tag{8-15}$$

with the dimensionless number $M_n^l$

$$M_n^l = \frac{(\Delta x)^2}{\Delta t} \cdot \frac{1}{D_n^l} \tag{8-16}$$

and the diffusivity $D^l$

$$D^l = \exp\left(\frac{-A^l}{C^p + C^l \cdot \alpha^l} - B^l\right) \tag{8-17}$$

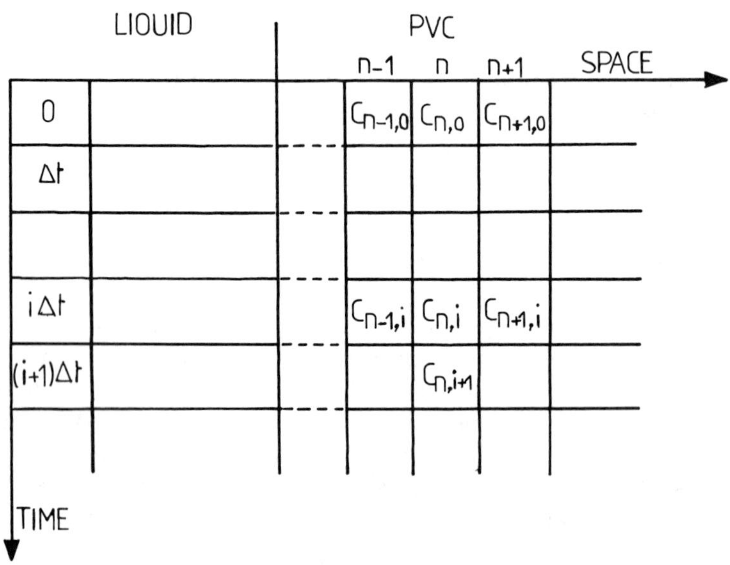


**Figure 8–12** Space-time diagram for calculation of concentrations of matter (liquid, plasticizer) through the PVC thickness.

For the plasticizer

$$CN_n^p = \frac{1}{M_n^p}\,[C_{n-1}^p + (M_n^p - 2)C_n^p + C_{n+1}^p] \tag{8-15'}$$

with the dimensionless number $M^p$

$$M_n^p = \frac{(\Delta x)^2}{\Delta t}\cdot\frac{1}{D_n^p} \tag{8-16'}$$

and diffusivity $D^p$

$$D^p = \exp\left(\frac{-A^p}{C^p + C^l\cdot\alpha^p} - B^p\right) \tag{8-17'}$$

***PVC Faces.*** The matter balance is calculated in the slice of thickness $\Delta x/2$ next to the PVC surface, by considering the coefficient of matter transport on the PVC surface. The new concentration $CN_o$ on the surface is obtained as a function of the concentration obtained at the previous time on the surface and plane 1, as well as of the concentration at equilibrium $C_{eq}$.

For the liquid

$$CN_o^1 = \frac{1}{M_o^l}\,[2N^1C_{eq}^1 + (M_o^1 - 2N^1 - 2)C_o^1 + 2C_1^1] \tag{8-18}$$

with the dimensionless numbers $M^1$ shown previously and $N^1$

$$N^1 = \frac{h^1\cdot\Delta x}{D_o^1} \tag{8-20}$$

where $h$ is the coefficient of matter transport through the liquid-PVC interface.

For the plasticizer

$$CN_o^p = \frac{1}{M_o^p}\,[2N^pC_{eq}^p + (M_o^p - 2N^p)C_o^p + 2C_1^p] \tag{8-18'}$$

with the dimensionless numbers $M^p$ and $n^p$

$$N^p = \frac{h^p\cdot\Delta x}{D_o^p} \tag{8-20'}$$

The coefficients $A$ and $B$ are constant, and the coefficient $\alpha$ evaluates the contribution of the liquid according to the expression:

$$\alpha_n = \frac{V_n^1}{V_n^1 + V_n^p} \tag{8-21}$$

where $V_n^1$ and $V_n^p$ are the volumes of liquid and plasticizer located in plane $n$.

The concentration at equilibrium of the liquid, which is used for calculating the concentration on the PVC surface, is expressed in terms of the concentrations of the liquid and plasticizer on the PVC surface.

$$C_{eq}^1 = \frac{C_o^1}{C_o^1 + C_o^p} C \tag{8-22}$$

where $C$ is a constant.

The coefficient of matter transport $h$ is determined as follows:

$$h = \frac{1}{S \cdot \Delta C} \cdot \frac{dM}{dt} \tag{8-23}$$

where $S$ is the surface of PVC in contact with liquid and $\Delta C$ is the difference between the concentration at equilibrium and the concentration on the surface.

***Total Amount of Matter Located in PVC.*** The total amount of liquid and plasticizer located in the PVC sheet is obtained by integrating the respective concentrations with respect to space.

***Conditions of Stability for Calculation.*** A condition of stability must be considered in calculation for the liquid and plasticizer.

$$M - 2N - 2 > 0 \tag{8-19'}$$

### 8-3-4 Results

**Data obtained from short tests and long tests.** Experiments are carried out at 30°C with the well-stirred liquid, by using various PVC discs with a plasticizer concentration ranging from 20 to 50 wt %.

Diffusivities of liquid and plasticizer are calculated from the slopes of the straight lines obtained by plotting the amount of matter transported versus the square root of time. As the logarithm of diffusivity for the liquid and plasticizer is found to vary linearly with the reciprocal of concentration of plasticizer, the coefficients $A$ and $B$ can be calculated (Figure 8–8). These coefficients $A$ and $B$ appearing in the equations of diffusivity are shown in Table 8–7.

The coefficient of matter transport $h$ is also determined from short-test measurements when the difference in the concentration $\Delta C$ is well known.

The value of diffusivity and coefficient of matter transport are higher for the liquid than for the plasticizer. Moreover, these values are increased when the plasticizer concentration in PVC is increased.

**TABLE 8–7** Data for matter transports

| | % DEHP | $A$ | $B$ | $h.10^6$ | $C^1_{eq}$ |
|---|---|---|---|---|---|
| Plasticizer | 20 | 234 | 9.7 | 8 | |
| | 25 | 234 | 9.7 | 18 | |
| | 30 | 234 | 9.7 | 30 | |
| Liquid | 20 | 160 | 10.9 | 42 | 0.77 |
| | 25 | 160 | 10.9 | 75 | 0.80 |
| | 30 | 160 | 10.9 | 89 | 0.85 |

**Validity of the model.** The validity of the model as well as the accuracy of the data concerned with diffusivity and amount of matter transported at equilibrium are tested by comparing the experimental kinetics of transport with the calculated ones. The kinetics of plasticizer and n-heptane transported in long real tests are drawn in the following figures, at 30°C:

- Figure 8–13 with an initial plasticizer concentration of 20 wt %.
- Figure 8–14 with an initial plasticizer concentration of 25 wt %.
- Figure 8–15 with an initial plasticizer concentration of 30 wt %.

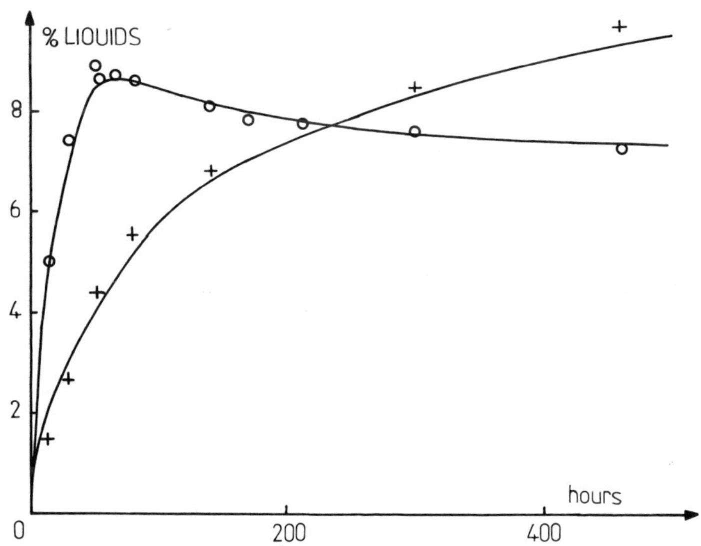


**Figure 8–13** Kinetics of transport of plasticizer and n-heptane. Experiments: plasticizer (+), liquid (O). Calculated: full line – 30°C – Initial wt % plasticizer = 20. (Reprinted from *J. Appl. Polym. Sci.*, 31, (1986), 111–122, J. L. Taverdet and J. M. Vergnaud, ref. 74, by courtesy of John Wiley and Sons, Inc.)

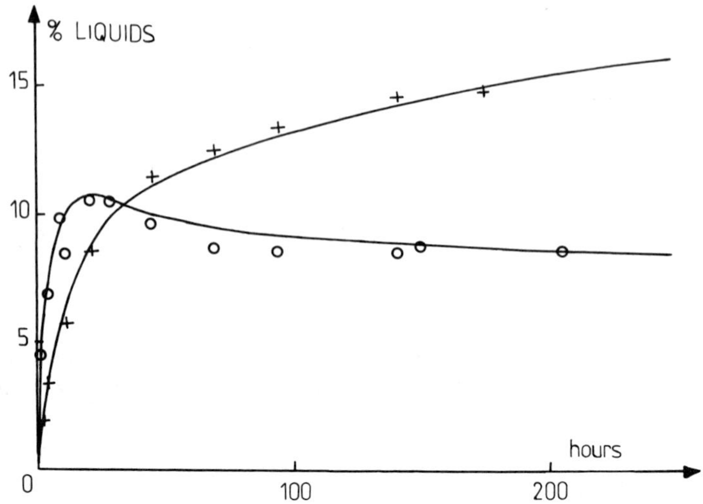


**Figure 8–14** Kinetics of transport of plasticizer and n-heptane. Experiments: plasticizer (+), liquid (○). Calculated: full line – 30°C – Initial wt % plasticizer = 25. (Reprinted from *J. Appl. Polym. Sci.*, 31, (1986), 111–122, J. L. Taverdet and J. M. Vergnaud, ref. 74, by courtesy of John Wiley and Sons, inc.)

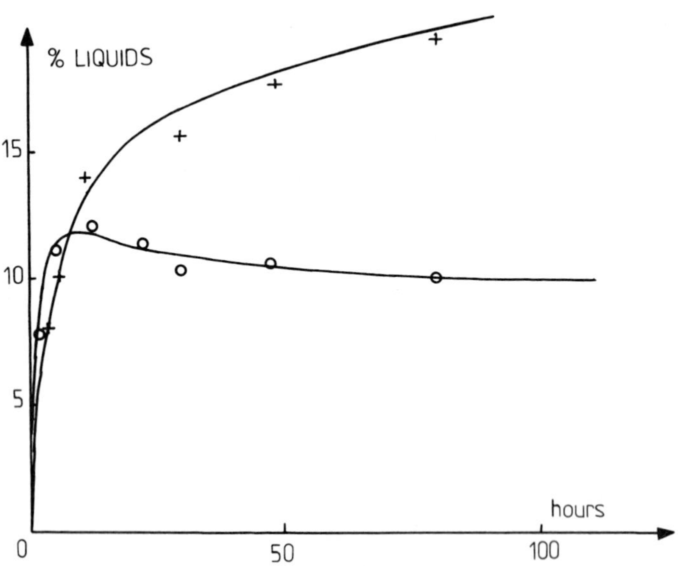


**Figure 8–15** Kinetics of transport of plasticizer and n-heptane. Experiments: plasticizer (+), liquid (○) – Calculated: full line – 30°C – Initial wt % plasticizer = 30. (Reprinted from *J. Appl. Polym. Sci.*, 31, (1986), 111–122, J. L. Taverdet and J. M. Vergnaud, ref. 74, by courtesy of John Wiley and Sons, Inc.)

As shown in these three figures, the experimental and calculated kinetics are very well superimposed, proving the quality of the model and data. The following results of interest are worth noting.

1. The behaviors of the liquid and plasticizer are quite different. The plasticizer is continuously extracted from the PVC and transported into the liquid until equilibrium is reached after a long time. A complex transport is shown for the liquid, the amount of liquid located in the polymer passing through a maximum.
2. The effect of the value of the initial concentration of plasticizer in PVC on the maximum reached by the amount of liquid in PVC is of importance. The higher the percentage of plasticizer in PVC, the shorter the time at which the maximum is observed.
3. A swelling takes place at the beginning of the process, the rate of transport being higher for the liquid than for the plasticizer. The extent of the swelling depends on the initial concentration of plasticizer in PVC with the statement: the lower the percentage of plasticizer, the larger the swelling.
4. The swelling extent also passes through a maximum, and then decreases. After the two kinetic curves of plasticizer and liquid intersect, the swelling is followed by a shrinkage when the amount of plasticizer transported is higher than the amount of liquid.
5. The time at which the two kinetic curves intersect varies with the initial plasticizer concentration in PVC: the higher the initial concentration of plasticizer, the longer this time.
6. At the end of the experiment, the PVC sheets are shrunk. This shrinkage extent is increased when the initial percentage of plasticizer in PVC is increased.

The profiles of concentration of all liquids can be calculated with the model, as they are developed through the thickness of the PVC sheet at various times. Two profiles are drawn: the one in Figure 8–16 in the case of a PVC sheet with the initial weight percentage of plasticizer of 20, and the other in Figure 8–17 with the initial value of 25 wt %. These curves afford further insight into the nature of phenomena taking place within the PVC sheets:

1. The validity of these profiles is proved by the agreement obtained between the experimental and calculated kinetics of transport, these kinetics being obtained by integrating these profiles of concentration with respect to space.
2. A continuous decrease in the concentration of plasticizer can be appreciated with respect to time and position in PVC sheets.

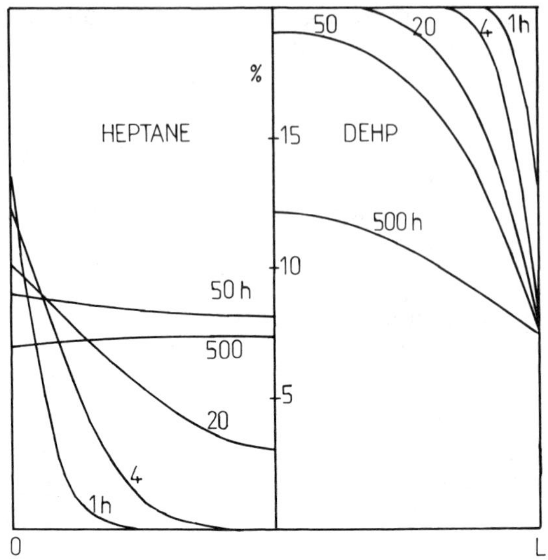


**Figure 8–16** Profiles of concentration of liquids expanded through the PVC thickness at 30°C, with initial wt % plasticizer of 20. Left: n-heptane. Right: plasticizer. (Reprinted from *J. Appl. Polym. Sci.*, 31, (1986), 111–122, J. L. Taverdet and J. M. Vergnaud, ref. 74, by courtesy of John Wiley and Sons, Inc.)

3. A complex transport is observed for n-heptane, and the gradients of concentration are quite different before and after the maximum of liquid located in PVC.
4. The maximum of liquid located in PVC corresponds with a change in the slope of these gradients of concentration from negative to positive at about 50 h for 20 wt % plasticizer and 10 h for 30 wt %.

**Conclusions.** The process of matter transport taking place when plasticized PVC is in contact with liquids is very complex for the following reasons [74]:

1. Two matter transports are observed: the liquid entering the PVC, while the plasticizer leaves the polymer.
2. Both these transports are controlled by transient diffusion, and the diffusivities are expressed in terms of the concentration of the liquid and plasticizer.
3. The rate of transport is higher for the liquid than for the plasticizer at the beginning of the process, provoking a swelling of the PVC.
4. As the amount of the liquid located in the polymer passes through a maximum, and the amount of plasticizer leaving the polymer in-

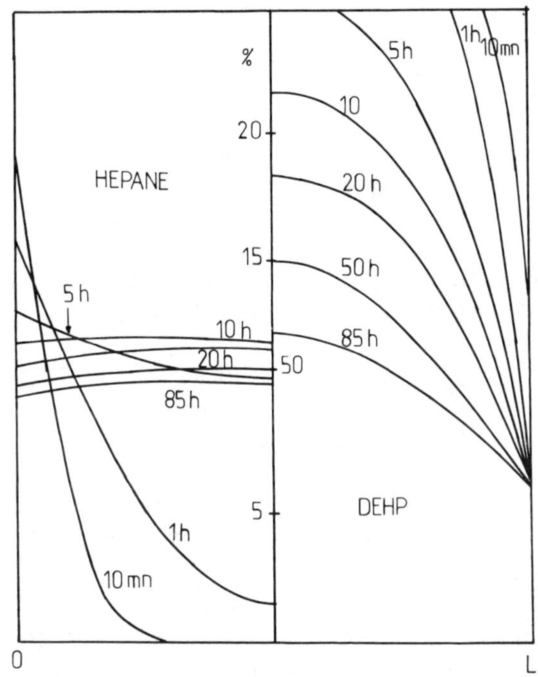


**Figure 8–17** Profiles of concentration of liquids expanded through the PVC thickness at 30°C, with initial wt % plasticizer of 25. Left: n-heptane. Right: plasticizer. (Reprinted from *J. Appl. Polym. Sci.*, 31, (1986), 111–122, J. L. Taverdet and J. M. Vergnaud, ref. 74, by courtesy of John Wiley and Sons, Inc.)

creases continuously, the two kinetic curves intersect, and a shrinkage of the PVC sheet is observed.

Although these transport phenomena are very complicated, the mathematical model is proved to describe the process very well from the beginning to end.

Another point is worth noting because it is the main characteristic in the process of matter transport in plasticized PVC. When plasticized PVC is immersed into a liquid and extracted before the time of equilibrium, two types of profiles of concentration are expanded through the PVC thickness.

1. A gradient of concentration of plasticizer with lower concentration of plasticizer on the PVC surface.
2. A gradient of concentration of liquid with a higher concentration of liquid on the PVC surface, when the time of immersion is shorter than the time required to attain the maximum of liquid transported.

This last result is the basis for bringing forward a theoretical explanation of two phenomena which appear in case of plasticized PVC:

1. The high difficulty to evaporate the liquid which has entered the PVC.
2. The preparation of plasticized PVC with perfect flexibility and low matter transports when in contact with liquids.

## 8-4 DRYING OF PLASTICIZED PVC PREVIOUSLY IMMERSED INTO LIQUID: MODELING AND EXPERIMENTS

Plasticized PVC is very often used in the preparation of packagings for liquid foods, blood, hydrocarbons, and other kinds of liquids [19] with some advantages and some drawbacks.

1. The advantages of PVC result certainly from the low price of the components (PVC and plasticizer) and also from the flexibility of the final material which can be adjusted to the desired level by addition of the right extent of plasticizer.
2. The main drawback is due to the presence of plasticizer in PVC. Generally two matter transports take place when the PVC comes into contact with the liquid. The liquid enters the PVC, while the plasticizer leaves the polymer.

Both these matter transports are controlled by transient diffusion with concentration-dependent diffusivities. Following these transports, two kinds of profiles of concentration are expanded through the thickness of PVC sheets. One is concerned with the plasticizer with a low concentration on the PVC surface, the other with the liquid and a high concentration on the polymer surface [22–26, 29, 30, 59, 68–70, 73, 74].

An application of interest is the preparation of soft and flexible plasticized PVC sheet able to release the plasticizer with a very low rate when the polymer comes into contact with a liquid. This preparation consists of the following two-step process.

1. Immersion of the plasticized PVC sheet into the liquid during a short time.
2. Extraction of the sheet out of the liquid and evaporation of the liquid out of the PVC sheet.

During the first step, the two previously described profiles of concentration are created through the PVC sheet, with high gradients next to the surface as short tests are considered.

The second step involves the evaporation of the liquid, and this process is of such importance for the quality of the final plasticized PVC that a fundamental study is of interest.

The process of drying in the case of PVC is not simple because the rate of drying is controlled not only by the rate of evaporation of the liquid on the PVC surface, but also by diffusion of the liquid through the thickness of the plasticized PVC sheet. The rate of drying is unusually low because the diffusivity of the liquid depends largely on the concentration of the plasticizer and liquid, and because of the presence of steep gradients of concentration of plasticizer.

### 8-4-1 Experimental

**Materials.** Sheets of plasticized PVC with the following characteristics are used:

- Thickness = 0.060 cm    Area = $S$ = 9 $cm^2$/side.
- Percent plasticizer = 30 by weight (30 g plasticizer mixed with 100 g pure PVC, and 2 g additives).
- Plasticizer = DEHP (diethylhexylphthalate).

The liquid chosen is n-hexane, as this liquid has been selected by the Food and Drug Administration for simulating liquid foods.

**Apparatus for experiments.** The thin PVC sheet, rectangular in shape (3 × 3 cm) with a weight of around 0.65 g, is immersed in 100 ml of liquid at 20°C in a flask with controlled stirring. The PVC sample is weighed at intervals, while the concentration of plasticizer in the liquid is determined by UV-spectrometry (U-1100 Hitachi).

Two types of experiments are carried out for absorption.

- One conducted during a long time in order to determine the kinetics of transport of the liquid and plasticizer, and also the amount of matter transported at equilibrium (after infinite time).
- The second performed during a very short time (for instance, 15 min).

The desorption process is studied by evaporating the liquid from the PVC sheet previously immersed in the liquid for 15 min. The desorption is done under controlled conditions of temperature until equilibrium is attained, and the sample is weighed at intervals. Various values are chosen for the temperature: 20°C, 85°C, and 120°C.

### 8-4-2 Theoretical

**Assumptions.** The following assumptions are made for the steps of absorption and desorption of liquid.

#### ABSORPTION

1. Two matter transports take place during the first step of immersion of a PVC sheet into the liquid. The liquid enters the PVC while the plasticizer leaves the polymer.
2. These two matter transports are controlled by transient diffusion. Constant diffusivity for each transfer is taken to simplify the problem, this test being very short.
3. During this step of immersion, the concentration of each matter (liquid, plasticizer) on the PVC surface reaches equilibrium as soon as the process starts. Special attention is given to the equilibrium concentration of the liquid.

#### DESORPTION

1. The desorption of liquid is controlled by diffusion through the PVC sheet and by evaporation on the surface.
2. The evaporation rate is proportional to the concentration of the liquid on the PVC surface.
3. The diffusivity of the liquid is expressed in terms of the concentration of the liquid and plasticizer in the PVC.
4. The plasticizer does not diffuse during the desorption step. This assumption is correct when desorption is conducted at low temperature for a short time.

**Mathematical treatment.** The equation of one-direction diffusion is:

$$\frac{\partial C}{\partial t} = \frac{\partial}{\partial x}\left(D \cdot \frac{\partial C}{\partial x}\right) \tag{8-4}$$

where $D$ is the diffusivity.

***Absorption.*** The initial and boundary conditions are:

| | | | | | |
|---|---|---|---|---|---|
| Initial | $t = 0$ | $0 < x < L$ | $C = 0$ | PVC | (8-24) |
| Boundary | $t > 0$ | $x = 0, x = L$ | $C = C_{eq}$ | surface | (8-25) |

When the diffusivity is constant, the concentration of the liquid and plasticizer is given by the trigonometrical series in terms of time and space.

$$C_{x,t} = \frac{4\, C_{eq}}{\pi} \sum_{n=0}^{\infty} \frac{1}{2n + 1} \cdot \sin\left(\frac{(2n + 1)\,\pi x}{L}\right) \cdot \exp\left[-\frac{(2n + 1)^2\, \pi^2}{L^2}\, Dt\right] \quad (8\text{-}26)$$

with $L$ the thickness of the PVC sheet.

The amount of matter transported (liquid, plasticizer), $M_t$, is expressed as a fraction of the matter transported after infinite time, $M_\infty$.

$$\frac{M_\infty - M_t}{M_\infty} = \frac{8}{\pi^2} \cdot \sum_{n=0}^{\infty} \frac{1}{(2n + 1)} \exp\left[-\frac{(2n + 1)^2\, \pi^2}{L^2}\, Dt\right] \quad (8\text{-}27)$$

For very short times, the value of the matter transported at time $t$ is given by the simple equation:

$$\frac{M_t}{M_\infty} = \frac{4}{L}\left(\frac{D \cdot t}{\pi}\right)^{0.5} \quad (8\text{-}8)$$

***Desorption.*** The initial and boundary conditions are for the liquid (n-heptane), since the polymer is assumed to be motionless.

| | | | | | |
|---|---|---|---|---|---|
| Initial | $t = 0$ | $0 < x < L$ | $C = C_{in}$ | PVC | (8-28) |
| Boundary | $t > 0$ | $D\dfrac{\partial C}{\partial x} = -F_o \dfrac{C_o}{\rho}$ | | surface | (8-29) |

where $C_o$ is the actual concentration of liquid on the PVC surface, $F_o$ is the evaporation rate of the pure liquid, and $\rho$ is the density of the liquid,

As the diffusivity of the liquid is a function of the concentrations of the liquid and plasticizer, no analytical solution can be found, and the problem must be solved by using a numerical method with finite differences.

**Numerical analysis for desorption.** The cross section of the PVC sheet of thickness $L$ is divided into $N$ equal finite slices of thickness $\Delta x$, by concentration-time reference planes, $n$ and $i$ being integers characterizing the position and time (Figure 8–18).

***Within the PVC.*** The balance of the liquid transported in plane $n$ during the increment of time $\Delta t$ leads to the relation between the new concentration at position $n$ after the elapse of time $\Delta t$, $CN_n$, and the concentrations at the previous time $i \cdot \Delta t$ at the same position, $C_n$, and adjacent planes.

$$CN_n = \frac{1}{M_n}\left[C_{n-1} + (M_n - 2)C_n + C_{n+1}\right] \quad (8\text{-}30)$$

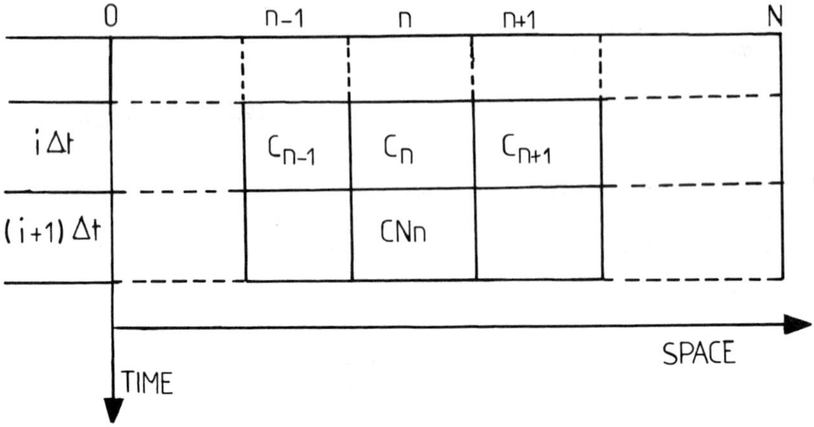


**Figure 8–18** Space-time diagram for numerical analysis.

with the dimensionless number $M_n$

$$M_n = \frac{(\Delta x)^2}{\Delta t} \cdot \frac{1}{D_n} \tag{8-31}$$

with the diffusivity expressed in terms of the concentrations of the liquid and plasticizer.

$$D_n = D_0 \cdot \exp \frac{-A}{C_n^l + C_n^p} \tag{8-32}$$

where $C_n^l$ and $C_n^p$ are the concentration of liquid and plasticizer at position $n$ and time $i \cdot \Delta t$.

***PVC Surface.*** The basic boundary condition can be written, by considering the slice of thickness $\Delta x/2$ next to the PVC surface.

$$D \cdot \left(\frac{\partial C}{\partial x}\right)_{0.5} - F_0 \cdot \frac{C_0}{\rho} = \left(\frac{\partial C}{\partial x}\right)_{0.5} \cdot \frac{\Delta x}{2} \tag{8-33}$$

By assuming that

$$CN_{0.25} - C_{0.25} = CN_0 - C_0 \tag{8-34}$$

and using finite differences, the boundary condition becomes:

$$CN_o = C_o + 2 \frac{\Delta t}{(\Delta x)^2} D_{0.5} \cdot (C_1 - C_o) - 2 \frac{F_o \cdot C_o}{\rho} \cdot \frac{\Delta t}{\Delta x} \tag{8-35}$$

where $F_o$ is expressed in g/cm$^2$·s and $D_{0.5}$ is the diffusivity of the liquid at position 0.5 and time $i \cdot \Delta t$.

The diffusivity $D_{0.5}$ is calculated with the equation of diffusivity expressed in terms of concentrations of liquid and plasticizer. The concentrations of liquid and plasticizer at position 0.5 are the mean value of concentration taken at position 0 and 1.

***Amount of Liquid Evaporated.*** The amount of liquid evaporated during the process of desorption is determined in two ways. One is by integrating the rate of evaporation with respect to time.

$$M_t = \sum_{i=0}^{i} S \frac{F_o \cdot C_o}{\rho} \Delta t \tag{8-36}$$

The other is by considering the amount of liquid remaining in the PVC sheet, which is obtained by integrating the concentration of liquid with respect to space.

### 8-4-3 Results

The whole process consists of two steps, with each step considered successively.

1. The immersion of plasticized PVC sheet in n-hexane for a short time.
2. The evaporation and diffusion of the liquid which is located in the PVC at the end of the first step.

**Immersion of the PVC sample in the liquid.** Two series of PVC samples, with the same dimension (3 × 3 × 0.6 cm) and a weight around 0.65 mg, are immersed in n-hexane (100 ml for each PVC sample). The former series of PVC samples is used for determining the kinetics of matter transports, and the latter series is used for preparing the PVC samples necessary for desorption studies.

***Kinetics of Matter Transports During Absorption.*** The PVC sample is immersed in the liquid for a long time, up to equilibrium of matter transports, and it is weighed at intervals while the concentration of plasticizer in the liquid is measured.

Two matter transports take place simultaneously, the one with the liquid entering the PVC, the other with the plasticizer leaving the PVC. The kinetics of these transports are drawn (Figure 8–19) and the following facts are worth noting.

1. The rate of transport is higher for the liquid than for the plasticizer, at the beginning of the process for times shorter than 1 h.
2. Both these transports are controlled by diffusion, as indicated by the square root of time dependence of the amount of matter transported (Figure 8–20).

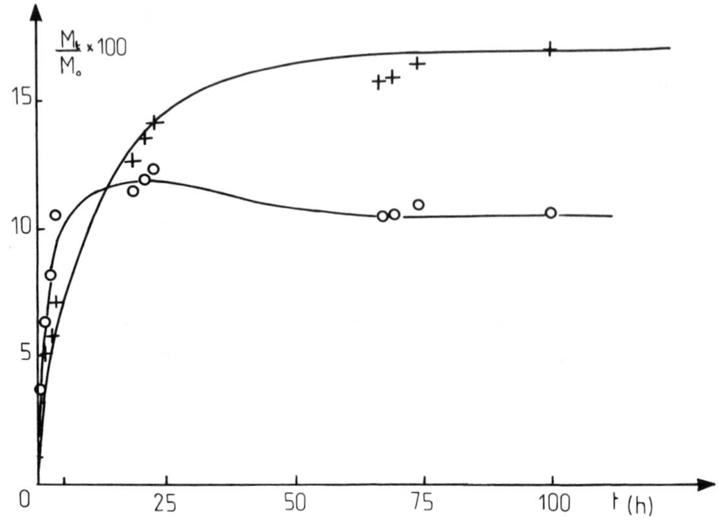


**Figure 8–19** Kinetics of transports of liquid during the absorption process at 20°C. Plasticizer (+)–n-heptane (○). (Reprinted from *Europ. Polym. J.*, 26, 1990, 285–91, A. Aboutaybi, J. Bouzon and J. M. Vergnaud, ref. 70, by courtesy of Pergamon Press.)

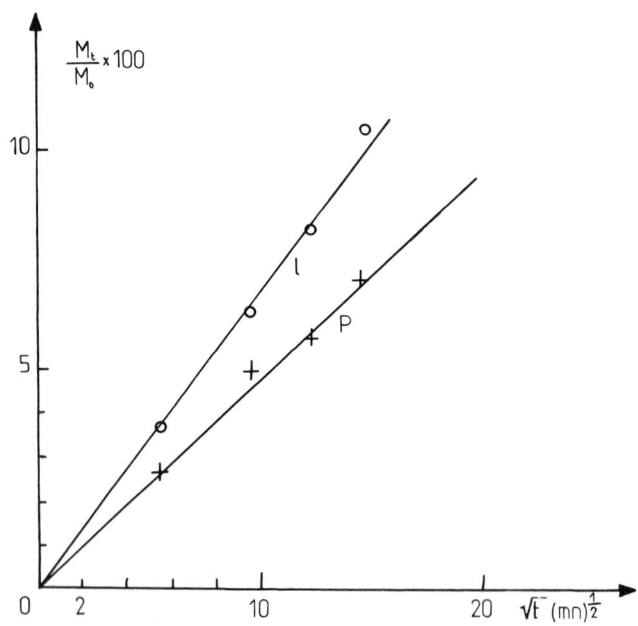


**Figure 8–20** Square root of time dependence of the amount of matter transported during the absorption step. (Reprinted from *Europ. Polym. J.*, 26, 1990, 285–91, A. Aboutaybi, J. Bouzon and J. M. Vergnaud, ref. 70, by courtesy of Pergamon Press.)

3. The amount of liquid transported in PVC passes through a maximum at around 15 h.

From the slopes obtained in Figure 8–20, the diffusivities are calculated for each matter, by using the amount of matter transported at equilibrium $M_\infty$ and equation (8-8). This amount $M_\infty$ is easy to determine for the plasticizer, but it is not so simple for the liquid. Three values can be considered for $M_\infty$ of the liquid:

1. the value obtained after long contact, after 100 h.
2. the value attained at the maximum.
3. the value obtained by extrapolating the amounts measured before 4 h of immersion.

The third method is chosen for determining the amount of n-hexane transported at equilibrium, $M_\infty$.

The value of diffusivity and amount at equilibrium are shown in Table 8–8.

**Preparation of PVC samples by immersion.** PVC samples (3 × 3 × 0.06 cm) are immersed in n-hexane (100 ml) for 15 min. Then, the samples are weighed and the amount of plasticizer extracted is determined by UV-measurement in the liquid.

The amounts of matter transported after this short immersion are compared with those calculated by using constant diffusivities and trigonometrical series. The values obtained for n-hexane are plotted in Figure 8–21, showing a good agreement.

The profiles of concentration of liquid and plasticizer expanded through the PVC thickness are calculated at the end of the absorption process, by using the mathematical model with constant diffusivities (Figure 8–22). The following conclusions are of interest.

1. Steep gradients of concentration are shown for the plasticizer, with a low concentration on the PVC surface. This concentration on the PVC surface is not zero as shown previously [59].
2. Another steep gradient of concentration is obtained for the liquid, with a high concentration on the PVC surface.

**TABLE 8–8** Diffusivities and amount at equilibrium during the absorption step.

| Matter | $D$(cm$^2$/s) | $M_\infty$% |
|---|---|---|
| n-hexane | $5 \times 10^{-8}$ | 12 |
| plasticizer | $7 \times 10^{-9}$ | 17 |

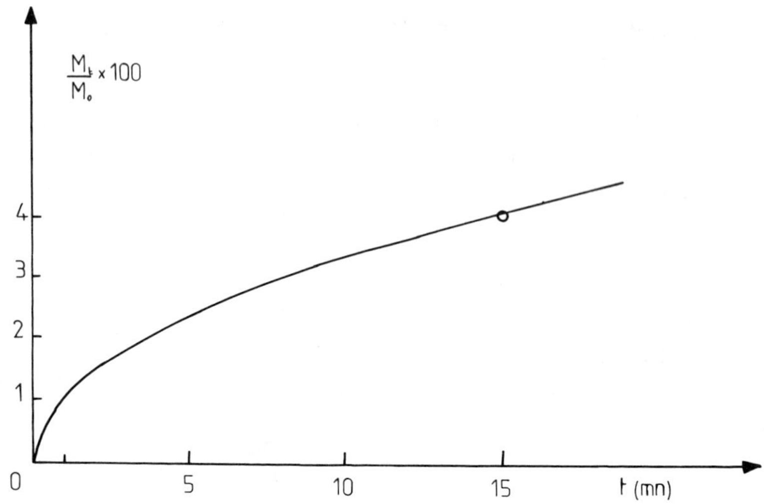


**Figure 8–21** Kinetics of n-hexane transported in PVC during the absorption step. Calculated (full line). Experiment (+). (Reprinted from *Europ. Polym. J.*, 26, 1990, 285–91, A. Aboutaybi, J. Bouzon and J. M. Vergnaud, ref. 70, by courtesy of Pergamon Press.)

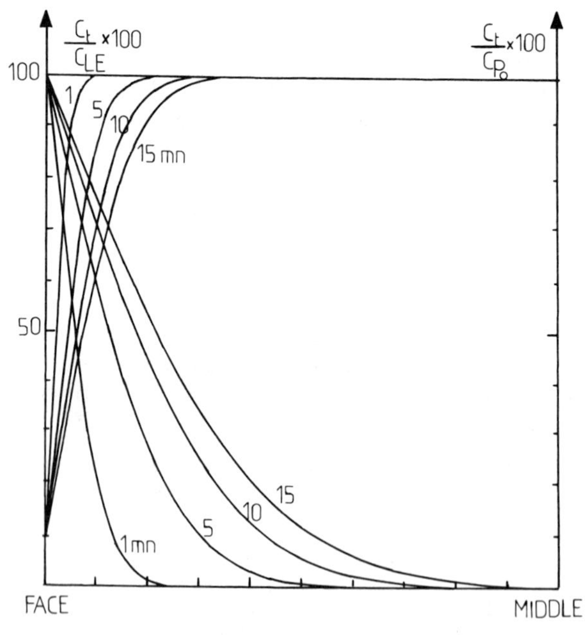


**Figure 8–22** Profiles of concentration of n-hexane and plasticizer developed through the PVC sheet at the end of the absorption step (calculated values). (Reprinted from *Europ. Polym. J.*, 26, 1990, 285–91, A. Aboutaybi, J. Bouzon and J. M. Vergnaud, ref. 70, by courtesy of Pergamon Press.)

The value of the concentration of plasticizer on the PVC surface is of high importance. It can be controlled by selecting the ratio of the volumes of liquid and PVC: the higher this ratio, the lower the concentration of plasticizer on the PVC surface.

**Drying step of the PVC sample.** After immersion of 15 min, the PVC sample is extracted and the process of drying begins. The operation of drying is performed at 22°C in motionless air, with the PVC sheet being vertical. The sample is weighed at intervals up to equilibrium, or rather when a negligible rate of evaporation is obtained.

As the operation at room temperature is not sufficient to evaporate the liquid completely, the sample is dried at 85°C and then at 120°C. The kinetics of drying obtained in the process using these three temperatures in succession are drawn (Figure 8–23). Some conclusions are given from these experiments.

1. It is impossible, or at least very difficult, to completely dry the PVC sample at 22°C.
2. The operation of drying at 85°C does not lead to complete dryness since 7.5% of the liquid remains in the PVC sample.

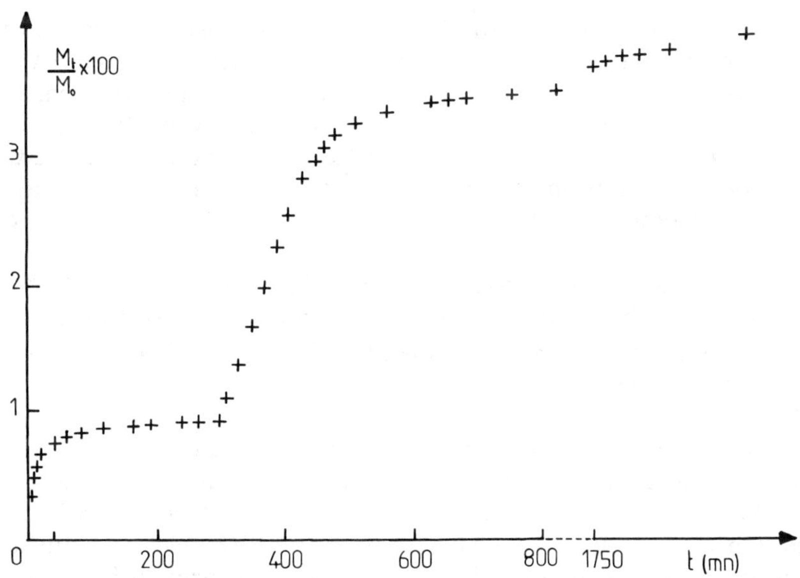


**Figure 8–23** Kinetics of drying of the PVC sheet (previously immersed in n-hexane for 15 min) at 22, 85 and 120°C, in succession. (Reprinted from *Europ. Polym. J.*, 26, 1990, 285–91, A. Aboutaybi, J. Bouzon and J. M. Vergnaud, ref. 70, by courtesy of Pergamon Press.)

3. The third stage of drying at 120°C is then necessary to evaporate all the liquid.
4. The kinetics of drying at 85°C are quite different from the kinetics determined at 22°C.

An attempt to describe the process with the help of the mathematical model is made, by considering the diffusion of the liquid through the PVC and the evaporation on the surface, and by assuming that the plasticizer is motionless. The diffusivity of plasticizer, as obtained from experiments, is expressed in terms of the concentration of the liquid and plasticizer.

$$D^l = 2.7 \times 10^{-7} \times \exp \frac{-0.435}{C^p + C^l} \, (cm^2/s) \tag{8-37}$$

The coefficients in the expression of diffusivity are calculated by taking these facts into account.

1. The liquid and plasticizer play the same role for the transport of liquid [74, 75].
2. The value of the diffusivity in equation (8-37) is the same as that found for the liquid entering the polymer at the beginning of the process. At this time, the initial concentration of the plasticizer is about equal to the sum of the concentrations of liquid and plasticizer.

As the profiles of concentration of liquid and plasticizer are very steep next to the PVC surface, it is necessary to divide the thickness of PVC into a large number of slices. The time of calculation is proportional to the number of slices and to the reciprocal of $\Delta t$, and $\Delta t$ is proportional to $(\Delta x)^2$ as shown in the dimensionless number $M$. Then the time of calculation is proportional to the (number of slices)$^3$. After several attempts, the most convenient way is:

1. Twenty slices are considered through half the thickness of the PVC sheet (half the thickness is studied because of the symmetry).
2. Each of the two slices located next to the PVC surface is divided into 40 small slices.

The validity of the model is tested by comparing the kinetics of drying of the PVC sheet at 22°C obtained from experiments and calculation (Figure 8–24). The complementary information is given with the profiles of concentration of n-hexane developed through the PVC thickness during the drying process. The calculated profiles are drawn through half the thickness of PVC (Figure 8–25), and half the slice located next to the PVC surface (Figure 8–26).

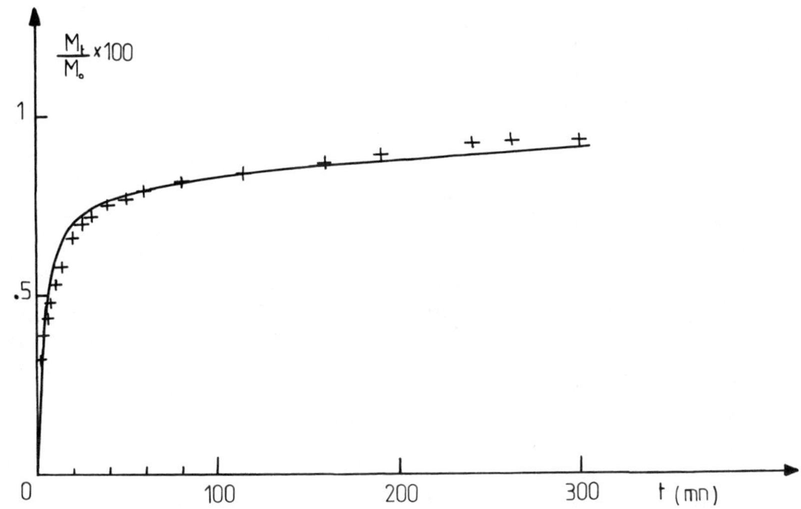


**Figure 8–24** Validity of the model. Kinetics of drying at 22°C of PVC sheet. Calculated (full line). Experiments (+). (Reprinted from *Europ. Polym. J.*, 26, 1990, 285–91, A. Aboutaybi, J. Bouzon and J. M. Vergnaud, ref. 70, by courtesy of Pergamon Press.)

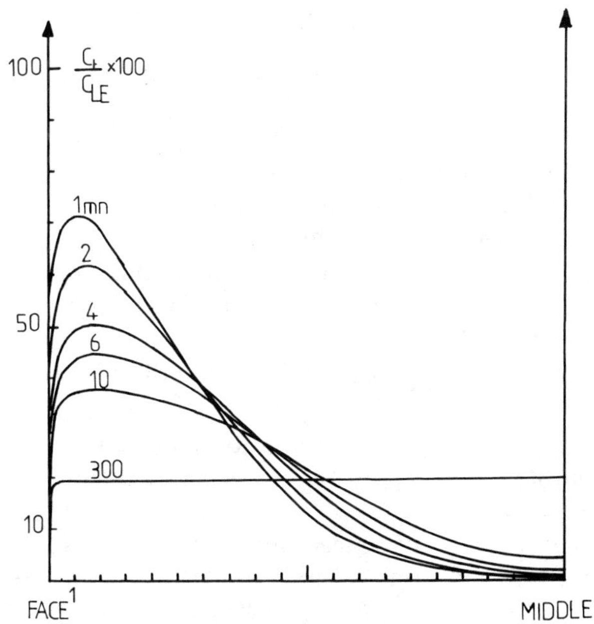


**Figure 8–25** Profiles of concentration of liquid expanded through the PVC thickness during the drying at 22°C, at various times. (Reprinted from *Europ. Polym. J.*, 26, 1990, 285–91, A. Aboutaybi, J. Bouzon and J. M. Vergnaud, ref. 70, by courtesy of Pergamon Press.)

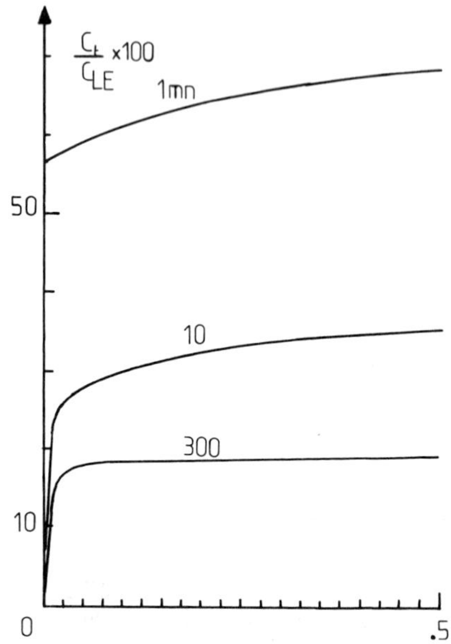


**Figure 8–26** Profiles of concentration of liquid developed through half the slice located next to the PVC surface, during the drying at 22°C. (Reprinted from *Europ. Polym. J.*, 26, 1990, 285–91, A. Aboutaybi, J. Bouzon and J. M. Vergnaud, ref. 70, by courtesy of Pergamon Press.)

The following conclusions can be drawn from these results.

1. A good agreement is observed for the kinetics of drying obtained from experiment and calculation, proving the quality of the model.
2. The concentration of liquid on the PVC surface decreases very quickly because of the evaporation of the liquid.
3. The value of the concentration of liquid on the PVC surface is not zero during the drying at 22°C, because of the liquid transported by diffusion to the surface.
4. A slight transport of liquid toward the middle of PVC sheet can be noticed.
5. Very high gradients of concentration of liquid are expanded next to the PVC surface. This fact explains the method chosen for dividing the PVC sheet into slices.
6. The very high gradient of concentration of liquid, associated with the other high gradient of concentration of plasticizer, with the low value of concentration of liquid and plasticizer on the PVC surface, are responsible for the very low rate of diffusion of liquid next to the surface.

**Conclusions.** Generally, the process of drying of polymers is controlled by diffusion of the liquid through the polymer and by evaporation from the solid surface. When no change in the polymer occurs during the previous step of absorption of liquid and the step of desorption, the polymer can be dried to completion at a temperature lower than the boiling temperature of the liquid, even if a high swelling and following shrinkage are observed during the step of absorption and desorption [75–79].

In this case of plasticized PVC, the problem is more complex, and the liquid previously absorbed by the solid cannot be evaporated at temperatures lower than the boiling point of the liquid. Moreover, very high temperatures, ranging from 85°C to 120°C, are necessary to evaporate all the liquid located in PVC sheets [69, 70, 80].

The process is explained with the help of the model and experiments used to test the model. During the first step of liquid absorption by the PVC, two matter transports are observed with the following production of steep gradients of concentration for the liquid with a high concentration on the solid surface and for the plasticizer with a low concentration on the surface.

As the diffusivities of the liquid and plasticizer are expressed in terms of the concentration of the liquid and plasticizer, it is clear that the PVC sheet obtained at the end of the process of absorption is quite different from the initial polymer, as far as the transport properties are concerned. Then, during the step of evaporation, the liquid can diffuse at an appreciable rate as long as the concentration of the liquid on the surface and within the polymer next to the surface is not too low. When most of the liquid located next to the PVC surface is evaporated, the diffusivity of the liquid decreases to a very low value.

The validity of the model is tested at the temperature of 22°C because the data concerned with diffusivities can be obtained under these conditions. The model could be used to describe the following process with evaporation at temperatures of 85°C and 120°C, if the data were available.

## 8-5 PREPARATION OF PLASTICIZED PVC SAMPLES WITH VERY LOW MATTER TRANSPORTS

Plasticized PVC is often used in the packaging industry for liquid foods [18, 19] and in many biomedical applications as blood bags and other blood-handling devices [2, 10–14]. In the latter case, the solid consists of less than 70 percent PVC, the balance being made up of additives and plasticizer which are potential migrants. In most applications, the plasticized PVC is in contact with the liquid, and the plasticizer may migrate with the following results.

1. Because of loss of plasticizer, a deterioration in mechanical properties of the polymer occurs.
2. The liquid is contaminated [2, 10–14, 18, 19, 31–37].

The problem is set in a rather complex way, because of the following facts:

1. There is a simultaneous transport of the liquid into the PVC and the plasticizer out of it [22–24, 73, 74].
2. Both these transports are controlled by transient diffusion, with concentration-dependent diffusivities.

Many factors may affect the migration process [19], such as the type of the liquid, the nature and amount of plasticizer, the quality of polymer and the process of plasticization, besides the conditions of the migration process, such as stirring, temperature, and time. Any studies on these matter transports between plasticized PVC and liquids are of interest if they help to reduce these transports. Previous basic works have paved the way for the preparation of flexible and soft plasticized PVC with low matter transports.

1. Short-test techniques [73] have been used successfully because they are suitable for the kinetic studies of the two simultaneous matter transports, the concentration of the plasticizer and liquid remaining about constant during the short time of experimentation. Experiments performed with PVC samples of various concentrations of plasticizer can then lead to the expression of the diffusivity in terms of the concentrations of plasticizer and liquid.
2. Calculations, done by using numerical methods with finite differences, are able to take into account all the facts and then to describe the process.

The model provides insight into the nature of these transports, not only by simulating the process and then giving the kinetics of transports, but also by calculating the profiles of concentration of each of the two matters as they are developed through the PVC sample. From this knowledge, a method of preparation of flexible plasticized PVC with very low matter transports can be elaborated. This method consists of a three-step process.

1. Immersion of plasticized PVC sample, in order to create two types of gradients of concentration for the liquids. The former with a high

concentration of liquid on the PVC surface, and the latter with a low concentration of plasticizer on this surface.

2. Evaporation of the liquid located in the PVC. At the end of this step, two gradients of concentration exist within the PVC, with a low concentration on the PVC surface for either the liquid or the plasticizer.
3. The plasticized PVC prepared in this way is then able to release the plasticizer with a very low rate when it is in contact with any liquid.

Of course, various parameters are of interest, such as the conditions of immersion and desorption.

### 8-5-1 Experimental

**Materials.** A commercial PVC resin in the form of white powder, Lucovyl (Rhone-Poulenc), and DEHP (diethylhexyl phthalate) are mixed. The compounds are pressed into sheets in a steel mold at 150°C operated by a press under a pressure of 50 bars. Discs of 18 mm in diameter and 3 mm thick are then cut from these sheets.

**Apparatus for matter transports.** Short tests and long real tests are done with 20 PVC discs in 200 ml of liquid in a 500-ml flask with a controlled rate of stirring. At intervals, one PVC disc is extracted for weighing as well as 10 ml of liquid for analysis, so that the PVC-liquid ratio is kept constant. The concentration of plasticizer in the liquid is determined by gas chromatography.

From short-test experiments carried out with samples of various plasticizer content, linear relationships between the amount of liquid transported and the square root of time are obtained, and the diffusivity can be expressed in terms of the concentration of the plasticizer in PVC.

From long real tests, the amount of matter transported is obtained after infinite time, as well as the kinetics of transport during the whole process.

**Preparation of PVC samples with low matter transport.** Plasticized PVC samples are soaked in the liquid (n-heptane) for a definite time. The discs are then extracted and weighed, while the amount of plasticizer transported is determined by gas chromatography.

These PVC discs are put in contact with air at various temperatures, in order to evaporate the liquid previously absorbed. The discs are weighed at intervals to follow the process. These PVC discs, prepared in this way, are then tested by putting them into contact with liquid and following the kinetics of transport by weighing the PVC and analyzing the amount of plasticizer transported.

### 8-5-2 Theoretical

**Assumptions.** The following assumptions are made for the first step of absorption.

1. Two simultaneous transports take place, with the liquid entering the PVC and the plasticizer leaving the polymers.
2. Both these transports are controlled by transient diffusion, with concentration-dependent diffusivities.
3. The concentrations of the plasticizer and liquid on the PVC surface are equal to those at equilibrium as soon as the PVC sample is immersed in the liquid.

**Mathematical treatment for absorption.** Fick's second equation of transient diffusion in one dimension:

$$\frac{\partial C}{\partial t} = \frac{\partial}{\partial x}\left(D \cdot \frac{\partial C}{\partial x}\right) \tag{8-4}$$

cannot be solved with the preceding conditions.

However, for short tests, the amount of matter transported $M_t$ is so low at time $t$, that the concentrations of plasticizer and liquid in the PVC are constant during the test. The value of $M_t$, as a fraction of the amount transported after infinite time $M_\infty$, is expressed in terms of the square root of time by:

$$\frac{M_t}{M_\infty} = \frac{4}{L}\left(\frac{D \cdot t}{\pi}\right)^{0.5} \tag{8-8}$$

where $L$ is the thickness of the sheet and $D$ is the diffusivity. This equation is useful for determining the diffusivity corresponding to the concentration of plasticizer in the PVC.

**Numerical analysis for long tests.** The problem is solved by a numerical method with finite differences. The thickness of the PVC sheet is divided into equal finite slices of thickness $\Delta x$ by concentration-time reference planes $(n, i)$, $n$ and $i$ being integers which characterize the space and time (Figure 8–27).

The balance of matter transports done on the slice centered on plane $n$ leads to the new concentration after the elapse of time $\Delta t$:

$$CN_n = \frac{1}{M_n}\left[C_{n-1} + (M_n - 2)C_n + C_{n+1}\right] \tag{8-30}$$

where $C_n$ is the concentration of matter in the slice $n$ at the previous time

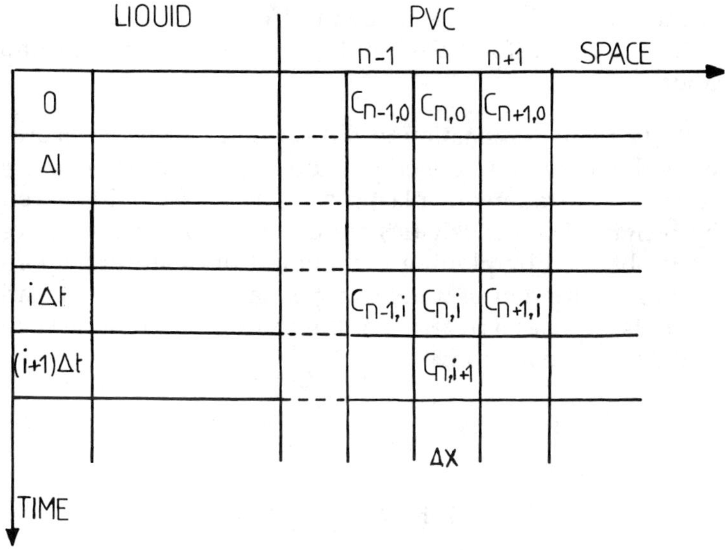


**Figure 8–27** Space-time diagram for numerical analysis.

$i \cdot \Delta t$ and the dimensionless number $M_n$ is:

$$M_n = \frac{(\Delta x)^2}{\Delta t} \cdot \frac{1}{D_n} \tag{8-31}$$

The diffusivities are expressed as a function of the concentration of plasticizer and liquid.

$$D^p = \exp\left(\frac{-A^p}{C^p + C^l \cdot \alpha} - B^p\right) \tag{8-32$'$}$$

$$D^l = \exp\left(\frac{-A^l}{C^p + C^l \cdot \alpha} - B^l\right) \tag{8-32$''$}$$

The amount of matter located in the PVC sample is obtained by integrating the concentration with respect to space.

### 8-5-3 Results

Three steps are considered in succession.

- the absorption, with two matter transports
- the desorption, with evaporation of a part of the liquid
- the experiment of absorption done with the new PVC sample

**Absorption of liquid.** The following two points are of interest: the determination of the diffusivities, and the calculation of the profiles of concentration.

**Short tests and determination of diffusivities.** Short tests are carried out following the method described previously [73], by using PVC samples with various concentrations of plasticizer ranging from 20 percent to 50 percent by weight. The diffusivities are calculated from the slopes of the straight lines obtained by plotting the amount of matter transported as a function of the square root of time, using equation (8-8). The logarithm of the diffusivity is expressed in terms of the concentration of plasticizer and liquid in PVC by the equation:

$$D^p = \exp\left(\frac{-234}{C^p + C^1\cdot\alpha} - 9.7\right) \tag{8-38}$$

$$D^l = \exp\left(\frac{-160}{C^p + C^l\cdot\alpha} - 10.9\right) \tag{8-39}$$

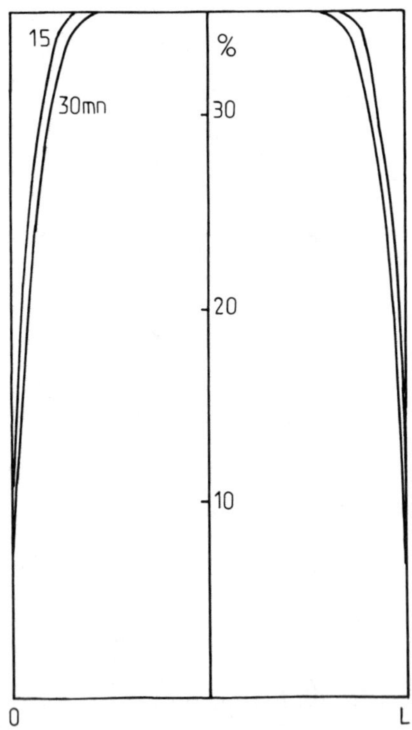


**Figure 8–28** Profiles of concentration of the plasticizer developed through the PVC thickness, after 15 and 30 min of absorption in n-hexane. (Reprinted from *Europ. Polym. J.*, 12, (1986), 959–962, J. L. Taverdet and J. M. Vergnaud, ref. 68, by courtesy of Pergamon Press.)

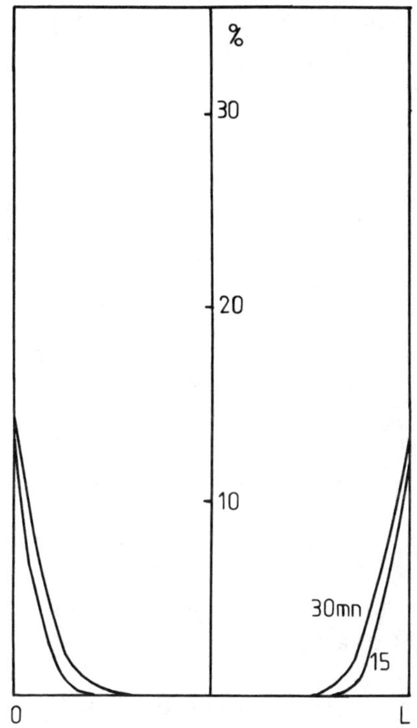


**Figure 8–29** Profiles of concentration of liquid developed through the PVC thickness, after 15 and 30 min of absorption in n-hexane. (Reprinted from *Europ. Polym. J.*, 12, (1986), 959–962, J. L. Taverdet and J. M. Vergnaud, ref. 68, by courtesy of Pergamon Press.)

**Profiles of concentrations through the PVC thickness.** PVC samples with the same initial concentrations of plasticizer (35.5 wt %) are used for the purpose. The profiles of concentration of plasticizer (Figure 8–28) and liquid (Figure 8–29) developed through the thickness of the PVC sheet are calculated by using the model at various times (15 min and 30 min).

As shown in these figures, high gradients of concentration are expanded through the PVC, and especially next to the PVC surface. Of course, the higher value of concentration on the surface is for the liquid, and the lower for the plasticizer.

**Desorption of the liquid.** After their removal from the liquid, the PVC samples are dried either at room temperature or at 80°C. The conditions of interest, essentially the time of absorption and desorption and temperature of drying, as well as the amount of plasticizer extracted during the step of absorption, are noticed in Table 8–9.

**TABLE 8–9** Conditions of preparation of PVC samples.

| Sample | Soaking Time (min) | Loss in Weight Plasticizer (%) | Drying Time (min) | Drying Temperature (°C) |
|---|---|---|---|---|
| 1 | 16 | 2.4 | 30 | 25 |
| 2 | 30 | 3.2 | 30 | 25 |
| 3 | 36 | 3.6 | 30 | 25 |
| 4 | 36 | 3.6 | 30 | 80 |

**Long real test for the PVC samples.** The PVC samples prepared by the absorption-desorption process are then immersed in n-hexane at 30°C for long real tests. The kinetics of transport of plasticizer are determined by measuring the concentration of plasticizer in the liquid at intervals [68, 75].

Three curves are drawn in various conditions of preparation of PVC samples, and for various liquids. In each case, the kinetics of release of plasticizer obtained with the pretreated PVC samples are compared with PVC samples without treatment.

- Figure 8–30, for long real test in n-heptane by using the original PVC sheet with 35.5 wt % plasticizer, the pretreated PVC sheet with 33.1 wt % plasticizer (sample 1, soaked for 16 min), and the pretreated PVC sheet with 32.3 wt % plasticizer (sample 2, soaked for 30 min).
- Figure 8–31, for long real test in arachis oil by using the original PVC sheet with 35.5 wt % plasticizer, and the pretreated PVC sheet with 31.9 wt % plasticizer (sample 3, soaked for 36 min).
- Figure 8–32, for long real test in n-heptane by using the original PVC sheet with 35.5 wt % plasticizer, and the pretreated PVC sheet with 33.1 wt % plasticizer (sample soaked for 16 min and dried at 80°C).

**Conclusions.** The method consisting of pretreating a plasticized PVC sheet by using absorption and desorption is of interest to reduce the rate of release of plasticizer of these samples. The process is very complex, and overall data are difficult to obtain. However, studies are under way, and the models can take into account all the facts. The following qualitative results are of interest, when it is desirable to reduce the release of plasticizer.

***Absorption.*** A low value of the concentration of plasticizer is required on the PVC surface. This objective can be attained by selecting the time of soaking which must be very low, and by increasing the ratio of volumes of liquid and plasticized PVC.

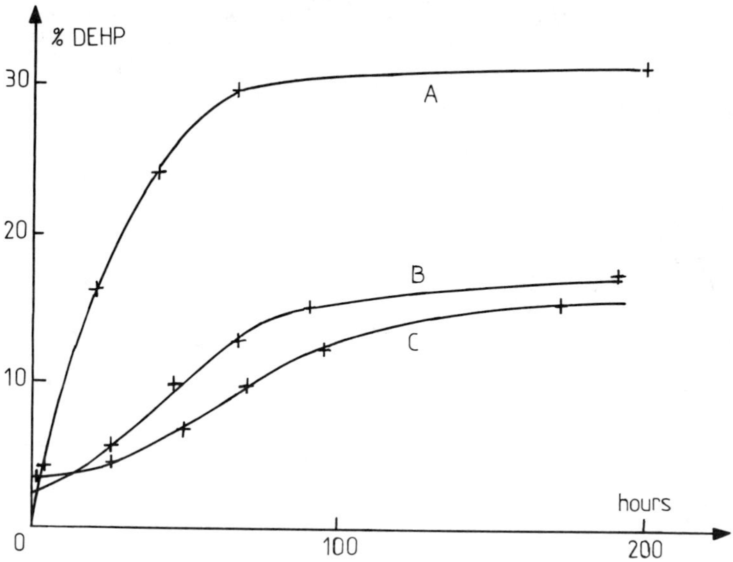


**Figure 8–30** Kinetics of plasticizer release in n-heptane at 20°C. (A) original PVC sheet with 35.5 wt % plasticizer. (B) pretreated PVC sheet with 33.1 wt % plasticizer (16 min soaking, 30 min drying at 20°C). (C) pretreated PVC sheet with 32.1 wt % plasticizer (30 min soaking, 30 min drying at 20°C). (Reprinted from *Europ. Polym. J.*, 12, (1986), 959–962, J. L. Taverdet and J. M. Vergnaud, ref. 68, by courtesy of Pergamon Press.)

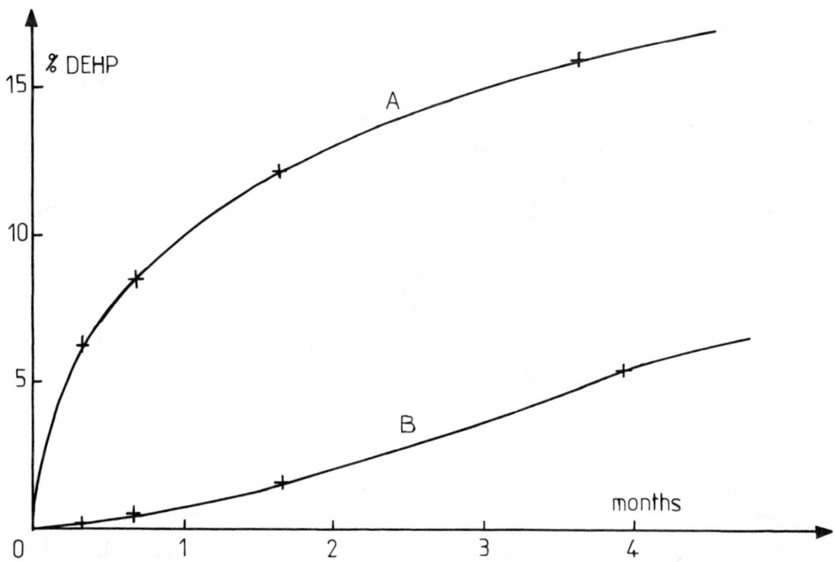


**Figure 8–31** Kinetics of plasticizer release in arachis oil at 20°C. (A) original PVC sheet with 35.5 wt % plasticizer. (B) pretreated PVC sheet with 31.9 wt % plasticizer (36 min soaking in n-heptane, 30 min drying at 20°C). (Reprinted from *Europ. Polym. J.*, 12, (1986), 959–962, J. L. Taverdet and J. M. Vergnaud, ref. 68, by courtesy of Pergamon Press.)

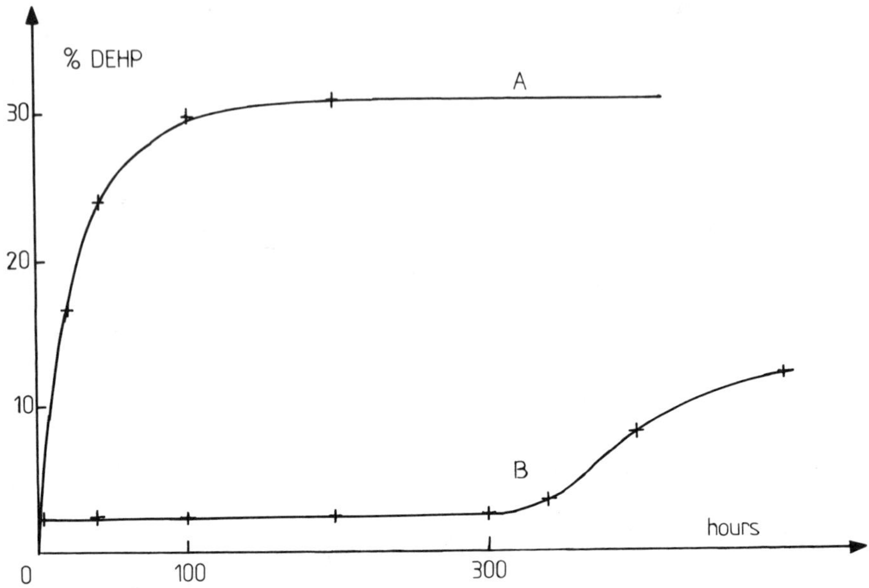


**Figure 8–32** Kinetics of plasticizer release in n-heptane at 20°C. (A) original PVC sheet with 35.5 wt % plasticizer. (B) pretreated PVC sheet with 33.1 wt % plasticizer (16 min soaking, 30 min drying at 80°C). (Reprinted from *Europ. Polym. J.*, 12, (1986), 959–962, J. L. Taverdet and J. M. Vergnaud, ref. 68, by courtesy of Pergamon Press.)

***Desorption.*** The purpose is to evaporate the liquid without any transport of plasticizer, so as to obtain on the PVC surface a low concentration for the plasticizer and for the liquid on the PVC surface. Following the results already shown, a desorption at high temperature for a short time seems to be the best way.

The following conclusions can be drawn from these experiments.

1. The pretreated PVC samples exhibit a much lower release of plasticizer as compared with the original PVC sheet.
2. The conditions of pretreatment are of interest for the retardation in the kinetics of plasticizer release.
3. Only a low amount of plasticizer is extracted from the PVC sheets during the pretreatment.
4. The time of immersion in the liquid, during the first step, is of importance.
5. The temperature of desorption is also of interest. Finally, the desorption must be done at a high temperature for a short time.

To resolve the problem of reducing the matter transport from plasticized PVC, the presence of the plasticizer in the polymer and the formation of gradients of concentration in the PVC when contacted with a liquid has been taken advantage of. This way is quite different from the other which has been laid in the case of release applications in the field of agriculture. In this latter example, the process of release of the active agent is controlled by transient diffusion with the help of a rather complex device [81–85]. The device consists of a core and shell made of various polymers, the core exhibiting a large capacity of absorption for the liquid agent, while the shell regulates the flux.

## REFERENCES


1. B. F. Greek, "Demand for Architectural Products Helps Boost PVC Capacity," *C and Engings. News*, November 28, 1988, pp. 15–17.
2. P. Blais, "DEHP in blood bags and medical plastics: their limitations," *Can. Research*, (June/July 1984), pp. 13–25.
3. R. F. Van Der Heide, "The safety of health of plastics food packaging materials, principles and chemical methods," Kemink En Zoon N.V.-Utrecht 1964 (dissertation).
4. C. G. Vom Bruck, W. R. Eckert, and F. B. Rudolph, "Migration von Vinylchlorid aus PVC-Packungen," *Fette-Seifen Anstrichmittel*, 78, (1976), 334–337.
5. K. Figge and J. Piater, Migration von Hilfsstoffen der Kunststoffverarbeitung aus Folien, VII. Mitteilung: Methodik und Ergebnisse der Extraktions und Migrationsversuche mit festen Nahrungsfetten bzw. Simulantien; Eindringtiefen der Hilfsstoffe in feste Fette," *Dtsch. Lebensmittel-Rdsch.*, 67, (1971), 235–242.
6. K. Figge, "Migration of Additives from Plastics Films into Edible Oils and Fat Simulants," *Fd. Cosmet. Toxicol.*, 10, (1972), 815–828.
7. K. Figge, "Radioananalytische Verfahren zur Bestimmung des Ubertritts von Packstoffbestandteilen in Nahrungsmittel," *J. Radioananal. Chem.*, 32, (1976), 315–343.
8. K. Figge and A. Zeman, "Zur Migration zinnhaltiger Stabilisatoren aus Hart-PVC-Folien in fetthaltige Nahrungsmittel, Rötgenphoto-elektronenspektroskopische und radioanalytische Untersuchungen an PVC-Oberflächen," *Kunststoffe*, 63, (1973), 543–550.
9. K. Figge and H. Piater, "Migration von Hilfsstoffen der Kunststoffverarbeitung aus Folien in flüssige und feste Fette bzw. Simulantien, X. Mitteilung: Additivübertritt in n-Alkane und Di-n-alkyläther unterschieldlicher Kettenlänge sowie in andere organische Flüssigkeiten," *Dtsch. Lebensmittel. Rdsch.*, 68, (1972), 313–319.
10. D. Goullet and M. Fusselier, "Evaluation des risques toxicologiques engendrés par la migration de deux plastifiants: diéthyl hexyl phthalate et diethyl hexyl adipate à partir de materiel en chlorure de polyvinyle utilisé en médecine," *Labo. Phar.*, 304, (1980), 939–948.
11. J. Bommer, E. Ritz, and K. Andrassy, "Side effects due to materials used in hemodialysis equipment," *Adv. Nephr.*, 14, (1985), 409–438.
12. J. P. Gibson, W. A. Briggs, and B. J. Boone, "Delivery of di-2-ethylhexyl phthalate to patients during hemodialysis," *J. Lab. Clin. Med.*, 87, (1976), 519–524.

13. A. E. Ganning, U. Brunk, and G. Dallner, "Phthalate esters and their effect on the liver," *Hepatol.*, 4, 3, (1984), 541–547.
14. G. M. Pollack, J. F. Buchanan, R. L. Slaughter, R. K. Kohli, and D. D. Shen, "Circulating concentrations of di (2-ethylhexyl) phthalate and its de-esterified phthalic acid products following plasticizer exposure in patients receiving hemodialysis," *Toxicol. Apl. Pharmacol.*, 79, (1985), 257–267.
15. J. Simonik and J. Brexler, "Interaction of PVC with linear and branched phthalates," private communication.
16. D. Messadi, J. L. Taverdet, and J. M. Vergnaud," Plasticizer migration from plasticized PVC into liquids. Effect of several parameters on the transfer," *I and EC Prod. Res. Dev.*, 22, 1983, pp. 142–146.
17. K. Figge, "Migration from plastics-packaging materials into packed goods—Test method and diffusion models," *Prog. Polym. Sci.*, 6, (1980), 187–252.
18. G. Haesen and A. Schwarze, "Migration phenomena in food packaging," Commission of the European Communities, 1978.
19. J. M. Vergnaud, "Scientific aspects of plasticizer migration from plasticized PVC into liquids," *Polym-Plast. Technol. Eng.*, 20, (1983), 1–22.
20. E. M. Kampouris, "The migration of plasticizers into petroleum oils," *Europ. Polym. J.*, 11, (1975), 705–710.
21. E. M. Kampouris, F. Regas, S. Rokotas, S. Polychronakis, and M. Pantazoglou, "Migration of PVC plasticizers into alcohols," *Polymer*, 16, (1975), 840–844.
22. D. Messadi, M. Hivert, and J. M. Vergnaud, "A new approach for the study of plasticizer migration from PVC into methanol," *J. Appl. Polym. Sci.*, 26, (1981), 667–677.
23. D. Messadi and J. M. Vergnaud, "Simultaneous diffusion of benzyl alcohol into plasticized PVC and of plasticizer from polymer into liquids," *J. Appl. Polym, Sci.*, 26, (1981), 2315–2324.
24. D. Messadi and J. M. Vergnaud, "Plasticizer transfer from plasticized PVC into ethanol-water mixture," *J. Appl. Polym. Sci.*, 27, (1982), 3945–3955.
25. A. Grigorakakis, J. L. Taverdet, and J. M. Vergnaud, "Transfer of liquids between plasticized PVC and acetic acid," *Europ. Polym. J.*, 11, (1985), 967–972.
26. A. Grigorakakis, J. L. Taverdet, and J. M. Vergnaud, "Study of PVC packaging of interest for vinegar and acetic acid solutions," in *The Shelf Life of Foods and Beverages*, ed. G. Charalambous, Elsevier Sci. Publ., (1986), 525–533.
27. K. Figge and J. Koch, "Effect of some variables on the migration of additives from plastics into edible fats," *Food. Cosmet. Toxicol.*, 11, (1973), 975–988.
28. C. G. Vom Bruck, F. B. Rudolph, K. Figge, and W. R. Eckert, "Application of the diffusion theory to migration of plastics components into packed foods: survey of recent migration studies," *Food Cosmet. Toxicol.*, 17, (1979), 153–157.
29. D. Messadi, J. L. Taverdet, and J. M. Vergnaud, "Pollution of liquid food by PVC container," in *The Quality of Foods and Beverages, Chemistry and Technology*, eds. G. Charalambous and G. Inglete, Academic Press, (1981), 113–125.
30. J. L. Taverdet and J. M. Vergnaud, "Modelisation et simulation des transferts de matière entre le PVC plastifié et les liquides: n-heptane et huile d'arachide," *J. Chim. Phys.*, 82, (1985), 643–645.
31. R. W. Baker, "Diethylhexyl phthalate as a factor in blood transfusion and hemodialysis," *Toxicol.*, 9, (1978), 319.
32. J. H. Corley, T. E. Needham, E. D. Summer, and R. Mikeal, "Effect of various

factors on the amount of plasticizer in intravenous solutions packaged in flexible bags," *Am. J. Hosp. Pharm.*, 34, (1977), 264.

33. M. Horioka, T. Aoyama, and H. Karasawa, "Particles of DEHP in intravenous infusion fluids migrating from PVC bags," *Chem. Pharm. Bull.*, 25, (1977), 1971.
34. J. T. B. Gray, K. R. Butterworth, I. F. Gaunt, P. Grasso, and S. D. Gangolli, "Short-term toxicity of DEHP in rats," *Food Cosmet. Toxicol.*, 15, (1977), 389.
35. F. P. Bell and D. J. Nazir, "Effect of dietary DEHP on liquid biosynthesis in selected tissues from the rat, in vitro," *Lipids*, 11, (1976), 216.
36. A. E. Jones, R. H. Kahn, J. T. Groves, and E. A. Napier, "Phthalate ester toxicity in human cell culture," *Toxicol. Appl. Pharm.*, 31, (1975), 283.
37. G. Rock, V. E. Secours, C. A. Franklin, J. Chiu, and D. C. Villeneuve, "The accumulation of mono-2-ethylhexylphthalate during storage of whole blood and plasma," *Transfusion*, 18, (1978), 553.
38. Second Report on Toxicity by the British Plastics Federation (1962), (BPF), 6.
39. Commision of the European Communities, document 3003/VI/74-D Orig, Brussels, September 1, 1974.
40. D. Messadi and J. M. Vergnaud, "Etude de la diffusion du DEHP du PVC plastifié trempé dans du n-heptane," *J. Chim. Phys.*, 77, (1980), 935–939.
41. K. Figge and H. Piater, "Migration von Hilfsstofen der Kunststoffverarbeitung aus Folien in flüssige und feste Fette bzw. Simulantien, V. Mitteilung: Ergebnisse der Extraktions-oder Kurzzeitversuche," *Dtsch. Lebensmittel-Rdsch.*, 67, (1971), 100–119.
42. H. Piater and K. Figge, "Migration von Hilfsstoffen der Kunststoffverarbeitung aud Folien in Flüssige und feste Fette vzw. Simulantien, VIII. Mitteilung: Vergleich der gravimetrisch bestimmten Rückstände der Extractionslösungen mit den tatsächlich extrahierten Additivmengen," *Dtsch. Lebensmittel Rdsch.*, 67, (1971), 265–269.
43. K. Figge and J. Koch, "Effect of some variables on the migration of additives from plastics into edibles fats," *Food. Cosmet. Toxicol.*, 11, (1973), 975–988.
44. K. Figge, S. R. Eder, and H. Piater, "Migration von Hilfsstoffen der Kunststoffverarbeitung aus Folien in flüssige und feste Fette bzw. Simulantien, XI. Mitteilung Ein synthetisches Triglyceridgemisch als universelles Fettsimulans," *Dtsch. Lebensmittel-Rdsch.*, 68, (1972), 359–367.
45. H. Woggon, D. Jehle, and W. J. Uhde, "Beitrag zur Prüfung von Bedarfsgegenständen aus Plasten. Zur Migration von PVC. Stabilisatoren in Speiseöl," *Die Nahrung*, 13, (1969), 343–350.
46. W. Pfab, "Diffusionserscheinungen in System Kunststoff-Fett," *Dtsch. Lebensmittel-Rdsch.*, 69, (1973), 151–162.
47. H. Woggon, W. J. Uhde, and G. Zydek, "Beitrag zur Prüfung von Bedarfsgegenständen aus Plasten. Zur Migration von 2,6 Di-ter. butyl-p-kresol (BHT) in speiseöl, Z.," *Lebensmitt.-Untersuch. u. Forsch.*, 138, (1968), 169–177.
48. K. Figge and H. Piater, "Migration von Hilfsstoffen der Kunststoffverarbeitung aus Folien in flüssige und feste Fette bzw. Simulantien, VI. Mitteilung: Ergebnisse der Migrations- oder Langzeitversuche mit flüssigen Fetten bzw. Simulantien," *Dtsch. Lebensmittel-Rdsch.*, 67, (1971), 154–161.
49. D. Messadi and J. M. Vergnaud, "Quick identification and analysis of plasticizers in PVC by programmed T gas chromatography using the best stationary phases," *J. Appl. Polym. Sci.*, 24, (1979), 1215–1225.

50. M. P. Petitjean and J. M. Vergnaud, "Analysis of peanut oil in plasticized PVC and plasticizer in peanut oil," in *Instrumental Analysis of Foods*, 1, Academic Press, (1983), 379–390.
51. W. J. Koros and H. B. Hopfenberg, "Scientific aspects of migration of indirect additives from plastics to food," *Food Technol.*, 4, (1979), 56–60.
52. A. Kishimoto and Y. Enda, "Diffusion of benzene in polyacrylates," *J. Polym. Sci.*, 1979, A, 1.
53. P. Meares, "Diffusion of gases through poly (vinyl acetate)," *J. Amer. Chem. Soc.*, 16, (1954), 2171.
54. V. Stannet, "Simple gases," in *Diffusion in Polymers*, eds. J. Crank and G. S. Park, Academic Press, (1968), p. 45.
55. H. Fujita, "Organic vapors above the glass transition temperature," in *Diffusion in Polymers*, eds. J. Crank and G. S. Park, Academic Press, (1968), p. 75.
56. G. Park, in *Diffusion in Polymers*, eds. J. Crank and G. S. Park, Academic Press, (1968), p. 5.
57. J. Crank, *The Mathematics of Diffusion*, 2nd ed. (Oxford: Clarendon Press, 1976).
58. H. L. Frisch, "Simultaneous diffusion of a solvent and organic penetrant in a polymer," *J. Polym. Sci., Phys. ed.*, 16, (1978), 1651–1664.
59. J. L. Taverdet and J. M. Vergnaud, "Surface analysis of plasticized PVC packagings by attenuaded total reflectance," in *Instrumental Analysis of Foods*, 1, Academic Press, (1983), 367–378.
60. M. Kessler, "Bio-compatibilité à long terme en hémodialyse," *Néphroligie*, 6, (1985), 231–234.
61. Y. Ishikawa, K. Honda, S. Sasakawa, K. Hatada, and H. Kobayashi, "Prevention of leakage of di-(2-ethylhexyl) phthalate from blood bags by glow discharge treatment and its effect on aggregability of stored platelets," *Vox. San.*, 45, (1983), 68–76.
62. C. D. Papaspyrides, "Flexible poly (vinyl chloride) sheets: interrelation between ultra-violet irradiation and plasticizer migration into alcohols," *Polymer*, 27, (1986), 1967–1970.
63. C. Baquey, B. Basse-Chatalinat, L. Bordenave, J. Cayx, A. J. Brendel, and D. Ducassou, "Topographic and dynamic radiosotopic investigations of interactions between biomaterials and blood components," *Artif. Organs.*, 10, 6, (1986), 481–488.
64. L. Ljunggren, "Plasticizer migration from blood lines in hemodialysis," *Artif. Organs.*, 8, 1, (1984), 99–102.
65. H. Igarashi, S. Kioi, F. Gejko, and M. Arakawa, "Physiologic approach to dialysis-induced hypoxemia: effects of dialyser material and dialysate composition," *Nephon*, 41, (1985), 62–69.
66. D. Barron, S. Harbottle, N. A. Hoenich, A. R. Morley, D. Appleton, and J. F. McCabe, "Particle spallation induced by blood pumps in hemodialysis tubing sets," *Artif. Organs.*, 10, 3, (1986), 226–235.
67. M. Garreau, Thesis, Contribution à l'amélioration de la biocompatibilité des circuits extracorporels pour hemodialyse, Compiègne, June 1989.
68. J. L. Taverdet and J. M. Vergnaud, "Preparation of plasticized PVC samples with very low matter transfers," *Europ. Polym. J.*, 12, (1986), 959–962.
69. A. Aboutaybi, B. Touchard, J. Bouzon, and J. M. Vergnaud, "Drying of plas-

molar volume of liquid, $\mu$ and $\rho$ are a constant characterizing the solvent-rubber interaction, and the density of the network.

Several standardized test procedures [3–6] are concerned with the measurement of change in mechanical properties such as tensile strength, hardness, and change in dimensions or in volume. Because swelling is relatively simple to measure, changes in volume or rather changes in mass are easily followed when the test piece is immersed in the test liquid at the selected temperature.

The rate of uptake of liquid is considered to be the result of a diffusion process with either a constant diffusivity or a concentration-dependent diffusivity [7]. This case of absorption is sometimes easy to solve when the sheet of rubber, initially free of liquid, is immersed in the liquid and when the concentration of liquid on the surface attains the concentration corresponding to the equilibrium uptake as soon as the process starts. Analytical solutions are found when the diffusivity is constant. Generally the problem is more complex, because of concentration-dependent diffusivity or of a double matter transport. The extraction of plasticizer and additives from plasticized PVC is explained by simultaneous diffusion of the solvent into and plasticizer out of the plasticized PVC, when plasticized PVC sheets are in contact with liquids [8–10]. In this last case, no analytical solution can be given and modeling using numerical analysis is the only possibility [11].

Desorption process is the complementary problem in which the solute is at first uniformly distributed through the solid or not, and diffuses out of the rubber. If free and immobilized solute is considered to be initially in equilibrium everywhere in the solid of simple shape, the analytical solution presented for absorption also describes desorption, provided a few conditions are accepted: the diffusivity is constant, the concentration of the solute is zero on the rubber faces as soon as desorption starts. Another more real case of desorption is diffusion-controlled evaporation of the liquid initially located in the rubber. The process of evaporation is controlled by diffusion in the sense that the rate of evaporation depends largely on the rate at which solvent is supplied to the evaporating surface by internal transport by diffusion. In fact, it is possible to express the kinetics of the loss in weight of liquid in terms of a simple expression only in these two cases:

1. The surface of the rubber sheet is assumed to reach equilibrium with external atmosphere when evaporation commences, and thus the concentration on the surface falls immediately to zero when the atmosphere is free of vapor (or to the concentration at equilibrium with the vapor pressure of vapor in external atmosphere).
2. The rate of evaporation from the surface equals the rate of transfer by diffusion to this surface. The concentration of liquid on the surface

which is at equilibrium with the vapor pressure in external atmosphere must be known at the temperature considered.

The first purpose in this chapter is to show that the process of absorption (and desorption) cannot be perfectly defined by only one measurement of the uptake of liquid at one time. It is preferable to take several readings to ensure that the full curve of absorption is recorded. In fact, two independent parameters at least are necessary to describe the whole process of absorption: the diffusivity which can be either constant or concentration dependent, and the capacity of absorption (or the amount of liquid absorbed at infinite time, when equilibrium is reached).

For the process of desorption, these two parameters, diffusivity and capacity of desorption (the amount of liquid desorbed at infinite time) are needed, as well as the rate of evaporation. It is clear that using a single time for characterizing the absorption process and the resistance of rubber to the liquid may lead to a definite mistake, as shown in Figure 9–1. The amount of liquid observed at the time *t* can correspond with various curves of absorption with different values for diffusivity and capacity of absorption.

The second main purpose of this chapter is concerned with modeling

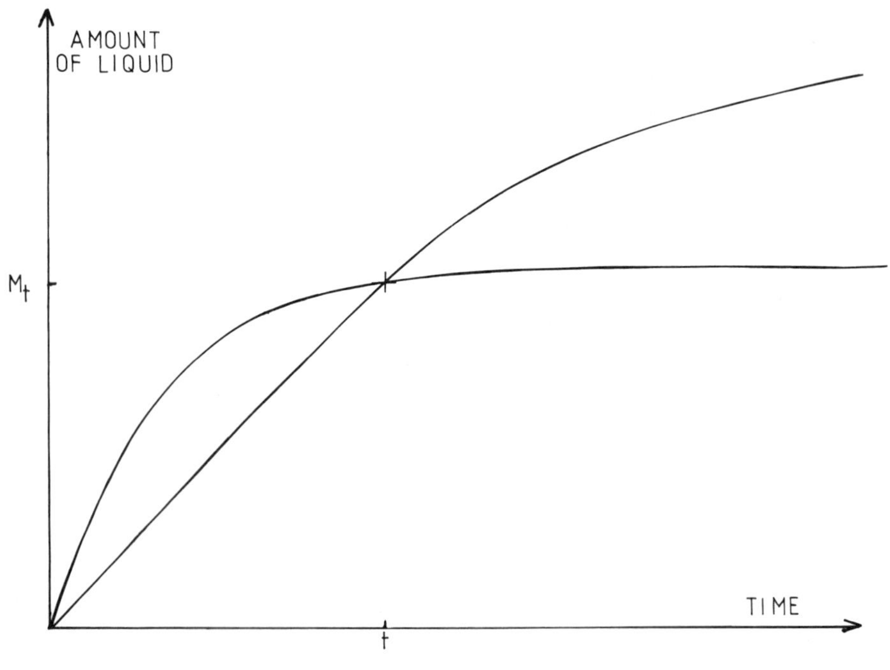


**Figure 9–1** Absorption-time histories of rubber which can be supposed with simple test and one reading of absorption.

of the process of absorption and desorption for various shapes of rubber samples, as making experiments with real large real devices is costly, time consuming, or even impossible for the following reasons:

1. A large system for experimenting absorption is needed.
2. Experiments can be highly time consuming because of the time dependence with the square of the main dimension (thickness for sheet, radius for sphere or long cylinder, small dimension for parallelepipede).
3. During the absorption process, some additives are generally extracted, and so one cannot speak of a nondestructive test.

Modeling of the process is of interest if the following steps are studied in succession:

1. Experiments carried out in order to get a good knowledge of the process (reversibility in absorption-desorption, change in properties, extraction of additives), and to get data such as the diffusivity which can be constant or concentration dependent, capacity of absorption, and rate of evaporation.
2. Modeling the process in the case of samples with defined common shapes: thin sheets, parallelepipede, cylinder.
3. Experiments performed by using these samples so as to test the validity of the model.

As a result, any real rubber device with a special shape can be studied, as far as the absorption of liquid is concerned, without having to experiment with it. Only a small aliquot sample is needed, the best being a thin sheet representative of the large device, for measuring the previously cited parameters.

## 9-2 LIQUID SORPTION AND EVAPORATION WITH RUBBER SHEET: MODELS AND EXPERIMENTS

The first purpose of this chapter is to study the process of absorption of liquid by rubber sheets, and evaporation of liquid out of the same sheets. These sheets are rather thin for two reasons: to reduce the time of experiments, and to provide a negligible amount of liquid transported through the lateral surface.

The second objective is to elaborate a method for determining all data of interest which are able to characterize the process:

- diffusivity and its concentration dependency

- capacity of absorption and desorption
- rate of evaporation of liquid out of the rubber surface
- testing the accuracy of these data

The third goal in this study is to build mathematical models based on explicit numerical methods with finite differences by considering the simple case of thin sheets of rubber, and to establish the validity of these models [12].

### 9-2-1 Experimental

**Rubber sheets.** Rubber sheets of various thicknesses ranging from 0.15 cm to 0.48 cm are prepared by curing rubber in a steel mold operated by a power press under a pressure of 50 bars. Operational conditions for the cure are determined from earlier studies [13–17] in order to make sure that vulcanizates have a high state of cure. Several discs (3.8 cm in diameter) are cut from these sheets.

**Test conditions for absorption.** Experiments are performed with two series of rubber discs immersed in about 300 ml of toluene in a flask (500-ml capacity), using either motionless liquid or liquid with a controlled rate of stirring. Test samples are weighed at intervals, after removal of the liquid in excess on the surface.

**Test for evaporation.** The samples previously saturated by toluene are exposed either to motionless air or to air with forced convection (stirring). The sheets are in a vertical position and the temperature is constant at 20°C. The weight of the samples is measured at intervals.

### 9-2-2 Theoretical

#### Assumptions

A few assumptions are made:

1. A thin plane sheet is used, and the one-dimensional transport is considered.

**TABLE 9–1** Cure Conditions for Rubber Samples.

| Thickness (cm) | Mold Temperature | Time (min) |
|---|---|---|
| 1.5 | 170°C | 10 |
| 4.8 | 170°C | 20 |

2. The process (absorption, desorption) is controlled by transient diffusion.
3. During the step of absorption when the sample is in contact with liquid, the concentration of liquid on the surface reaches the equilibrium value as soon as the process starts.
4. The step of desorption is controlled by diffusion-evaporation. The rate of evaporation is either very high (with a zero concentration of liquid on the surface) or proportional to the difference of liquid concentration: the present one and its value at equilibrium with the external atmosphere.

### Mathematical treatment

The equation of one-dimensional diffusion is described by Fick's law

$$\frac{\partial C}{\partial t} = \frac{\partial}{\partial x}\left(D\cdot\frac{\partial C}{\partial x}\right) \tag{9-1}$$

Initial and boundary conditions are:

| | | | | | |
|---|---|---|---|---|---|
| Initial | $t = 0$ | $0 < x < L$ | $C = 0$ | absorption | (9-2) |
| | | | $C = C_\infty$ | desorption | |
| Boundary | $t = 0$ | $0 < x < L$ | $C = C_{t.x}$ | absorption-desorption | (9-3) |
| | $x = 0$ and $x = L$ | | $C = C_{eq}$ | absorption | |

In the case of constant diffusivity, an analytical solution exists for the process of absorption and desorption. The kinetics are then described by the well-known series, as well as the profiles of concentration developed through the thickness.

### Absorption

$$\frac{M_\infty - M_t}{M_\infty} = \frac{8}{\pi^2}\cdot\sum_{m=0}^{\infty}\frac{1}{(2m+1)^2}\cdot\exp\left[\frac{-(2m+1)^2\pi^2}{L^2}Dt\right] \tag{9-4}$$

where $M_t$ and $M_\infty$ are the amount of liquid absorbed at time $t$ and after infinite time.

$$\frac{C_\infty - C_{tx}}{C_\infty}\cdot\frac{\pi}{4} = \sum_{m=0}^{\infty}\frac{1}{2m+1}\cdot\sin\frac{(2m+1)\pi x}{L}\cdot\exp\left[\frac{-(2m+1)^2\pi}{L^2}Dt\right] \tag{9-5}$$

where $C_{tx}$ and $C_\infty$ are the concentration at position $x$, and time $t$ and infinite time, respectively, and $L$ is the thickness of the sample.

**Desorption.** The rate of loss of liquid by evaporation is represented by the surface condition:

$$-D\cdot\left(\frac{\partial C}{\partial x}\right)_{x=0} = F_0\cdot(C_0 - C_{eq}) \tag{9-6}$$

where $F_0$ is the rate of evaporation of the pure liquid, $C_0$ is the actual concentration on the surface, and $C_{eq}$ is the concentration required to maintain equilibrium with the vapor pressure in surrounding atmosphere.

When the diffusivity is constant, the initial concentration $C_{in}$ is uniform through the sheet, the profiles of concentration through the thickness $2L$ of the sheet, as well as the kinetics of desorption are then given:

$$\frac{C_{eq} - C_{tx}}{C_{eq} - C_{in}} = \sum_{n=1}^{\infty} \frac{2M\cos\dfrac{\beta_n x}{L}}{(\beta_n^2 + M^2 + M)\cos\beta_n} \exp\left(-\frac{\beta_n^2}{L^2}Dt\right) \tag{9-7}$$

$$\frac{M_\infty - M_t}{M_\infty} = \sum_{n=1}^{\infty} \frac{2L^2}{\beta_n^2(\beta_n^2 + M^2 + M)}\cdot\exp\left(-\frac{\beta_n^2 Dt}{L^2}\right) \tag{9-8}$$

where the $\beta_n$s are the positive roots of

$$\beta\cdot\tan\beta = M \tag{9-9}$$

and the dimensionless number $M$ is:

$$M = \frac{L\cdot F_0}{D} \tag{9-10}$$

**Numerical analysis.** The sheet occupying the space $0 < x < L$ is divided into $N$ equal intervals of thickness $\Delta x$, and the time into intervals $\Delta t$. The space-time region is covered by a grid of rectangles of sides $\Delta x$ and $\Delta t$. The coordinate of a point is $i\cdot\Delta t$ and $n\cdot\Delta x$ where $i$ and $n$ are integers (Figure 9–2).

The matter balance is calculated during the increment of time $\Delta t$ in the slice located at position $n$, and then the new concentration after elapse of time is expressed in terms of concentrations at the previous time.

**Inside the rubber sheet.** The new concentration at time $(i + 1)\Delta t$ is as follows:

$$CN_n = \frac{1}{M_n}[C_{n+1} + (M_n - 2)C_n + C_{n+1}] \tag{9-11}$$

where $C_n$ is the concentration at time $i\cdot\Delta t$ and position $n$, and the dimensionless number $M_n$ is:

$$M_n = \frac{(\Delta x)^2}{\Delta t}\cdot\frac{1}{D_n} \tag{9-12}$$

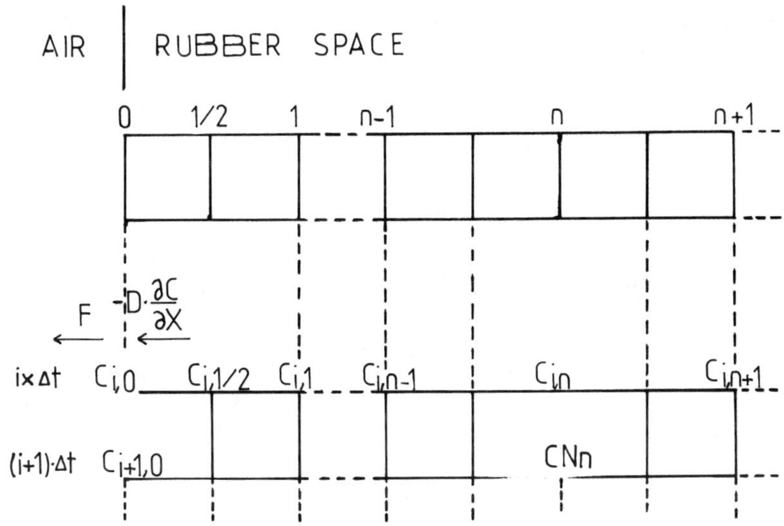


**Figure 9–2** Time-space diagram for numerical analysis.

**Rubber sheet faces.** The matter balance is calculated in the rubber slice of thickness $\Delta x/2$ next to the face, by considering the transport by diffusion and evaporation on each surface.

$$\left[D\cdot\left(\frac{\partial C}{\partial x}\right)_{0.5} - F_0\,(C_0 - C_{eq})\right]\Delta t = \frac{\Delta x}{2}\,[CN_{0.25} - C_{0.25}] \qquad (9\text{-}13)$$

By making the approximation, for the concentration at position 0.25:

$$CN_{0.25} - C_{0.25} = CN_0 - C_0 \qquad (9\text{-}14)$$

and for the gradient of concentration at position 0.5, the new concentration on the surface, $CN_o$, is obtained in terms of the concentrations at the previous time.

$$CN_0 = \frac{1}{M_0}\,[2C_1 + (M_0 - 2 - 2N)\cdot C_0 + 2N\cdot C_{eq}] \qquad (9\text{-}15)$$

$M_o$ and $N$ being dimensionless numbers:

$$N = \frac{F_0\cdot\Delta x}{D} \qquad (9\text{-}16)$$

**Amount of liquid located in rubber sheet.** The amount of liquid located in the rubber sheet is calculated by integrating the concentration with respect to the thickness of the sheet.

### 9-2-3 Results

**Diffusivity and evaporation rate.** Three various ways are used for determining the value of diffusivity. The first way is by using the short-test technique [11] and equation.

$$\frac{M_t}{M_\infty} = \frac{4}{L}\left(\frac{D \cdot t}{\pi}\right)^{0.5} \tag{9-17}$$

where $L$ is thickness of the sheet. This method can be easily used in the case of absorption of liquid, at the beginning of the process when the amount of liquid located in rubber is nearly zero.

A second way to determine the value of diffusivity is use of half-time of process. The diffusivity can be calculated from the time necessary for the amount of liquid to reach half the value at infinite time ($Mt/M_\infty = 0.5$), by using the relation (with an error of about 0.001 percent). The value of this half-time is determined easily from the kinetic curve.

$$D = 0.049 \cdot \left(\frac{L^2}{t_{0.5}}\right) \tag{9-18}$$

A third method to determine the value of diffusivity is the use of final rates of the process. In the later stages of diffusion in the sheet, only the first term of the series can be retained.

$$\frac{M_\infty - M_t}{M_\infty} = \frac{8}{\pi^2} \cdot \exp\left(-\frac{\pi^2}{L^2} Dt\right) \tag{9-19}$$

The values of diffusivity obtained by using these methods are given in Table 9–2.

The values of diffusivity obtained by using the half-time and long-time techniques are about the same. In all these calculations, the value of the thickness considered is that of initial thickness of the unswollen sample.

The amount of liquid absorbed at equilibrium is expressed in weight percent of the initial rubber for absorption and desorption. It is seen that the whole amount of liquid previously absorbed is completely desorbed.

The rate of evaporation is determined by using the initial rate of de-

**TABLE 9–2** Diffusivity and Amount of Liquid at Equilibrium

| Process | Thickness (cm) | $D \cdot 10^{-7}$ ($cm^2/s$) | Equation | $M_{eq}$% |
|---|---|---|---|---|
| Sorption | 0.15 cm | 5 | 18 | 338 |
| Sorption | 0.15 | 5 | 19 | 338 |
| Sorption | 0.48 | 5.3 | 18 | 320 |
| Desorption | 0.15 | 5.1 | 19 | 338 |

**TABLE 9–3** Rate of Evaporation ($F_o \cdot C_o$)

| Conditions | $F_o \cdot C_o$(cm/s) | $C_o$ |
|---|---|---|
| Motionless air | $10.4 \times 10^{-5}$ | 0.77 |
| Stirred air | $38.4 \times 10^{-5}$ | 0.77 |

sorption, when the process may be assumed to be controlled by evaporation.

$$F_0 \cdot C_0 = \frac{dM_t}{dt} \quad \text{for} \quad t \to o \tag{9-20}$$

where $C_0$ is the concentration of liquid on the surface when the rubber is saturated with liquid.

The values of this rate of evaporation depend largely on the air velocity and stirring, as shown in Table 9–3.

**Validity of the model and data.** The kinetics of absorption and desorption are calculated either by using the model with the numerical

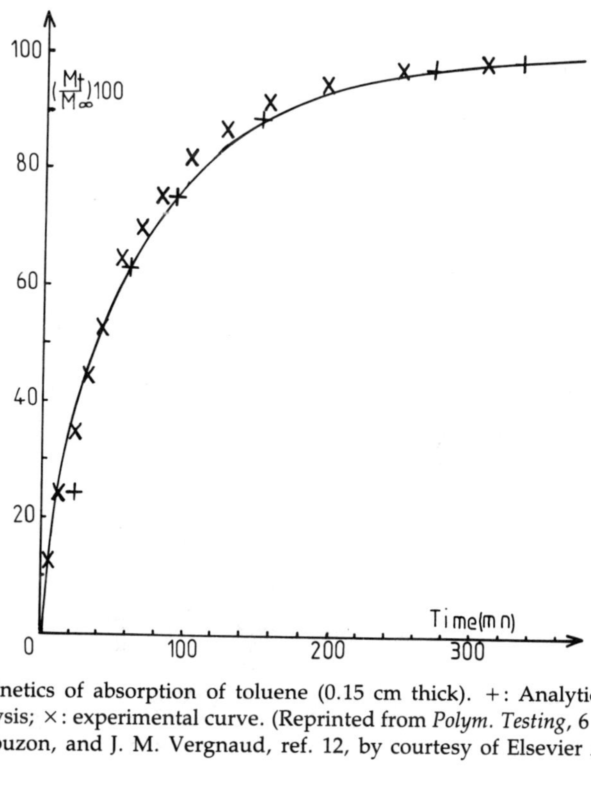


**Figure 9–3** Kinetics of absorption of toluene (0.15 cm thick). +: Analytical solution; —: numerical analysis; ×: experimental curve. (Reprinted from *Polym. Testing*, 6 (1986), 253–265, Y. Khatir, J. Bouzon, and J. M. Vergnaud, ref. 12, by courtesy of Elsevier Applied Science Publishers)

or by using the analytical solution. These theoretical kinetics are then compared with the experimental ones.

The validity of the models as well as the constance of diffusivity are illustrated in the case of absorption (Figure 9–3) with a thin rubber sheet (0.15 cm). A good agreement is also obtained in the case of thicker sheets (0.48 cm) [12].

For the process of desorption, two cases are of interest, following the agitation of air. In the case of motionless air (Figure 9–4), a good agreement exists between the experimental kinetics and the kinetics obtained with the model, while the analytical solution gives too high a value for the amount of liquid evaporated, especially at the beginning of the process. This difference is due to the fact that the evaporation rate is rather low, and so the concentration on the surface is higher than the concentration at equilibrium.

When air is strongly stirred, the kinetics calculated with the help of the analytical solution and the model are superimposed on the experimental kinetics (Figure 9–5). With the higher value of the rate of evaporation, the liquid concentration on the rubber surface falls more quickly to zero.

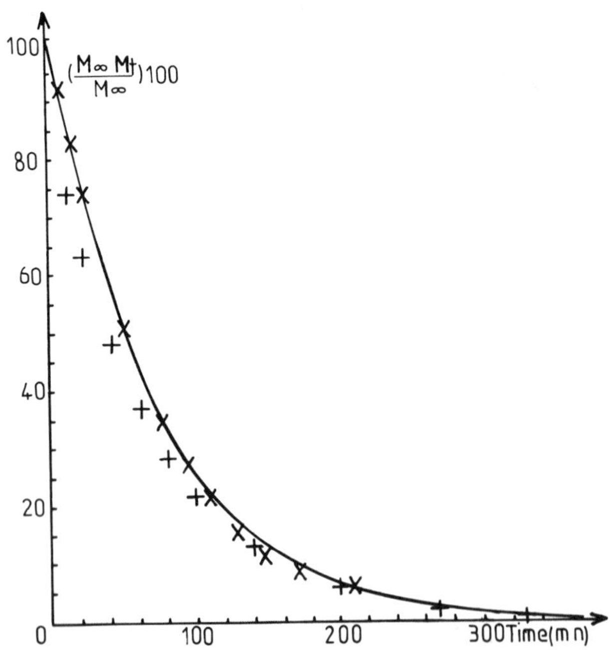


**Figure 9–4** Kinetics of desorption of toluene in motionless air (0.15 cm thick). + Analytical solution; − model; ×: experimental curves. (Reprinted from *Polym. Testing*, 6 (1986), 253–265, Y. Khatir, J. Bouzon, and J. M. Vergnaud, ref. 12, by courtesy of Elsevier Applied Science Publishers)

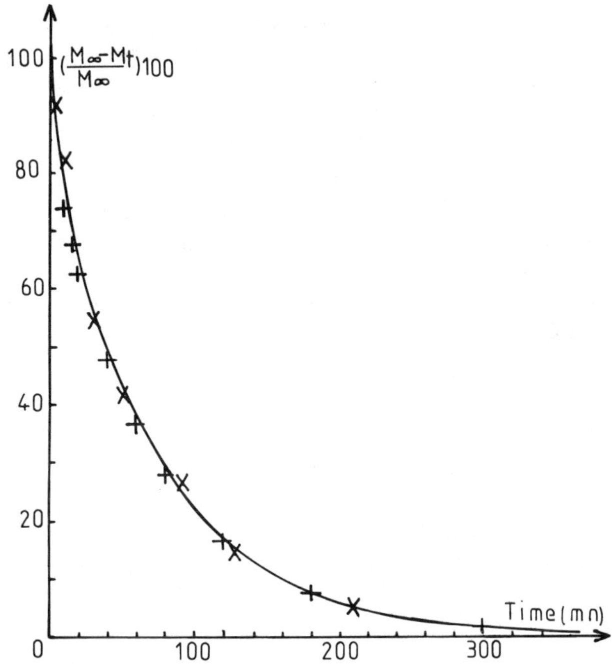


**Figure 9–5** Kinetics of desorption of toluene in stirred air (0.15 cm thick). + Analytical solution; – model; ×: experimental curves. (Reprinted from *Polym. Testing*, 6 (1986), 253–265, Y. Khatir, J. Bouzon, and J. M. Vergnaud, ref. 12, by courtesy of Elsevier Applied Science Publishers)

**Conclusions.** The model based on an explicit numerical method with finite differences can describe the process of liquid transport in rubber either for absorption when the rubber is in contact with liquid or for desorption when the liquid is evaporated.

In the case of desorption, the effect of stirring of air on the kinetics is of high importance. All data obtained by using various methods, especially diffusivity, are thus very accurate.

## 9-3 LIQUID ABSORPTION AND EVAPORATION WITH CYLINDERS OF FINITE LENGTH: MODELS AND EXPERIMENTS

As already noticed (Introduction), real tests for measuring the kinetics of absorption and desorption of liquid by real rubber devices are not only time consuming and costly but also destructive for the sample. So the prediction of the process of liquid absorption and desorption, by using simple and short tests with an aliquot sample of the real piece, and by calculation with a model, can be a way of approaching the problem.

The first purpose of this chapter is to show that mathematical models based on a numerical method with finite differences can describe these processes in the case of a cylinder of finite length. An analytical solution is obtained for cylinders of finite length when the diffusivity is constant, and the ratio of the height and diameter is of importance. When the diffusivity is concentration dependent, mathematical models must be used to solve the problem.

The second purpose is to give an idea of the effect of the dimensions of the sample on the process of absorption and desorption, by especially considering the ratio of the height and diameter of the cylinder.

The parameters which are worth attention during the step of absorption are the diffusivity expressed in terms of concentration of liquid, and the capacity of absorption. During the step of desorption, the rate of evaporation, in addition to the preceding two parameters, is important to consider. These parameters are determined by using the short-test technique and a thin sheet made of the same material as that of the real device.

### 9-3-1 Experimental

**Rubber samples.** Rubber cylinders of various sizes (thickness and diameter) are cured in a steel mold operated by a press under a pressure of 50 bars at a temperature of 170°C. The conditions of cure are previously determined by using a model taking into account not only the heat transported through the rubber but also the kinetics of heat evolved from the cure [13–15]. The kinetics of cure are determined from calorimeter measurements in scanning mode [16, 17]. Thus, the time of cure is obtained in terms of dimensions of each sample, to make sure that all vulcanizates have a high value of the state of cure.

**Operational conditions for absorption.** The rubber samples are immersed in toluene with a controlled rate of stirring. At intervals, the samples are weighed after extraction of the excess of liquid on the surface, in order to determine the kinetics of absorption. All experiments are performed at 20°C.

**TABLE 9–4** Cure Conditions of Rubber Samples.

| Diameter (cm) | Thickness (cm) | Tmold (°C) | Time (min) |
|---|---|---|---|
| 0.47 | 0.73 | Calorimeter | |
| 3.8 | 0.92 | 170 | 25 |
| 2.2 | 0.49 | 170 | 15 |

**Operational conditions for desorption.** The rubber samples presaturated with toluene are exposed either to motionless air or to air with a forced convection. The samples are kept vertical at a temperature of 20°C. The weight is measured at various times for the kinetics of desorption. As shown previously (Introduction), the gravimetric method consisting of weighing the sample in air either during the process of absorption or of desorption, is preferably used [1].

### 9-3-2 Theoretical

**Assumptions.** The following assumptions are made.

1. A cylinder of finite length is considered, with an isotropic material.
2. The transport of liquid is controlled by transient diffusion, the diffusivity being concentration dependent as found from experiments.
3. In the case of absorption, the concentration of liquid on the rubber surface reaches equilibrium, as soon as the process starts.
4. In the case of desorption, evaporation takes place on the rubber surface. The rate of evaporation is proportional to the difference between the concentration on the surface and the concentration which is at equilibrium with the vapor pressure in the atmosphere.
5. In spite of the change of volume, a frame of reference is considered for calculation with the initial dimensions of the rubber cylinder free of liquid.

**Mathematical treatment.** The rate of variation of the concentration of liquid is defined by Fick's second law with radial and longitudinal directions of diffusion.

$$\frac{\partial C}{\partial t} = \frac{1}{r} \cdot \frac{\partial}{\partial r}\left(D \cdot r \cdot \frac{\partial C}{\partial r}\right) + \frac{\partial}{\partial z}\left(D \cdot \frac{\partial C}{\partial z}\right) \tag{9-21}$$

where $D$ is the concentration-dependent diffusivity and the dimensions of cylinder $r$ and $z$ are:

$$0 \leq r \leq R \qquad 0 \leq z \leq h \tag{9-22}$$

$h$ and $R$ being the length and radius of the cylinder.

In the case of absorption, the initial and boundary conditions are:

$$t = 0 \quad 0 < r < R,\ 0 < z < h \qquad C_{r,z} = 0 \quad \text{sample} \tag{9-23}$$

$$t > 0 \quad r = R,\ z = 0 \text{ and } z = h \qquad C = C_\infty \quad \text{surface} \tag{9-24}$$

In the case of desorption, the initial and boundary conditions are:

$$t = 0 \quad 0 < r < R \quad C_{r,z} = C_\infty \quad \text{sample}$$
$$0 < z < h \tag{9-25}$$

$$t > 0 \quad r = R$$
$$z = 0 \text{ and } z = h \quad \text{surface condition} \tag{9-26}$$

$$-D\left(\frac{\partial C}{\partial x}\right) = F_0\cdot(C_0 - C_{eq}) \tag{9-6}$$

where $C_o$ is the concentration on the surface, $C_{eq}$ is the concentration on the surface in equilibrium with the vapor pressure in external atmosphere, and $F_o$ is the rate of evaporation of the pure liquid.

**Numerical analysis in the case of absorption.** As no analytical solution can be found with concentration-dependent diffusivity, the problem is solved by using a numerical method with finite differences. As shown in the concentration-time diagram (Figure 9–6) the cylinder is divided into various annuli of constant thickness $\Delta r$ and constant length $\Delta z$, while $\Delta t$ is the increment of time.

The increments of space are $\Delta r$ and $\Delta z$, $N_r$ and $2N_h$ being the number of space intervals. The integers $i$ and $j$ are used to define the position, with:

$$r = i\cdot\Delta r \quad \text{and} \quad R = N_r\cdot\Delta r \tag{9-27}$$

$$z = j\cdot\Delta z \quad \text{and} \quad h = 2N_z\cdot\Delta z \tag{9-28}$$

(Because of the symmetry, only the part located between 0 and $R$ and 0 and $z$ is considered, 2z being the length of the cylinder.)

The balance written in the small annulus of radius $r$ and thickness $\Delta r$ and $\Delta z$ gives:

$$CN_{r,z} - C_{r,z} = \frac{D\cdot\Delta t}{r(\Delta r)^2}\left[\left(r + \frac{\Delta r}{2}\right)\cdot C_{r+\Delta r,z} - 2r\cdot C_{r,z} + \left(r - \frac{\Delta r}{2}\right)\cdot C_{r-\Delta r,z}\right] + \frac{D\cdot\Delta t}{(\Delta z)^2}\cdot[C_{r,z+\Delta z} - 2C_{r,z} + C_{r,z-\Delta z}] \tag{9-29}$$

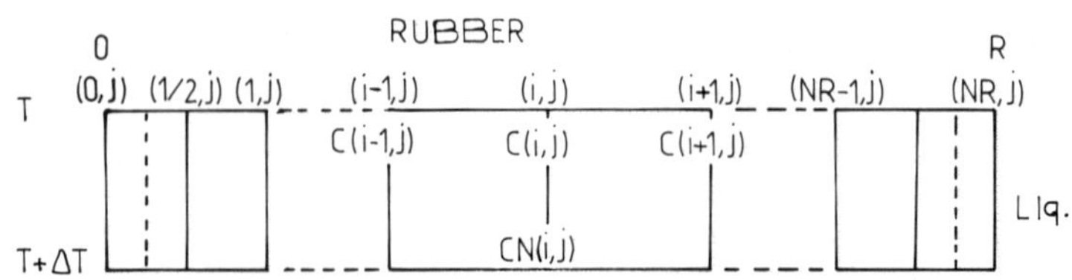


**Figure 9–6** Diagram for numerical analysis in the case of the cylinder of finite length.

***Inside the Cylinder* (r,z).** The preceding equation can be rewritten, by considering the integers $i$ and $j$:

$$C_{i,j} = C_{i,j} + \frac{1}{i \cdot M_r}[G_{i+1,j} - G_{i,j}] + \frac{1}{M_h}[H_{i,j+1} - H_{i,j}] \tag{9-30}$$

and the functions $G$ and $H$

$$G_{i,j} = \left(i - \frac{1}{2}\right) \cdot [C_{i,j} - C_{i-1,j}] \tag{9-31}$$

$$H_{i,j} = C_{i,j} - C_{i,j-1} \tag{9-32}$$

as well as the dimensionless numbers $M_r$ and $M_h$

$$M_r = \frac{(\Delta r)^2}{\Delta t} \cdot \frac{1}{D} \tag{9-33}$$

$$M_h = \frac{(\Delta z)^2}{\Delta t} \cdot \frac{1}{D} \tag{9-34}$$

$D$ being the value of diffusivity at position $i,j$.

***Middle of the Cylinder* (r = *0*).** For the middle of the cylinder, the liquid balance is rewritten by considering the cylinder of thickness $\Delta r/2$, at position $z(j)$.

$$CN_{0,j} = C_{0,j} + \frac{4}{M_r}[C_{1,j} - C_{0,j}] + \frac{1}{M_h}[H_{0,j+1} - H_{0,j}] \tag{9-35}$$

***Surface of the Cylinder.*** Following the third assumption, the concentration of liquid on the surface reaches equilibrium as soon as the process starts:

$$C = C_\infty \qquad \text{surface} \tag{9-36}$$

***Amount of Liquid Located in the Solid.*** The amount of liquid located in the solid is obtained by integrating the concentration with respect to space.

**Numerical analysis in the case of desorption.** The balance of liquid is calculated in various annuli.

***Annulus on the Radial Surface.*** The annulus of thickness $\Delta r/2$ and $\Delta z$, at position $z$ is considered. The linear interpolation is used along the radial transfer.

$$C_{1/4,z} = \frac{3}{4} \cdot C_{0,z} + \frac{1}{4} \cdot C_{1,z} \tag{9-37}$$

The new concentration on the radial surface is then:

$$CN_{NR,j} = C_{NR,j} + \frac{1}{3}[C_{NR-1,j} - CN_{NR-1,j}] - \frac{8\left(R - \frac{\Delta r}{2}\right)}{3\left(R - \frac{\Delta r}{4}\right) M_r}[C_{NR,j} - C_{NR-1,j}] + \frac{1}{M_h}\left[K_{NR,j} + \frac{1}{3}K_{NR-1,j}\right] - \frac{8 \cdot R \cdot \Delta t \cdot F_0}{3\Delta r\left(R - \frac{\Delta r}{4}\right)} \tag{9-38}$$

with

$$K_{i,j} = C_{i,j+1} - 2 \cdot C_{i,j} + C_{i,j-1} \tag{9-39}$$

***Annulus on Both Plane Faces.*** The annulus of thickness $\Delta r$ and $\Delta z/2$, at position $r$, on each plane face, is considered:

$$CN_{i,0} = C_{i,0} - \frac{1}{3}[CN_{i,1} - C_{i,1}] + \frac{1}{i \cdot M_r}\left\{G_{i+1,0} - G_{i,0} + \frac{1}{3}[G_{i+1,1} - G_{i,1}]\right\} + \frac{8}{3M_h}[C_{i,1} - C_{i,0}] - \frac{8 \cdot \Delta t \cdot F_0}{3 \cdot \Delta z} \tag{9-40}$$

***Annulus at the Intersection of Radial Surface and Plane Surface.*** One of the two annuli, of thickness $\Delta r/2$ and $\Delta z/2$, is considered:

By assuming that

$$CN_{R-\Delta r/4, \Delta z/4} - C_{R-\Delta r/4, \Delta z/4} = CN_{R,0} - C_{R,0} \tag{9-41}$$

it is obtained:

$$CN_{R,0} = C_{R,0} - \frac{R - \frac{\Delta r}{2}}{M_r\left(2R - \frac{\Delta r}{2}\right)} \cdot \{3[C_{R,0} - C_{R-1,0}] + C_{R,1} - R_{R-1,1}\} + \frac{1}{2M_h}\{3[C_{R,1} - C_{R,0}] + C_{R-1,1} - C_{R-1,0}\} - \frac{4R \cdot \Delta t \cdot F_0}{\Delta r\left(2R - \frac{\Delta r}{2}\right)} - \frac{2\Delta t F_0}{\Delta z} \tag{9-42}$$

***Annulus on Both Plane Faces, at*** r ***= 0.*** These two annuli are located on each plane face at the position $r = 0$, and thus they are two cylinders of radius $\Delta r/2$ and thickness $\Delta z/2$. The assumption shown earlier for the concentration at position 0, $\Delta z/4$ is made:

$$CN_{0,\Delta z/4} - C_{0,\Delta z/4} = CN_{0,0} - C_{0,0} \tag{9-43}$$

The new concentration is then obtained:

$$CN_{0,0} = C_{0,0} + \frac{1}{3}[C_{0,1} - CN_{0,1}] + \frac{8}{3M_r}[C_{0,1} - C_{0,0}] + \frac{4}{M_r}\left\{C_{1,0} - C_{0,0} + \frac{1}{3}(C_{1,1} - C_{0,1})\right\} - \frac{8\Delta tF_0}{3\Delta z} \tag{9-44}$$

***Amount of Liquid Located in the Solid.*** The amount of liquid located in the cylinder is calculated by integrating the concentration with respect to space.

### 9-3-3 Results

**Diffusivity and evaporation rate.** The diffusivity is obtained from experiments done with thin sheets made of the same rubber as that of the cylinder. Thin sheets of thickness 0.15 cm are cured and tested for absorption of liquid [12]. The diffusivity is determined by using three various ways. The first way is by using the short-test technique [11, 12] and equation:

$$\frac{M_t}{M_\infty} = \frac{4}{L}\left(\frac{D\cdot t}{\pi}\right)^{0.5} \tag{9-17}$$

The second way to determine the diffusivity is by using the method with the time at half amount of liquid transported, and equation:

$$D = 0.049\cdot\left(\frac{L^2}{t_{0.5}}\right) \tag{9-18}$$

The third method to determine the diffusivity is by using the values of the amount of liquid transported at the end of the process, and equation:

$$\frac{M_\infty - M_t}{M_\infty} = \frac{8}{\pi^2}\cdot\exp\left(-\frac{\pi^2}{L^2}Dt\right) \tag{9-19}$$

The rate of evaporation at the beginning of the process of desorption, when the concentration is $C_{in}$, is determined as usual from the slope of the kinetics of desorption when $t = 0$.

$$F = \frac{dM}{dt} \qquad t = 0 \tag{9-20'}$$

The rate of evaporation of the liquid from the surface of the rubber cylinder is then proportional to the concentration of liquid on the surface, at any time.

$$F_t = F_0\cdot\frac{C_t}{C_{in}} \tag{9-45}$$

when $C_t$ and $C_{in}$ are the concentration of liquid on the surface at time $t$ and at the beginning of the desorption. Of course, the rate of evaporation depends on the motion of air, as shown previously.

The values of diffusivity are given in Table 9–5, as well as the value of the rate of evaporation.

As shown in this table, the diffusivity is found to be concentration dependent during the process of absorption, and about constant during desorption.

**Validity of the process in the case of absorption.** The experimental and calculated kinetics of absorption are drawn in Figure 9–7 for the cylinder with the following dimensions: diameter = 2.2 cm and height = 0.49 cm. The capacity of absorption (expressed in g. of liquid by $cm^3$ of the initial volume of rubber free of liquid) is 3.545. As shown in this figure, a good agreement is obtained, proving the validity of the model.

The profiles of concentration of liquid developed through the circular cross section of the cylinder taken at midheight are drawn in Figure 9–8.

**Validity of the process in the case of desorption.** The kinetics of desorption of liquid are compared in Figure 9–9, as they are obtained from experiment and calculation with the help of the model. Calculation is also made by using a more simple model built with an infinite value for the evaporation rate. This latter case is obtained when the concentration on the surface reaches the equilibrium value as soon as the desorption proceeds. In the present study, the concentration of liquid on the surface is then zero.

As already shown, this second model with $C = 0$ on the surface is not in good agreement with experiments, especially at the beginning of the process, when the concentration of liquid on the surface is far from being zero.

The profiles of concentration calculated by using the model are drawn through the circular cross section taken at the midheight of the cylinder in Figure 9–10.

**Conclusions.** Modeling the process of absorption and desorption of liquid accommodates all the facts in the case of cylinders of finite length, and success has attended calculation. The data, such as diffusivity, evaporation rate, and capacity of absorption, are determined by experiments done with thin sheets made of the same material.

**TABLE 9–5** Diffusivity ($cm^2/s$) and Rate of Evaporation.

| | |
|---|---|
| Absorption $D = 1.4 \times 10^{-7} \times \exp\left(1.186 \cdot \dfrac{C_{r,z}}{C_{eq}}\right)$ | |
| Desorption $D = 12.5 \times 10^{-7}$ | $F_0 = 10.4 \times 10^{-5}$ (cm/s) |

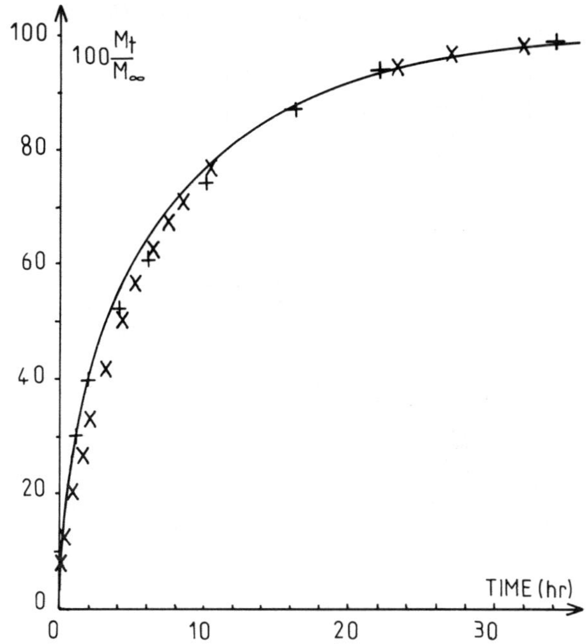


**Figure 9–7** Kinetics of absorption by cylinder of finite length; diameter = 2.2 cm, and h = 0.49 cm. +: calculated with $D = 1.4 \times 10^{-7} \times \exp(1.86\ C/C_\infty)$ and $C_\infty = 3.545$ g/cm$^3$; ×: experiment; full line: calculated with $D = 6.2 \times 10^{-7}$ (cm$^2$/s) and $C_\infty = 3.545$ g/cm$^3$. (Reprinted from *J. Polym. Engng.*, 7 (1987), 149–167, Y. Khatir, J. Bouzon, and J. M. Vergnaud, ref. 18, by courtesy of Freund Publishing House.)

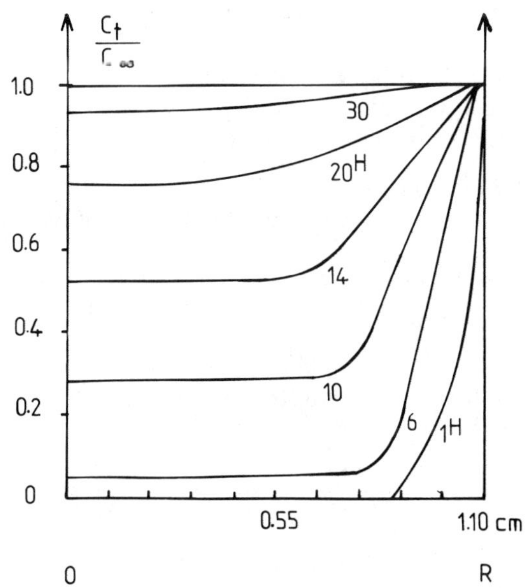


**Figure 9–8** Profiles of concentration of liquid during absorption through the cylinder of finite length (diameter 2.2 cm, h = 0.49 cm). (Reprinted from *J. Polym. Engng.*, 7 (1987), 149–167, Y. Khatir, J. Bouzon, and J. M. Vergnaud, ref. 18, by courtesy of Freund Publishing House)

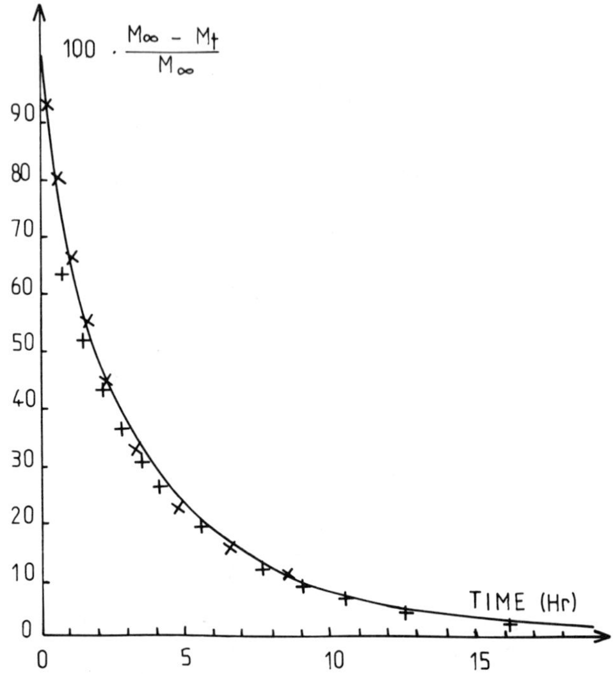


**Figure 9–9** Kinetics of desorption with cylinder of finite length (diameter = 2.2 cm, and h = 0.49 cm). ×: experiment; −model; +: analytical solution (infinite value of the evaporation rate). (Reprinted from *J. Polym. Engng.*, 7 (1987), 149–167, Y. Khatir, J. Bouzon, and J. M. Vergnaud, ref. 18, by courtesy of Freund Publishing House)

In the case of rubber, the process of the desorption is rather simple in the sense that all the liquid previously absorbed is totally desorbed and evaporated, when the atmosphere is free of liquid vapor.

***Remark 1. Other Equations.*** The preceding functions $G'$ and $H'$ can also be calculated, when the mean value of diffusivity is determined:

$$G'_{i,j} = \left(i - \frac{1}{2}\right) \cdot [D_{i-1,j} + D_{i,j}] \cdot [C_{i,j} - C_{i-1,j}] \cdot \qquad (9\text{-}31')$$

$$H'_{i,j} = [D_{i,j-1} + D_{i,j}] \cdot [C_{i,j} - C_{i,j-1}] \qquad (9\text{-}32')$$

Then, in equation (9-30), the terms $1/M_r$ and $1/M_r$ are replaced by $A$ and $B$ expressed as follows:

$$A = \frac{\Delta t}{2(\Delta r)^2} \quad \text{and} \quad B = \frac{\Delta t}{2(\Delta z)^2}$$

Equation (9-35) obtained for the middle of the cylinder (along the axis with $r = 0$) can be rewritten:

$$CN_{0,j} = C_{0,j} + 2A \cdot [D_{1,j} + D_{0,j}] \cdot [C_{1,j} - C_{0,j}] + B \cdot [H_{0,j+1} - H_{0,j}] \qquad (9\text{-}35)$$

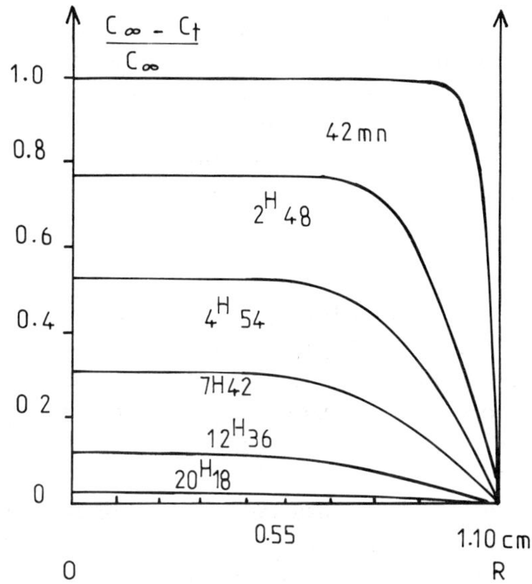


**Figure 9–10** Profiles of concentration of liquid during desorption, through the cross section at midheight (diameter = 2.2 cm, h = 0.49 cm). (Reprinted from *J. Polym. Engng.*, 7 (1987), 149–167, Y. Khatir, J. Bouzon, and J. M. Vergnaud, ref. 18, by courtesy of Freund Publishing House)

## 9-4 LIQUID ABSORPTION AND DESORPTION WITH RUBBER TUBINGS: MODELS AND EXPERIMENTS

Rubber tubings are widely used for liquid transport, and as new liquids are sometimes introduced on the market, the determination of the resistance of rubber to these liquids is of interest. These matter transports are controlled by diffusion of the liquid through the rubber in case of absorption, while the diffusion through the solid is followed by evaporation from the surface in case of desorption.

A special example is worth attention when the rubber tubing is in contact with liquid on the internal surface while it is allowed to evaporate from the external surface immersed in air. The liquid is transported under transient conditions at the beginning and as the transport is allowed to proceed, stationary conditions are then obtained in the final stage.

As the diffusivity is generally concentration dependent, no analytical solution can be found by mathematical treatment, and the problem is solved with the help of a numerical method with finite differences.

The main purpose is to set up a method of approach to the problem of liquid transport in the case of tubings when the diffusivity is concentration dependent. Short tests are used for measuring all data of interest such as diffusivity, capacity of absorption, and evaporation rate. An explicit numerical method with finite differences is developed in order to describe the process. It takes all the facts into account, and especially the radial diffusion of liquid with concentration-dependent diffusivity and the evaporation of liquid from the external surface. As usual, the evaporation rate is assumed to be proportional to the liquid concentration on the external surface of the tubing.

### 9-4-1 Experimental

**Rubber samples.** Rubber tubings of various lengths are prepared by curing the rubber in a steel mold operated by a press under a pressure of 30 bars. The operational conditions necessary for the tubing to get a high value of the state of cure are previously determined from experiments performed by calorimetry in scanning mode [13–17].

**Short-test conditions for measuring the parameters.** Rubber sheets of the same composition as the tubing are cured in a mold operated by a heated press. Operational conditions of cure are also determined.

The rubber sheets are immersed into the liquid and weighed at intervals during the process of absorption. The presaturated sheets are then exposed either to motionless air or to well-stirred atmosphere for studying the process of desorption.

**Test conditions for internal and external absorption.** The rubber tubing is immersed in toluene with a controlled rate of stirring. Great care is exercised in preventing the liquid from coming into contact with the plane faces of the tube. The test sample is weighed at intervals without altering the process.

**Test conditions for internal absorption and external desorption.** The liquid is continuously introduced into the internal free volume of the tubing, while the liquid is allowed to evaporate from its external surface. The external atmosphere is either motionless or well stirred. At intervals, the tubing is extracted and dried very quickly and weighed for determining the amount of liquid absorbed. Previous experiments have shown that these test conditions can therefore be used in these studies without leading us far astray.

### 9-4-2 Theoretical

**Assumptions.** The following assumptions are made.

1. The transport of liquid is only radial, as long rubber tubings are used without any longitudinal transport.

2. The transport is controlled by diffusion of liquid under transient conditions when both surfaces of the tubing are immersed into the liquid.
3. The transport is controlled by transient diffusion through the solid and evaporation on the external surface when the liquid is located in the internal part of the tubing.
4. The diffusivity can be concentration dependent, as determined from experiments.
5. The evaporation rate is proportional to the liquid concentration on the external surface.
6. In the case of absorption, the liquid concentration on the surface reaches equilibrium as soon as the process starts.
7. In spite of the change of volume, a frame of reference is taken for calculation, by considering the original dimension of the tubing free of liquid.

**Mathematical treatment.** As diffusion is only radial, the concentration of liquid is expressed in terms of radius $r$ and time $t$ by Fick's second law through the circular cross section.

$$\frac{\partial C}{\partial t} = \frac{1}{r} \cdot \frac{\partial}{\partial r}\left[r \cdot D \cdot \frac{\partial C}{\partial r}\right] \tag{9-46}$$

In the case of absorption of liquid on both surfaces, the initial and boundary conditions are:

$$t = 0 \qquad R_i < r < R_e \qquad C_{r,0} = 0 \qquad \text{tubing} \tag{9-47}$$

$$\begin{aligned} t > 0 \qquad & R_i < r < R_e && C_{r,t} && \text{tubing} \\ & r = R_i \text{ and } r = R_e && C = C_{eq} && \text{surface} \end{aligned} \tag{9-48}$$

In the case of absorption by the internal surface and evaporation on the external surface, the boundary conditions are:

$$t > 0 \qquad r = R_i \qquad C = C_{eq} \qquad \text{internal surface} \tag{9-49}$$

$$r = R_e \qquad F_0 \cdot C_{Re,t} \qquad \text{external surface} \tag{9-50}$$

**Numerical analysis in case of absorption of liquid on both surfaces.** As no analytical solution can be found when diffusivity is concentration dependent, a numerical method with finite differences is used.

The circular cross section of the tubing is divided into $N$ annuli of constant thickness $\Delta r$. The balance of the liquid is calculated for the annulus of radius $r$ and thickness $\Delta r$, during the increment of time $\Delta t$ (Figure 9–11).

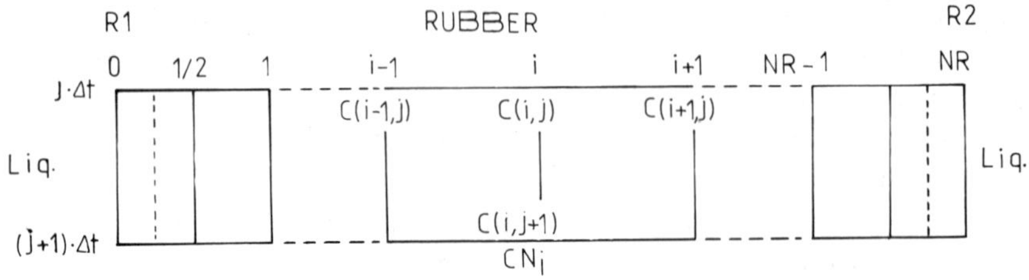


**Figure 9–11** Diagram for numerical analysis in the case of rubber tubing.

By putting the integer $i$, the radius $r$ of the annulus is defined by:

$$r = i \cdot \Delta r \tag{9-51}$$

with

$$N_i \leq i \leq N_e$$

and

$$R_i = N_i \cdot \Delta r \qquad R_e = N_e \cdot \Delta r \tag{9-52}$$

The new concentration after the elapse of time $\Delta t$ is obtained for the annulus of radius $r$ (and integer $i$) in terms of the concentrations at the same position and adjacent position.

$$CN_i = C_i + \frac{\Delta t}{i \cdot (\Delta r)^2} [I_{i-1/2} - I_{i+1/2}] \tag{9-53}$$

where the function $I$ is as follows:

$$I_{i-1/2} = \left(i - \frac{1}{2}\right) \cdot D_{i-1/2} \cdot (C_{i-1} - C_i) \tag{9-54}$$

The value of diffusivity at position $(i - \frac{1}{2})$ can be calculated by various ways. Here, a simple way is used by considering the mean diffusivity.

$$D_{i-1/2} = \frac{1}{2} (D_{i-1} + D_i) \tag{9-55}$$

Of course, when steep gradients of concentration are developed through the solid, or when the effect of concentration on diffusivity is high, other equations can be used for calculating the diffusivity at position $(i - \frac{1}{2})$.

***Concentration on Both Surfaces.*** The concentration on both surfaces of the tubing is at equilibrium.

$$C_{Ri} = C_{Re} = C_{eq} \tag{9-56}$$

The amount of liquid located in the rubber tubing per unit height is calculated by integrating the concentration of liquid with respect to space.

$$M_t = 2\pi \int_{Ri}^{Re} r \cdot C_r \cdot dr \tag{9-57}$$

which becomes:

$$M_t = \pi[N_i \cdot C_{Ni} + N_e \cdot C_{Ne}] \cdot (\Delta r)^2 + 2\pi \cdot \sum_{Ni+1}^{Nr-1} i \cdot (\Delta r)^2 \cdot C_i \tag{9-58}$$

Of course, in the preceding equation, slices having a thickness of $\Delta r/2$ are considered on both surfaces of the tubing.

**Numerical analysis in the case of absorption and evaporation.** The case with absorption of liquid by the internal surface and evaporation on the external surface is of high interest. Three positions of interest are the internal face, the external surface, and the part of the tubing located between them.

***Internal Surface* (R$_i$).** The concentration of liquid is at equilibrium.

$$C_{Ri} = C_{eq} \tag{9-56}$$

***Inside the Tubing* (Ri < r < R$_e$).** The concentration inside the tubing is given by the previous relation, equation (9-53).

***On the External Surface* (Re).** The balance of liquid is determined during the increment of time $\Delta t$ on the annulus of thickness $\Delta r/2$ located on the external surface, by considering the diffusion through the solid and evaporation from the external surface.

By making the assumption that

$$CN_{Re-\Delta r/4} - C_{Re-\Delta r/4} = CN_{Re} - C_{Re} \tag{9-59}$$

and considering the function $I$ in position $N_e - \frac{1}{2}$: (or $R_e - \Delta r/2$)

$$I_{Ne-1/2} = \left(Ne - \frac{1}{2}\right) \cdot D_{Ne-1/2}(C_{Ne-1} - C_{Ne}) \tag{9-54'}$$

the following equation is obtained for the concentration on $N_e$, the external surface, $CN_{Re}$, expressed in terms of the previous concentration $C_{Re}$:

$$CN_{Re} - C_{Re} = \frac{4\Delta t}{(\Delta r)^2 \left(Ne - \frac{1}{4}\right)} \cdot [F_0 \cdot C_{Re} \cdot N_e \cdot \Delta r - I_{Ne-1/2}] \tag{9-60}$$

***Amount of Liquid.*** The amount of liquid located in the tubing can be calculated by integrating the concentration with respect to space.

### 9-4-3 Results

**Diffusivity and evaporation rate.** All data, such as diffusivity, evaporation rate, and capacity of absorption, are determined from experiments carried out with rubber sheets. The three methods, already cited, are used to determine diffusivity. The first is the method with short tests, by plotting the amount of liquid transported as a function of the square root of time.

$$\frac{M_t}{M_\infty} = \frac{4}{L}\left(\frac{Dt}{\pi}\right)^{0.5} \tag{9-17}$$

The second method uses the time necessary for the half-sorption process and the following equation.

$$D = 0.049 \cdot \frac{L^2}{t_{0.5}} \tag{9-18}$$

The method of interest in the later stage of diffusion uses the first term of the series:

$$\frac{M_\infty - M_t}{M_\infty} = \frac{8}{\pi^2} \cdot \exp\left(-\frac{\pi^2}{L^2} Dt\right) \tag{9-19}$$

These three values obtained for the diffusivity are not the same, and the diffusivity is then concentration dependent. The expression shown in Table 9–6 is obtained.

The rate of evaporation is determined from the slope of the desorption curve plotted against time. Of course, the values of the evaporation rate depend on the stirring of air.

$$F = \frac{dM_t}{dt} \text{ for } t \to 0 \tag{9-20}$$

**Absorption of liquid on both surfaces.** Experiments for absorption of liquid on both surfaces are performed by completely immersing three rubber tubings of various lengths in the liquid.

length: 0.70 cm – 3.0 cm – infinite length

radius: internal = 0.98 cm – external = 1.66 cm

**TABLE 9–6** Diffusivity and Evaporation Rate.

| |
|---|
| $D = 0.4 \times 10^{-7} \times \exp\left(3.52 \frac{C}{C_{in}}\right)$ (cm²/s) |
| $F_0 = 10.4 \times 10^{-5}$ (g/cm²·s) pure liquid |
| $C_{eq} = 0.77$ |

The results of investigations for these three various tubings are represented in Figure 9–12, where the amount of liquid located in the tubing at time $t$ as a fraction of $M_\infty$, the amount at infinite time, is expressed as a function of time. Three results are worth pointing out:

1. The model gives kinetics in good agreement with the experimental kinetics, in the case of the tubings with infinite length.
2. The kinetics obtained with various annuli of different lengths (0.7 cm–3.0 cm) can also be compared with the kinetics of the tubing. In this case, the model used for calculation has been described in the previous chapter.
3. The kinetics of absorption are quite different for each of these rubber tubings. The effect of the longitudinal transport can be appreciated by comparing the kinetics. Of course, the larger the surface per unit

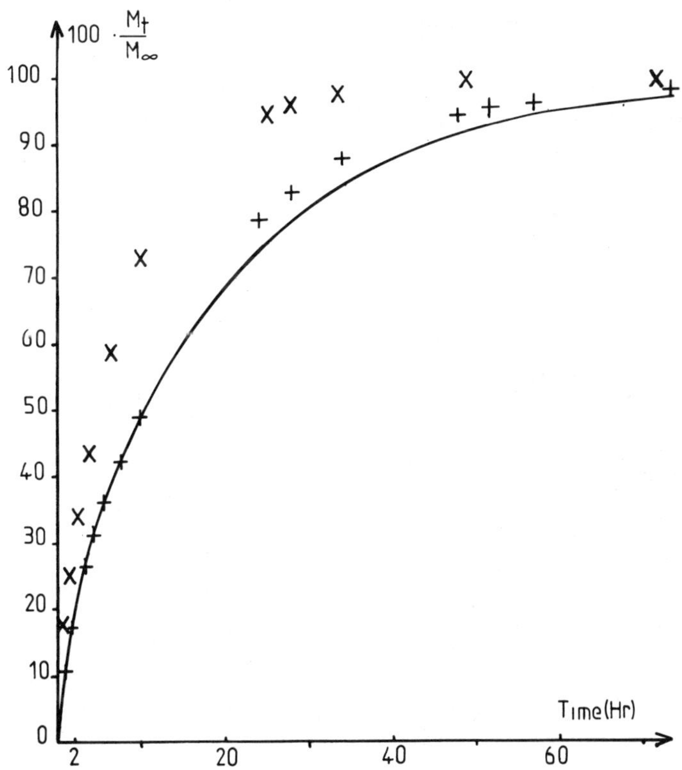


**Figure 9–12** Kinetics of absorption of liquid on both surfaces of rubber tubings: ×: length of 0.70 cm; +: length of 3.0 cm; −: infinite length (full line). (Reprinted from *J. Polym. Engng.*, 7 (1987), 275–299, Y. Khatir, J. Bouzon, and J. M. Vergnaud, ref. 19, by courtesy of Freund Publishing House)

mass of rubber in contact with the liquid, the higher the rate of absorption. The difference between the kinetics of absorption is especially significant for lower times of contact.

The profiles of concentration of the liquid located within the thickness are calculated at various times (Figure 9–13). As already said, it is difficult to do experiments providing one with this information [9]. However, the validity of these profiles can be evaluated, the kinetics of uptake of liquid being obtained by integrating these profiles of concentration with respect to space.

The progression of the liquid within the tubing is described in these two curves (Figures 9–12 and 9–13). Very deep gradients of concentration are exhibited at low time of absorption (1 h). These profiles are not exactly symmetrical with respect to the mean radius $\frac{1}{2}(R_i + R_e)$.

**Absorption on the internal surface and evaporation on the external face.** This problem is of high concern because it corresponds with the well-known use of tubing for transporting liquid. The liquid is in contact with the internal face, then it can diffuse within the rubber and evaporate on the external surface in contact with the external atmosphere.

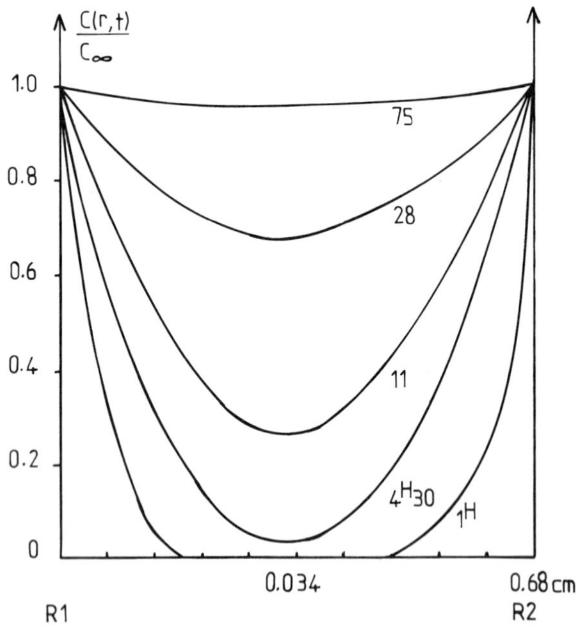


**Figure 9–13** Profiles of concentration of liquid within the tubing during the absorption of liquid on both surfaces. (Reprinted from *J. Polym. Engng.*, 7 (1987), 275–299, Y. Khatir, J. Bouzon, and J. M. Vergnaud, ref. 19, by courtesy of Freund Publishing House)

The kinetics of uptake of liquid are determined either from experiments or calculation. The amount of liquid located in the tubing as a fraction of the capacity of absorption of liquid is expressed as a function of time (Figure 9–14). The profiles of concentration developed through the thickness of the tubing are calculated at various times (Figure 9–15).

The following conclusions can be drawn.

1. The amount of liquid located in the tubing is far from being equal to the capacity of absorption, about half this value.
2. Steep gradients are obtained during the transient stage.
3. A rather long time is necessary to obtain the typical steady-state concentration distribution (113 h) with a linear gradient of concentration.

**Conclusion.** Modeling of the process is achieved in the following two cases concerned with rubber tubings.

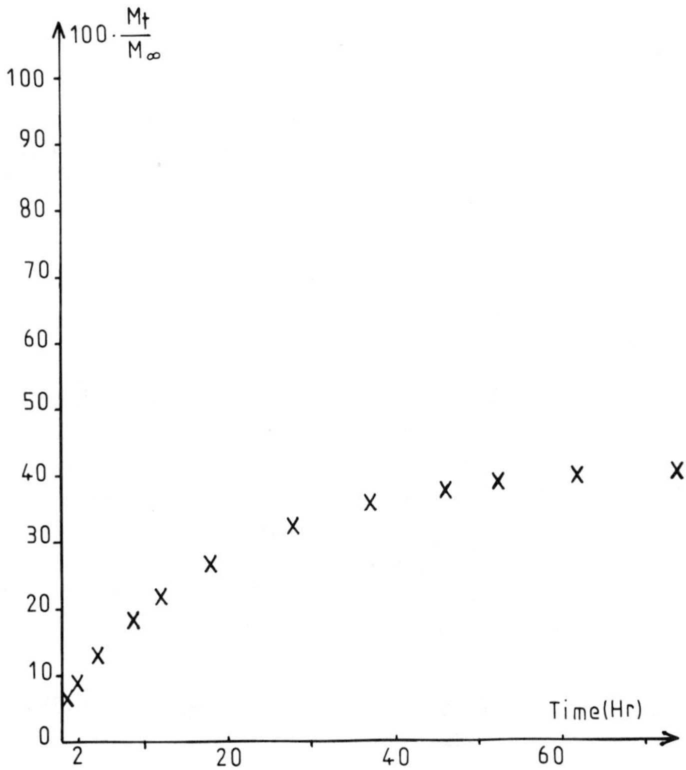


**Figure 9–14** Kinetics of absorption of liquid in the case of absorption on the internal surface and evaporation on the external surface. (Reprinted from *J. Polym. Engng.*, 7 (1987), 275–299, Y. Khatir, J. Bouzon, and J. M. Vergnaud, ref. 19, by courtesy of Freund Publishing House)

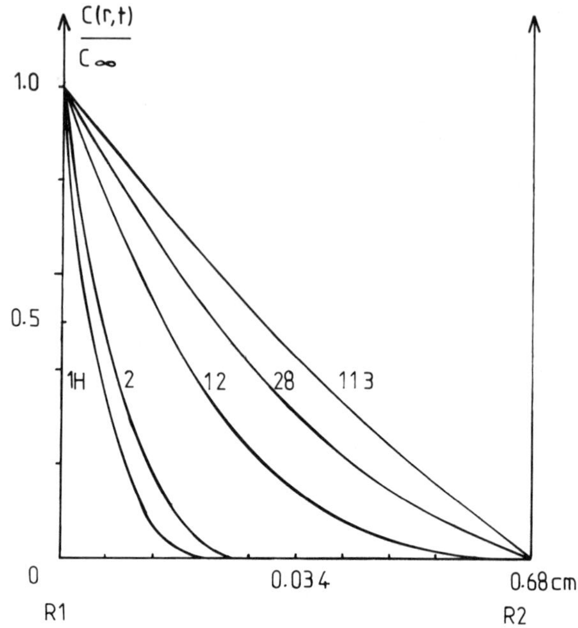


**Figure 9–15** Profiles of concentration of liquid through the thickness of the tubing in the case of absorption on the internal surface and evaporation on the external surface. (Reprinted from *J. Polym. Engng.*, 7 (1987), 275–299, Y. Khatir, J. Bouzon, and J. M. Vergnaud, ref. 19, by courtesy of Freund Publishing House)

1. The tubing is in contact with the liquid on both surfaces.
2. The liquid is in contact with the internal surface of the tubing, while the external surface is in contact with the external atmosphere.

Short tests conducted with thin rubber sheets are able to give all the information necessary for calculation: the diffusivity and rate of evaporation. As a result, the general method, consisting of determining the parameters of interest by using short tests and thin sheets and calculation with the help of the model, is a nondestructive technique for measuring the resistance of rubber tubings to liquids. Moreover, the method gives more information than experiments, with the profiles of concentration of liquid developed through the thickness of the tubing.

## 9-5 LIQUID ABSORPTION AND DESORPTION WITH RUBBER ANNULUS: MODELS AND EXPERIMENTS

Tests in which rubber pieces are exposed to liquids are called *swelling tests* because the resulting change in volume of the test piece is by far the most commonly used measure of the resistance of the rubber to the liquid [1]. These tests are also referred to as *oil aging*, as the most often specified liquids are standard grades of mineral oil. The degree of swelling is a good measure of the general resistance of a rubber to a given liquid, and it can be related to the state of cure of the rubber, the crosslink density being estimated by use of the Flory-Rehmer equation [1, 2].

Many rubber annuli of various shapes and dimensions are used in the way of sealing rings, and they can be in contact not only with the external atmosphere but also with various kinds of liquids, and especially oils and hydrocarbons.

The main purpose of this chapter is to describe a general method for studying the behavior of these rubber annuli when they are in contact with liquids. Toluene is selected for the liquid, as it is done in various tests [1, 3–6]. Although an annulus is circular (with the following symmetries), it may be a complicated two-dimensional system with radial and longitudinal transport of liquid.

The rate of uptake of liquid is generally controlled by diffusion [1, 7], the diffusivity being either constant or concentration dependent. In this first case with constant diffusivity, an analytical solution can be found, especially when the boundary conditions are simple (the concentration of liquid on the surface attains the value corresponding to the equilibrium uptake concentration and remains constant afterwards). In the case of concentration-dependent diffusivity, modeling of the process using a numerical method with finite differences is the only possibility.

The case of desorption is the complementary test, as the rubber piece is sometimes submitted to oil or hydrocarbon only from time to time. In the desorption process, the transport of liquid is generally done by diffusion through the solid and evaporation from the surface. Mathematical solutions can describe the desorption process when the diffusivity is constant, and for special boundary conditions [7]. When the diffusivity is concentration dependent, models must be built. The model takes into account the liquid transported by diffusion through the solid and the evaporation from the surface. No drastic assumption is imposed.

As data are necessary for calculation, the diffusivity and the rate of evaporation are determined experimentally, using thin sheets of the same rubber in absorption and desorption under the same conditions of temperature. Thin plates of rubber are chosen because there is only a one-dimensional diffusion and the tests are short (the time being proportional to the square of the thickness of the sheet). The short-test technique [11] is used for calculating the diffusivity at a defined value of the liquid concentration.

### 9-5-1 Experimental

**Rubber samples.** Rubber annuli are prepared by pressing the rubbery compound in a heated mold. The operational conditions (15 min) necessary for the cure to reach a value higher than 90 percent are determined by using calorimetric studies and a model [13–17]. The kinetics of heat evolved from the cure reaction is measured from the heat flux-temperature (or time) history obtained in scanning mode. The model takes into account not only the heat transported by conduction through the solid but also the internal heat generated by the cure reaction.

**Absorption and desorption.** In the case of absorption, the sample is immersed in the well-stirred liquid and kept at 20°C (pure toluene). The sample is weighed at intervals, after a proper drying of the surface able to extract the excess of liquid. Preliminary experiments are carried out on various rubber samples to make sure that the extraction of the sample during the weighing time does not alter the process of absorption.

For desorption, the sample previously presaturated by liquid is exposed to air at 20°C with defined conditions of stirring.

### 9-5-2 Theoretical

**Assumptions.** The following assumptions are made.

1. The transport of liquid is radial and longitudinal.
2. Both these transports are controlled by transient diffusion within the solid, either for absorption or for desorption.
3. Evaporation takes place on the surface in the case of desorption. The rate of evaporation is proportional to the liquid concentration on the surface.
4. In the case of absorption, the concentration of liquid on the surface reaches equilibrium as soon as the process starts.
5. A frame of reference, with the original dimension of the annulus free of liquid, is taken for calculation.
6. The cross section of the annulus is rectangular in shape.

**Mathematical treatment.** The rate of change in concentration of liquid in the annulus is expressed by Fick's second law by considering the radial and the longitudinal transports.

**TABLE 9–7** Condition of Cure of Rubber Annuli.

| Sample | $Ri$(cm) | $Re$(cm) | $H$(cm) | $t$(min) |
|---|---|---|---|---|
| 1 | 0.985 | 1.66 | 3 | 30 |
| 2 | 0.985 | 1.66 | 0.7 | 30 |
| 3 | 0.985 | 1.66 | 0.7 | 15 |

Temperature of mold = 160°C. Pressure = 30 bars.

$$\frac{\partial C}{\partial t} = \frac{1}{r} \cdot \frac{\partial}{\partial r}\left(rD \cdot \frac{\partial C}{\partial r}\right) + \frac{\partial}{\partial z}\left(D \cdot \frac{\partial C}{\partial z}\right) \tag{9-21}$$

where $D$ is the concentration-dependent diffusivity, and the dimensions of the annulus are:

$$R_i < r < R_e \qquad 0 \le z \le h \tag{9-61}$$

$R_i$ and $R_e$ being the internal and external radii of the annulus, and $h$ its height.

The rate of evaporation is proportional to the concentration of liquid on the surface.

$$F = F_o \cdot C_{surface} \tag{9-50}$$

**Numerical analysis in the case of absorption.** As no analytical solution is found with the concentration-dependent diffusivity, a numerical method with finite differences is used. The problem being symmetrical, only one half of the rectangular cross section is considered (Figure 9–16).

The rectangle of dimensions $(R_e - R_i)$ and $h/2$ is divided into small rectangles of dimensions $\Delta r$ and $\Delta z$. The number of rectangles is:

$$N_r = \frac{R_e - R_i}{\Delta r} \qquad N_h = \frac{h}{2 \cdot \Delta z} \tag{9-62}$$

By introducing the integer $i$ for radial position and $j$ for height, we have:

$$R_i \le r \le R_e \qquad \text{or} \qquad N_i \le i \le N_e \tag{9-63}$$

with:

$$r = i \cdot \Delta r \qquad R_i = N_i \cdot \Delta r \qquad R_e = N_e \cdot \Delta r$$
$$z = j \cdot \Delta z \tag{9-64}$$

***Within the Annulus.*** The balance of the liquid is written for the small annulus of thickness $\Delta r$ and height $\Delta z$, at position $r$ and $z$. The new con-

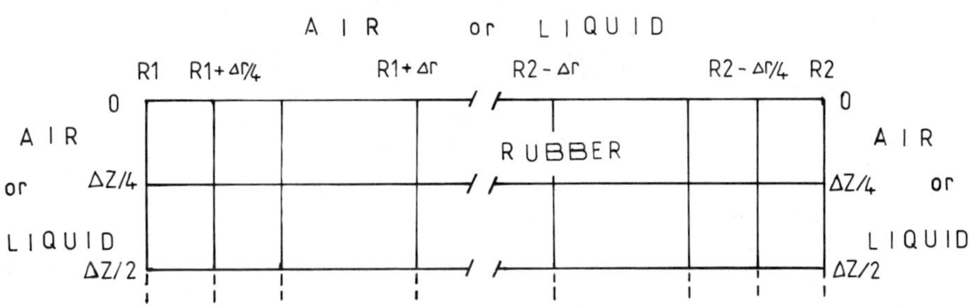


**Figure 9–16** Scheme of a part of the annulus (other parts can be obtained by symmetry).

centration at position $(r,z)$ after the elapse of time $\Delta t$ is given by:

$$CN_{i,j} = C_{i,j} + \frac{\Delta t}{i \cdot (\Delta r)^2}[J_{i-1/2,j} - J_{i+1/2,j}] + \frac{\Delta t}{(\Delta z)^2}[I_{i,j-1/2} - I_{i,j+1/2}] \tag{9-65}$$

where the functions $J$ and $I$ are:

$$J_{i-1/2,j} = \left(i - \frac{1}{2}\right) \cdot D_{i-1/2,j} \cdot (C_{i-1,j} - C_{i,j}) \tag{9-54}$$

$$I_{i,j+1/2} = D_{i,j+1/2} \cdot (C_{i,j} - C_{i,j+1}) \tag{9-66}$$

The diffusivity in functions $J$ and $I$ can be calculated in many ways.

***Surface of the Annulus.*** The concentration of liquid on the surface of the rubber annulus is equal to the concentration at equilibrium: $C_{eq}$.

***Amount of Liquid Located in the Rubber.*** The amount of liquid located in the rubber is obtained by integrating the concentration of liquid with respect to space.

**Numerical analysis in the case of desorption.** The rubber annulus, previously saturated with liquid, is exposed to motionless air. Various parts are successively considered.

***Annulus Next to the Internal Surface $R_i$ (thickness $\Delta r/2$ and $\Delta z$).*** This annulus is centered on the circle on radius $R_i + \Delta r/4$. By using the linear interpolation along the radius:

$$f\left(\frac{1}{4}\right) = \frac{3}{4} \cdot f(0) + \frac{1}{4} \cdot f(1) \tag{9-67}$$

the new concentration on the surface $R_i$ is:

$$CN_{Ni,j} = C_{Ni,j} - \frac{1}{3}[CN_{Ni+1,j} - C_{Ni+1,j}] + \frac{1}{M_h}\left[K_{Ni,j} + \frac{K_{Ni+1,j}}{3}\right] + \frac{1}{Mr}\frac{8\left(R_i + \frac{\Delta r}{2}\right)}{3\left(R_i + \frac{\Delta r}{4}\right)}[C_{Ni+1,j} - C_{Ni,j}] - \frac{8 \cdot R_i \cdot \Delta t}{3\left(R_i + \frac{\Delta r}{4}\right) \cdot \Delta r} \cdot F_0 \cdot C_{Ni,j} \tag{9-68}$$

where the function $K$ is defined as follows:

$$K_{i,j} = C_{i,j+1} - 2 \cdot C_{i,j} + C_{i,j-1} \tag{9-69}$$

and the dimensionless numbers:

$$M_r = \frac{(\Delta r)^2}{\Delta t} \cdot \frac{1}{D} \qquad M_h = \frac{(\Delta z)^2}{\Delta t} \cdot \frac{1}{D} \tag{9-70}$$

***Annulus Next to the External Surface* $\mathbf{R_e}$ *(Thickness* $\mathbf{\Delta r/2}$ *and* $\mathbf{\Delta z}$*).*** This annulus is centered on the circle of radius $R_e - \Delta r/4$. The calculation is done in the same way as for the annulus next to the internal surface. The new concentration is obtained:

$$CN_{Ne,j} = C_{Ne,j} - \frac{1}{3}[CN_{Ne-1,j} - C_{Ne-1,j}] + \frac{1}{M_h}\left[K_{Ne,j} - \frac{1}{3}K_{Ne-1,j}\right]$$
$$+ \frac{8\left(R_e - \frac{\Delta r}{2}\right)}{3M_r\left(R_e - \frac{\Delta r}{4}\right)}[C_{Ne-1,j} - C_{Ne,j}] - \frac{8R_e \cdot \Delta t}{3\left(R_e - \frac{\Delta r}{4}\right)\Delta r} \cdot Fo \cdot C_{Ne} \tag{9-71}$$

***Annulus Next to the Internal Face* ($\mathbf{R_i}$) *and Plane Face.*** This annulus of thickness $\Delta r/2$ and $\Delta z/2$ is centered on the circle of radius $R_i + \Delta r/4$ and at the position $\Delta z/4$. By asuming that the concentration of liquid at the center of this annulus is the same as that on the rubber face (with position $r = R_i$ and $z = 0$), the new concentration on the surface $(R_i, 0)$ is:

$$CN_{Ri,0} = C_{Ri,0} + \frac{2R_i + \Delta r}{M_r(4R_i + \Delta r)} \cdot [3(C_{Ri+1,0} - C_{Ri,0}) + C_{Ri+1,1} - C_{Ri,1}]$$
$$- \frac{2\Delta t}{\Delta z} Fo C_{Ri,0} - \frac{2R_i \cdot \Delta t}{\left(R_i + \frac{\Delta r}{4}\right)\Delta r} Fo C_{Ri,0} \tag{9-72}$$

***Annulus Next to the External Face* (Re) *and Plane Face.*** This annulus of thickness $\Delta r/2$ and $\Delta z/2$ is centered on the circle of radius $(R_e - \Delta r/4)$ and position $\Delta z/4$. By making the same assumption for the concentration at the center of this annulus and the surface, the following is obtained.

$$CN_{Ne,0} = C_{Ne,0} + \frac{(2R_e - \Delta r)}{M_r(4R_e - \Delta r)}[3(C_{Ne-1,0} - C_{Ne,0}) + C_{Ne-1,1}$$
$$- C_{Ne,1}] + \frac{1}{2M_h}[3(C_{Ne,1} - C_{Ne,0}) + C_{Ne-1,1} - C_{Ne-1,0}] \tag{9-73}$$
$$- \frac{2\Delta t}{\Delta a} Fo \cdot C_{Ne,0} - \frac{2R_e \cdot \Delta t}{\left(R_e - \frac{\Delta r}{4}\right)\Delta r} Fo \cdot C_{Ne,0}$$

***Annulus Next to the Plane Face* (z = *0*).** These annuli, of thickness $\Delta z/2$ and $\Delta r$, are centered on the radius ($r = i \cdot \Delta r$) and position $\Delta r/4$. The new concentration is then:

$$CN_{i,0} = C_{i,0} - \frac{1}{3}[CN_{i,1} - C_{i,1}] + \frac{1}{i \cdot M_r}\left[C_{i+1,0} - C_{i,0} + \frac{1}{3}(C_{i+1,1} - C_{i,1})\right] + \frac{8}{M_h}[C_{i,1} - C_{i,0}] - \frac{8\Delta t}{3\Delta z} \cdot F_0 \cdot C_{i,0} \tag{9-74}$$

***Annulus within the Rubber Annulus* (i,j).** The preceding equation obtained for the absorption can be used during the desorption.

***Amount of Liquid Located in the Rubber Annulus.*** The amount of liquid located in the rubber annulus is obtained by integrating the concentration with respect to space.

### 9-5-3 Results

**Determination of diffusivity and evaporation rate.** The diffusivity is obtained from experiments done with rubber sheets of 0.15 cm thickness which are properly vulcanized. Three methods are used for determining the diffusivity from the kinetics of absorption and desorption. One is the short-test method with a small amount of liquid absorbed (or desorbed) and the equation:

$$\frac{M_t}{M_\infty} = \frac{4}{L}\left(\frac{D \cdot t}{\pi}\right)^{0.5} \tag{9-17}$$

Second, the use of half-time of process of absorption and the equation:

$$D = 0.049 \cdot \left(\frac{L^2}{t_{0.5}}\right) \tag{9-18}$$

Third, the use of final rates of absorption and desorption:

$$\frac{M_\infty - M_t}{M_\infty} = \frac{8}{\pi^2} \cdot \exp\left(-\frac{\pi^2}{L^2} Dt\right) \tag{9-19}$$

For the rubber considered in this chapter, a constant diffusivity is obtained, as the values of the diffusivity calculated by all these methods are the same whether in the absorption or desorption (Table 9–8).

**TABLE 9–8** Diffusivity, Rate of Evaporation, and $M_\infty$.

| | Absorption | Desorption |
|---|---|---|
| $D.10^7$(cm²/s) | 6.5 | 14 |
| $F$(g/cm²·s) | | $10.4 \times 10^{-5}$ |
| $M_\infty/M_{in}$(%) | 289 | 289 |

The rate of evaporation is determined by using the initial rate of desorption, when the process is controlled by evaporation [12, 20]. The rate of evaporation depends largely on the agitation of air as well as the temperature. The rate of evaporation is proportional to the concentration of liquid on the surface.

$$F = \frac{dM_t}{dt} \quad \text{for } t \to 0 \tag{9-30}$$

The capacity of absorption is expressed by the weight of liquid absorbed by 100 g rubber. As shown in the table, all the liquid previously absorbed is completely desorbed.

**Absorption process by the swelling annulus.** The validity of the model is illustrated when the calculation is done with constant diffusivity in the case of the thin annulus (Figure 9–17) and in the case of the thick annulus

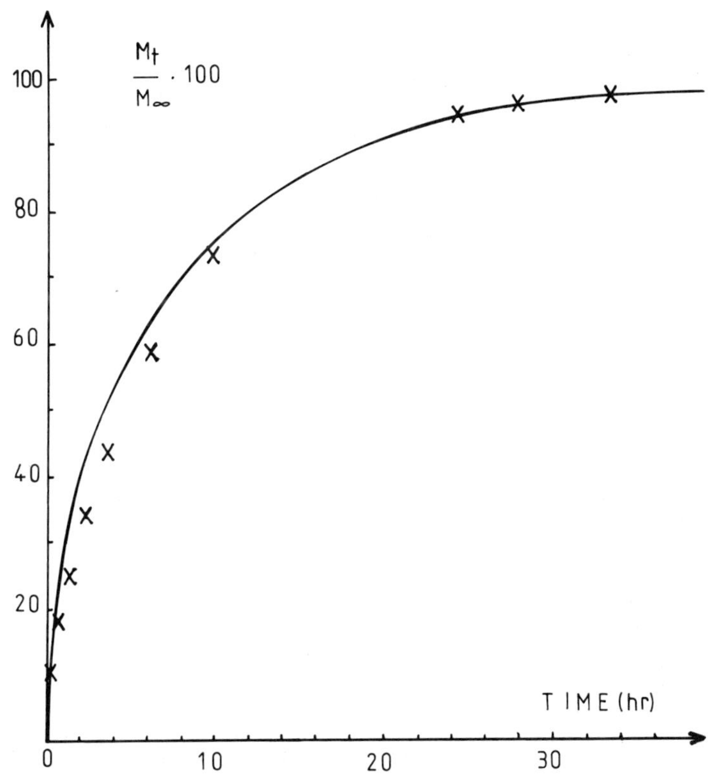


**Figure 9–17** Kinetics of absorption of toluene by the flat annulus ($Ri$ = 0.98 cm; $Re$ = 1.65 cm; $h$ = 0.70 cm). —calculated; ×: experimental. (Reprinted from *Plast. Rubber Proc. Applic.*, 9 (1988), 53–58, Y. Khatir, J. Bouzon, and J. M. Vergnaud, ref. 20, by courtesy of Elsevier Applied Science Publishers)

(Figure 9–18). A rather good agreement is obtained in both these cases, except at the beginning of the process. For short times of absorption, this slight difference between the curves appears when a change in shape of the annulus is observed due to the swelling.

The profiles of concentration expanded through the rubber annulus, in the cross section taken at midheight, are drawn in Figure 9–19 as they are calculated for each annulus.

**Desorption process from the rubber annulus.** A good agreement is also observed between the experimental (×) and calculated (−) kinetics obtained when swollen annuli are exposed to motionless air, for the flat (Figure 9–20) and the long annulus (Figure 9–21). The profiles of concentration are calculated through the cross section taken at midheight of the flat annulus, for various times.

**Conclusions.** The model, based on a numerical method with finite

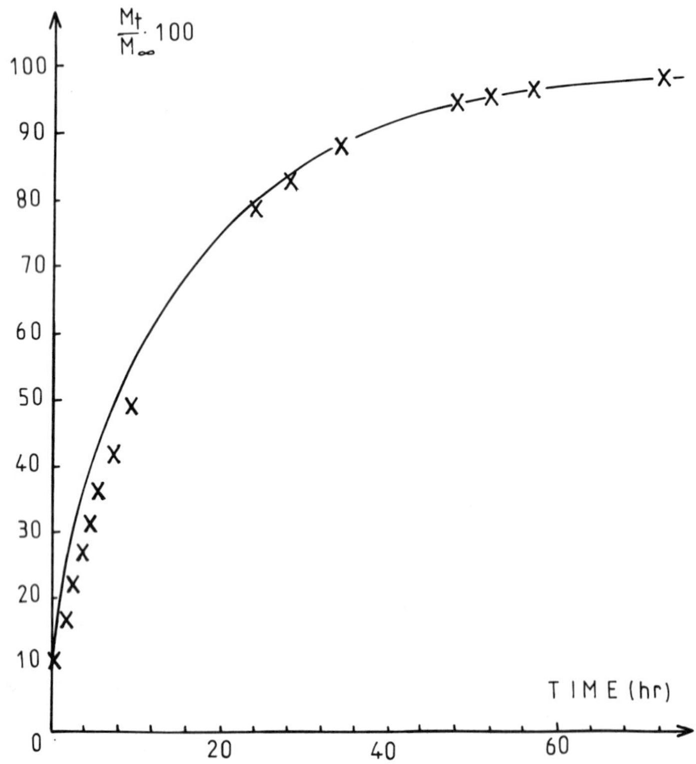


**Figure 9–18** Kinetics of absorption of toluene by the thick annulus ($Ri$ = 0.98 cm; $Re$ = 1.65 cm; $h$ = 3.0 cm). −calculated; ×: experimental. (Reprinted from *Plast. Rubber Proc. Applic.*, 9 (1988), 53–58, Y. Khatir, J. Bouzon, and J. M. Vergnaud, ref. 20, by courtesy of Elsevier Applied Science Publishers)

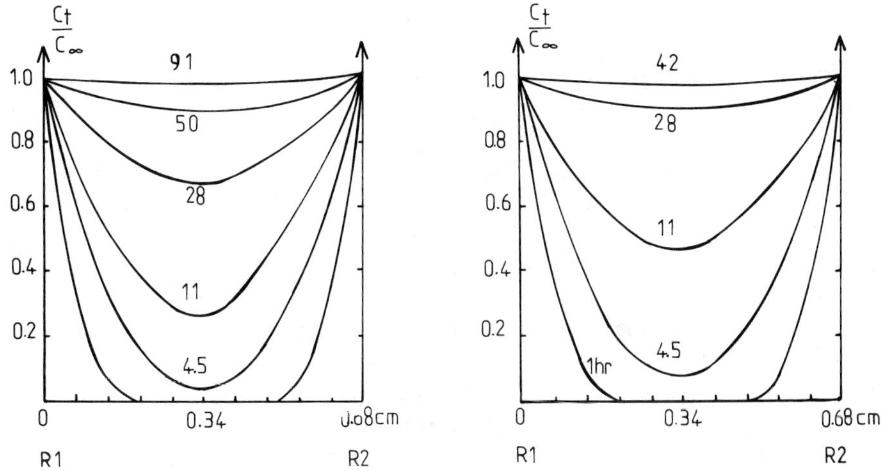


**Figure 9–19** Profiles of concentration of toluene through the rubber annulus at midheight; left: long annulus (3 cm); right: flat annulus (0.7 cm); $R_i$ = 0.98 cm; $R_e$ = 1.65 cm. (Reprinted from *Plast. Rubber Proc. Applic.*, 9 (1988), 53–58, Y. Khatir, J. Bouzon, and J. M. Vergnaud, ref. 20, by courtesy of Elsevier Applied Science Publishers)

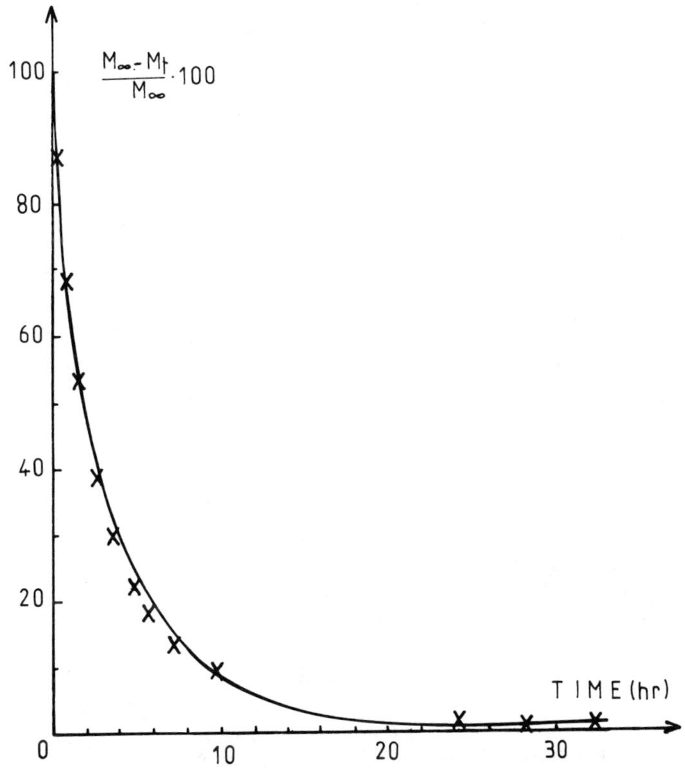


**Figure 9–20** Kinetics of desorption of toluene from the flat rubber annulus (*Ri*: 0.98 cm; *Re* = 1.65 cm; *h* = 0.70 cm); −calculated; ×: experimental. (Reprinted from *Plast. Rubber Proc. Applic.*, 9 (1988), 53–58, Y. Khatir, J. Bouzon, and J. M. Vergnaud, ref. 20, by courtesy of Elsevier Applied Science Publishers)

differences, describes the process of absorption and desorption in the difficult case of rubber annuli of various lengths. The model takes into account all the facts, especially the transient diffusion in the case of absorption and the transient diffusion coupled with evaporation in the case of desorption.

In the case of absorption of liquid, when the rubber sample is introduced in the liquid, a simple boundary condition can be used. The concentration of liquid on the surface reaches the equilibrium value as soon as the process starts.

During the desorption, the rate of evaporation is proportional to the actual concentration of liquid on the surface. So it is very fast at the beginning of the process, and then decreases strongly. In the case at hand with toluene and rubber, the diffusion through the solid plays the role of a limiting factor. These two parameters, the diffusivity which characterizes the rubber-liquid couple and essentially the rubber, and the rate of evap-

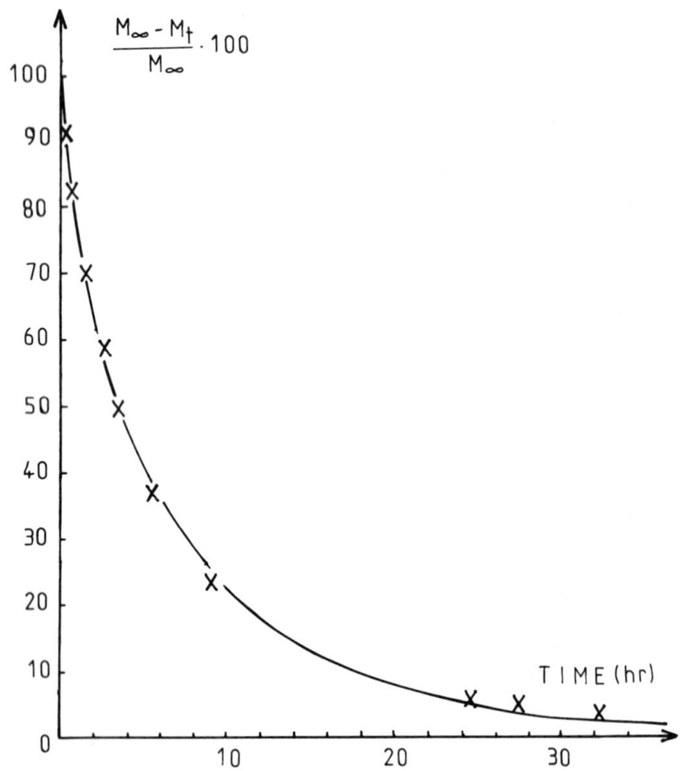


**Figure 9–21** Kinetics of desorption of toluene from the thick rubber annulus ($Ri$ = 0.98 cm; $Re$ = 1.65; $h$ = 0.70); – calculated; ×: experimental. (Reprinted from *Plast. Rubber Proc. Applic.*, 9 (1988), 53–58, Y. Khatir, J. Bouzon, and J. M. Vergnaud, ref. 20, by courtesy of Elsevier Applied Science Publishers)

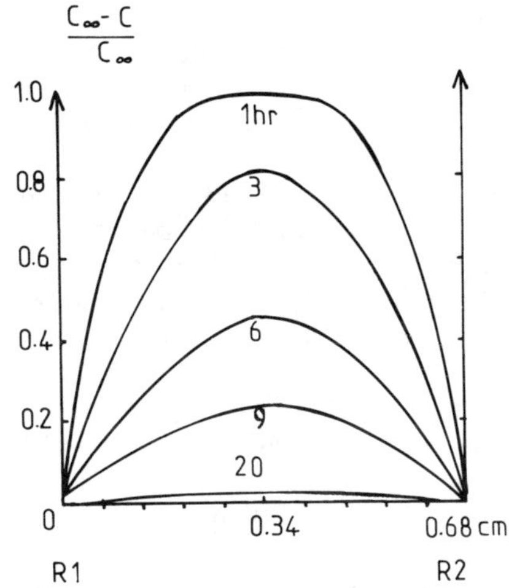


**Figure 9–22** Profiles of concentration of toluene through the cross section at midheight of the flat annulus ($Ri = 0.98$; $Re = 1.65$; $h = 0.70$). (Reprinted from *Plast. Rubber Proc. Applic.*, 9 (1988), 53–58, Y. Khatir, J. Bouzon, and J. M. Vergnaud, ref. 20, by courtesy of Elsevier Applied Science Publishers)

oration which depends on the volatility of the liquid are of interest, as well as the dimensions of the annulus. Of course, the model can be used for concentration-dependent diffusivity as well as for constant diffusivity.

## REFERENCES

1. R. P. Brown, "Physical testing of rubbers," *Applic. Sci. Publ.*, London (1979), pp. 268–293.
2. D. Hands and F. Horsfall, "A new method for simulating industrial cure processes," RAPRA Research Report 44 (1980).
3. ISO 1817, Resistance to Liquids. Methods of Test (1975).
4. BS 903, Part A16, The resistance of vulcanised rubber to liquids (1971).
5. ASTM D 471–75, Rubber Property. Effect of liquids.
6. ASTM D 1460, Change in length during liquid immersion.
7. J. Crank, *The Mathematics of Diffusion*, 2nd ed. (Oxford: Clarendon Press, 1975).
8. H. L. Frisch, "Simultaneous nonlinear diffusion of a solvent and organic penetrant in a polymer," *J. Polym. Sci.*, 16 (1978), 1651–1664.
9. D. Messadi and J. M. Vergnaud, "Simultaneous diffusion of benzyl alcohol into PVC and of plasticizer from polymer into liquid," *J. Applied. Sci.*, 26 (1981), 2315–2324.
10. J. M. Vergnaud, "Scientific aspects of plasticizer migration from plasticized PVC into liquids," *Polym. Plast. Technol. Eng.*, 20 (1983), 1–22.
11. J. L. Taverdet and J. M. Vergnaud, "Study of transfer process of liquid into and plasticizer out of plasticized PVC by using short tests," *J Applied. Polym. Sci.*, 29 (1984), 3391–3400.

12. Y. Khatir, J. Bouzon, and J. M. Vergnaud, "Liquid sorption by rubber sheets and evaporation; models and experiments," *Polym. Testing*, 6 (1986), 253–265.
13. A. Accetta, P. Le Parlouer and J. M. Vergnaud, "Kinetic parameters of the overall reaction of scrap rubber vulcanization by 1% sulfur," *Thermochim Acta*, 59 (1982), 149–156.
14. A. Sadr, R. Granger, and J. M. Vergnaud, "Calculation profiles of temperature and state of cure developed within the rubber mass in injection molding process," *J. Applied. Polym. Sci.*, 29 (1984), 955–963.
15. M. Abdul and J. M. Vergnaud, "Vulcanization progress in rubber sheets during cooling in motionless air after extraction from the mold," *Thermochim Acta*, 76 (1984), 161–170.
16. J. Y. Armand and J. M. Vergnaud, "Comparative study of a reaction of low enthalpy (rubber cure) by DC and DSC," *Thermochim Acta*, 116 (1987), 111–124.
17. J. Y. Armand, J. Bourgois, and J. M. Vergnaud, "Fundamental study of heating process in DSC. Effect of the cure enthalpy and heating rate," *Thermochim Acta*, 116 (1987), 301–316.
18. Y. Khatir, J. Bouzon, and J. M. Vergnaud, "Non-destructive testing of rubber for the sorption and desorption-evaporation of liquid by modeling: cylinders of finite length," *J. Polym. Engng.* 7 (1987), 149–167.
19. Y. Khatir, J. Bouzon, and J. M. Vergnaud, "Determination of processes of absorption and desorption of liquids with rubber tubings by using model and short tests," *J. Polym. Engng.*, 7 (1987), 275–299.
20. Y. Khatir, J. Bouzon, and J. M. Vergnaud, "The kinetics of absorption and desorption by a rubber annulus using a model and short-term tests, *Plast. Rubber Process. Applic.*, 9 (1988), 53–58.

# Chapter Ten

# Drug Delivery and Pharmaceutical Applications

## 10-1 INTRODUCTION

The main purpose of this introduction is to give general information on the problem of drug delivery from its dosage form, by following the drug through the body. Various terms generally used in medical science are also described. The final aim is to help dissemination of various sciences, by making scientists working out of the medical world capable of bringing their complementary competence to the field of drug delivery.

### 10-1-1 Position of the Problem of Drug Passing through the Body

**Path of the drug from the supply form to excretion.** A general diagram shows the path followed by the drug from the time it has been extracted from the dosage form (for example, by dissolution in gastric liquid, in the case of oral form), to the end of the process when the drug is eliminated (Figure 10–1). The following steps are of importance and are shortly described in other sections.

- Extraction of the drug from the dosage form
- Absorption of the drug into the organism, and problem of bioavailability

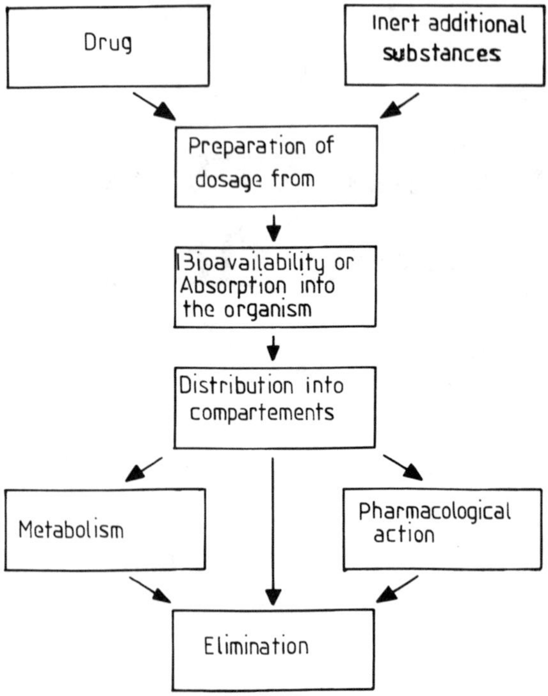


**Figure 10–1** Path followed by the drug from dosage form orally absorbed to excretion from the organism.

- Distribution of the drug into compartments
- Pharmaceutical action, metabolism
- Elimination of the drug out of the body

### Biopharmaceutics, pharmacokinetics, and pharmacodynamics

***Biopharmaceutics.*** The concept of biopharmaceutics was given in 1958 by Levy and variously interpreted as "the study of factors that affect the extent and speed of absorption or liberation of a drug from its various physicochemical forms . . . it deals essentially with the optimal or desired availability of a drug from the dosage form employed for the target site, i.e., the influence of drug formulation on the biological activity of the drug."*

The results of many works concerned with improvement of the therapeutic activity of the drug have shown that a pharmacologically active

* K. Heilmann, in *Therapeutic Systems: Rate-Controlled Drug-Delivery: Concept and Development*, Thieme Stratton Inc., New York (1984), p. 5.

substance is not necessarily an effective remedy for the sick. In other terms, the efficacy of the drug can be highly dependent on the method of delivery to the organism.

***Pharmacokinetics.*** Following Wagner's definition: "Pharmacokinetics is concerned primarily with the kinetics of absorption, distribution, metabolism, and excretion of drugs, metabolites produced by the body itself,"* by considering what is done to the drug by the body.

Pharmacokinetics essentially deal with the relationship between the concentration of drug in plasma and other body fluids, the amount of drug administered to the organism, and the change in drug concentration in body fluid with time.

Results worth noting are:

- The absorption and elimination of drugs generally follow a first-order reaction.
- Delivery of drugs from conventional dosage forms follows a first- or second-order kinetics.
- Therapeutic systems have to release the drug with a constant rate (meaning a zeroth-order reaction).

Multicompartiment models have been used in order to simulate pharmacokinetic processes. But the development will be possible by making pharmacokinetic measurements.

***Pharmacodynamics.*** Pharmacodynamics deal with the problem of what happens to the organism under the action of the drug.

**Release of drug from its supply form.** According to the World Health Organization, a drug is a biologically active substance used to modify physiological systems or pathological states for the benefit of the recipient. The drug (or active agent) is very often added to excipients which are auxiliary substances, as formulated and presented in dosage form. Galenic pharmacy is the science of the formulation of drugs.

When the dosage form has been administered, the drug is released from the form, during the first step of liberation. This initial phase is of great practical importance because of the relationship between the concentration of a drug (or its concentration-time history) and the efficacy.

The release of a drug is characterized by two factors: solubility (which is often pH dependent), and rate of dissolution, by considering solubility in either hydrophilic or lipophilic phases. Of course, the rate of dissolution

---

* K. Heilmann, in *Therapeutic Systems: Rate-Controlled Drug-Delivery: Concept and Development*, Thieme Stratton Inc., New York (1984), p. 6.

is the rate-limiting step in absorption, when it is slower than the rate of absorption. This is the very step which is considered with new dosage forms, by using either in vitro tests (with apparatus able to simulate the process of release) or in vivo tests.

### Absorption of drug—bioavailability

***Absorption of Drug.*** After its release from dosage form, the drug may be absorbed from the surface of the body in contact with it and transported into the blood circulation or the lymphatic system.

The drug must pass through complex barriers, or membranes, which delimitate morphological and functional entities. These membranes consist generally of a bimolecular layer of phospholipids, both of whose polar ends are covered by a layer of protein molecules. This structure in form of pleated sheet is responsible for the elasticity and plasticity of membranes.

The movement of the drug through the membrane is controlled by diffusion, and the rate of transport depends on the gradient of concentration, the size of the drug, and the thickness and the nature of the membrane.

***Bioavailability.*** The term *bioavailability* defines the percentage of an unchanged drug reaching the circulatory system after administration of a test dose introduced by using a standard dosage form. Bioavailability is determined by measuring the drug extent in blood or urine excretion.

***Absorption of Drug in the Stomach.*** Enteral absorption is obtained for the drug administered as oral formulation, through the membranes of the epithelial cells. The drug passes through the basement membrane, reaches the connective tissue and capillaries, and is transported to the portal vein in the bloodstream.

Of course, absorption of a drug may vary following the passage in the stomach and intestinal tract, and this depends on the pH, the degree of filling, enzymatic activation, and bacterial decomposition. In spite of this complex system, the penetration of epithelial cell membranes is considered as the rate-limiting step.

**Distribution of the drug in the body.** The organism consists of circulating fluid (intravascular compartment) and of cells with extracellular and intracellular compartments. The drug leaves the circulating fluid and is redistributed between extracellular and intracellular compartments.

Then, the ultimate distribution of the drug is obtained in the tissues, either in the case of drug binding the receptors with a local accumulation of drug on these sites, or in the case of drug without binding and evenly distributed between all fluid compartments.

**Elimination of the drug.** The drug is inactivated by various processes such as chemical alteration of the molecule, or metabolism, and by excretion by various organs.

***Metabolism.*** Metabolism generally takes place in tissues with the help of enzymes, and the metabolites formed are more soluble in water than the original drug, facilitating their excretion through the urine.

The drug molecule may be broken into simpler substances: Catabolism.

New molecules are prepared from the drug: Anabolism.

***Excretion via Organs.*** The major route of excretion for most substances is through the kidney in urine, except for gaseous anesthetics which are excreted in significant amounts through the lungs.

The rate and extent of renal excretion is controlled by glomerular filtration, tubular reabsorption, and tubular secretion. A lipid-soluble substance may be reabsorbed during passage through the kidney, and a slow excretion may thus be responsible for drug accumulation.

***Biological Half-Life.*** Derived from the radioactive decay in isotope chemistry, biological half-life is the time required for elimination of half the plasma content of the drug (by metabolism or by excretion). Since this time is subject to various factors which differ among individuals, a drug cannot be characterized by an absolute half-life.

The dosage regimen is the plan of administration of the drug. It is of great importance when a constant concentration of the drug in the tissues is desired for a long period.

**Therapeutic index.** The concept of the therapeutic index, introduced by Ehrlich, is related to the minimal curative dose and the maximum tolerated dose. In fact, it is the ratio of the median lethal dose (causing the death of 50 percent of animals experimented on) and the median effective dose (causing cure in 50 percent of cases). Of course, the higher the therapeutic index, the greater the safety margin.

### 10-1-2 Devices Used for Oral Drug Delivery

Various devices are examined, from the simple conventional forms to complex oral therapeutic systems. Special attention is given to devices where the drug is dispersed in polymer, this polymer playing the role of matrix. Advantages and drawbacks of each of these systems are considered in terms of the concentration of drug delivered by the device.

**Conventional dosage forms.** Conventional dosage forms consist of drug (or active agent) and auxiliary substances called excipients. Excipients which must be biologically inert (gelatin, lactose, starch, talc, paraffin) are commonly used for binding the drug in order to ensure a consistency and volume of supply form suitable for patient use. The release of the drug is controlled essentially by solubility of this drug and by solubility or degradability of excipients.

Generally the drug is very rapidly liberated from its dosage form because of the fast rate of solubility, and its concentration in gastric liquid quickly builds up to a high concentration. Then, because of the drug absorption through the gastric membrane, which follows a first-order kinetics, the drug concentration in the stomach falls exponentially until the next dose. As a result, there is an undulating concentration pattern for the drug in the stomach. The same pattern occurs for many drugs in the blood and tissues, the drug concentration reaching a therapeutic peak, and then rapidly falling below the minimum concentration for therapeutic activity because of their short biological half-life.

Then the optimal therapeutic level is attained only for a short time at any place in the body, and the time of cure is essentially occupied by overdosage and underdosage (Figure 10–2).

Another drawback for conventional dosage forms is that they necessitate high frequency of administration. And the problem of limited patient reliability must be considered with the following factors in noncompliance: wrong frequency, impermissible delay and omission, and execution in the wrong mode and at the wrong time.

Moreover, the fluctuating drug concentrations in blood and tissues

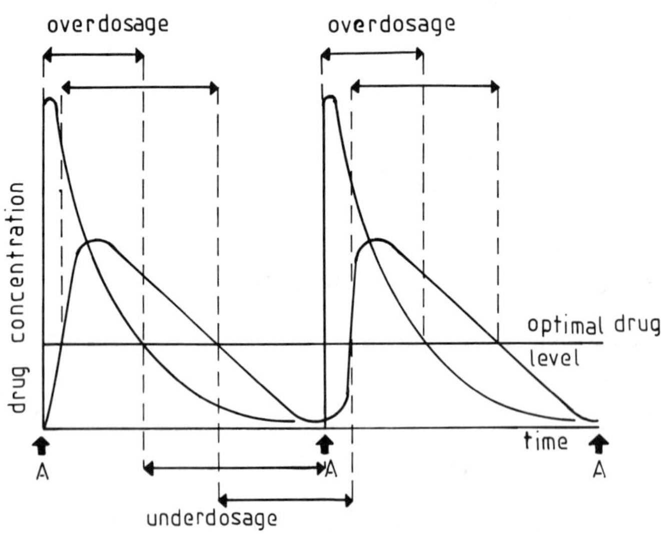


**Figure 10–2** Concentration of drug in the stomach and blood as a function of time, with conventional dosage form.

lead to an insufficient influence on the cure of disease and are responsible for an excessive use of the drug. These overdoses may produce disastrous side effects.

### Oral therapeutic systems

*Main Purpose.* Therapeutic systems may release the drug at a constant rate for a certain time period, and then offer important advantages over conventional dosage forms, especially in diseases requiring a constant drug concentration in the blood over prolonged duration of therapy. The drug is delivered to the target organ through the blood at a constant rate, and uniform concentrations of drug are attained. Thus, these dosage forms can decrease the total daily amount of drug administered, and in so doing decrease the importance of side effects. They also increase the patient compliance.

*Concept of Oral Therapeutic Systems.* A therapeutic system is a drug-containing preparation or dosage form that releases one or more drugs continuously in a predetermined pattern for a fixed period of time, either systemically or to a specified target organ [1].

Systemic treatment is often administered via the gastrointestinal tract because many dosage forms are easy to swallow.

The principle of retardation consists in slowing down the process of dissolution of the drug or delaying it for a period of time, by coating the drug with polymers with a low solubility in water, the retardation in drug delivery being increased by a higher thickness of the coat.

*Membrane-Controlled System OROS.* The OROS system (oral osmotic therapeutic system) consists of a drug reservoir surrounded by a semi-permeable membrane with an opening in it [2–5].

Osmotic pressure ensures that water from the gastrointestinal tract diffuses through the semi-permeable membrane. This water dilutes the drug (or dissolves it if the drug is in solid state). The influx of water causes an increase in hydrostatic pressure (osmotic pressure) which produces a release of the drug in solution through the delivery orifice in the membrane.

The rate of release is proportional to the square of the solubility of the drug. It follows that this simple osmotic system is only suitable for drugs having specific properties. Ideal substances are those with a rather high solubility, between 20 percent and 40 percent, which ensures an adequate osmotic pressure.

Only a few polymers are able to resist mechanical stress and the digestive liquids, thus ensuring a constant volume for the system. The most suitable up to now is cellulose acetate.

The delivery orifice has to comply with strict specification, for the accuracy of the pump system. $CO_2$ lasers are able to bore a hole (100–250 microns in diameter) through only one part of the membrane.

The OROS system is suitable as a therapeutic dosage form for many substances. The rate of drug release can be adapted to the required therapy by varying the surface, thickness, and specific permeability of the membrane. A few examples are given.

- Hypertension with beta-blockers, as metroprolos (LOPRESOR, from Ciba-Geigy) and oxprenolol (TRASICOR, Ciba-Geigy)
- Rheumatism, with indomethacin (INDOS, from Merck Sharp and Dohrme)
- Glaucoma, with acetazolamide

**Conclusions.** Of course, oral therapeutic systems of the OROS type are very useful. However, they can only work properly when the drug is rather soluble in gastric liquid. Moreover, they are not a simple dosage form like a pill, but a rather complex system.

### 10-1-3 Simple Monolithic Devices (Drug-Polymer)

**Main principles.** Various devices have been built up and studied in order to obtain a controlled release of the drug in gastric liquid. They can be divided into four categories based on the mechanism that controls the release of the drug from the delivery device [6–8]. These mechanisms are:

- Osmosis (already seen, with OROS systems).
- Polymer erosion. Two cases can be considered: devices where the drug is previously dispersed in the polymer, and devices surrounded by an erodible polymer.
- Diffusion and reaction with devices where the drug is bound to a polymer playing the role of a backbone.
- Diffusion with devices obtained by dispersing the drug in a non-erodible polymer, and pressing the mixture so that the polymer plays the role of a matrix.

It is clear, however, that for each device the release of the drug is controlled by more than one mechanism [8]. For instance, in the case of a device releasing the drug during polymer erosion, diffusion of the liquid and drug through the polymer generally takes place as well as polymer erosion.

**Monolithic devices (literature survey) [9].** Special attention has been given to finding a way for regulating the amount of drug released by means of monolithic devices where the drug is dispersed in an inert polymer matrix. The polymer is biocompatible and generally passes through the body without any change. Both biodegradable and nondegradable polymers have been used for the matrix [7, 10–12].

***Diffusional Transport of the Drug.*** Several theories have been put forward in an attempt to describe the release process of the drug from the inert matrix system. Sometimes they are built on a combination of hydrodynamic effects, the pores in the device playing a fundamental role [13], and diffusion effects.

Very often, experimental data have been studied only for short periods of time, and the diffusional process has been proved by the square root law of time dependence with the amount of drug transported.

$$\frac{M_t}{M_\infty} = \frac{4}{L}\left[\frac{D\cdot t}{\pi}\right]^{0.5} \tag{10-1}$$

*Mt* being the amount of drug delivered in time $t$, and $M_\infty$ the amount of drug released after infinite time. (This law can be used when $M_t/M_\infty < 0.5$.)

Various studies have been done on devices consisting of a soluble drug and a polymer.

- KCl dispersed in ethylcellulose [14].
- Various drugs and polymers, with degradable polymer [7].
- KCl dispersed in hydroxypropylmethylcellulose [15].
- Theophylline, acetaminophene, dyphylline dispersed in copolymers of polyvinyl acetate and polyvinyl chloride [11].
- Diphylline dispersed in various polyvinyl acetate [16].
- Theophylline dispersed in polyhydroxymethylmethacrylate and methyl methacrylate [17].
- Sodium salicylate dispersed in methylcellulose [12].
- Trinitrine dispersed in Eudragit (copolymer of dimethylaminoethylacrylate and ethylmethacrylate) [18].
- Drugs dispersed in biodegradable polymers (gelatin and polyethylacrylate) [19].

When long experiments have been carried out, the following equation (obtained with a sheet of thickness $L$) has been successfully used to describe the process of release.

$$\frac{M_t - M_\infty}{M_\infty} = \frac{8}{\pi^2}\cdot\sum_{n=0}^{\infty}\frac{1}{(2n+1)^2}\exp\left(-\frac{(2n+1)^2\pi^2}{L^2}Dt\right) \tag{10-2}$$

In all the preceding cases, two main assumptions have been made.

- The diffusivity of the drug is constant during the whole process.
- Only the transport of the drug is considered.

***Pseudo Steady State Conditions.*** A complex but interesting theory has been built up, based on the existence of pseudo steady state conditions for the drug transport. This fact occurs especially in some glassy polymers. A sharp linear concentration in front of the drug is formed between the extracted part and the untouched portion of the device. This drug front has been found to move at an almost constant rate [14, 20–22].

In all cases, drastic assumptions have been made:

- Only the drug transport is considered.
- The diffusivity of the drug is constant (and nonconcentration dependent).

In fact, this interesting observation could perhaps have been explained by the presence of liquid in the polymer. The liquid could diffuse in the polymer by two means: the one in glassy polymer, with a low rate of transport (low diffusivity) and low concentration of liquid; the other in the rubbery polymer with a high concentration of liquid, and a high rate of transport and diffusivity of the liquid and the drug. The polymer, initially under glassy state, becomes less and less glassy, and more and more rubbery, when the concentration of liquid increases.

***Release of Drug with a Constant Rate.*** A remarkable result has been shown for the transfer of the dyestuff Red-Sudan from polystyrene into n-hexane, when the dyed polymer is immersed in the liquid. The rate of release of the dye in liquid is constant during the greater part of the process. The rate of release is also a function of the initial dye percent in polymer and of the thickness of the polymer film [23]. Many attempts have been made in order to extend this example to various cases with a transport of drug [24].

Calculations, based on the pseudo steady state and on the rate of dissolution (especially when this rate of dissolution is low), have been made in order to find a zeroth-order kinetic [14]. However, the kinetics of drug release are found to follow a diffusional process, with perhaps a change in the value of diffusivity.

Various authors have shown in turn:

1. that it is impossible to obtain a constant rate of drug release by using devices consisting of a drug dispersed in a polymer [25]
2. that a constant rate for the release of drug could be attained from these simple devices after sufficient knowledge on the process of polymer swelling and diffusion [24].

***Swelling of Polymer Matrix.*** A few studies of interest have shown that a swelling of the polymer matrix may take place during the process of drug release, this swelling being due to the transport of liquid in the polymer with a higher rate than that of the drug release [17]. The rate of liquid transport has been determined in the case of KCl dispersed in hydroxypropylmethylcellulose [15], and follows a diffusional process. A dimensionless number, Deborah's number, has been introduced with interest [24].

**New principle with double transport.** More recent studies have shown that the process of drug release which takes place when the dosage form is in contact with the liquid is not simple. Experiments done with various materials, such as plasticized PVC, elastomers, and dosage forms, have proved that two matter transports are developed simultaneously, the one concerned with the liquid and the other with the drug (or plasticizer in case of plasticized PVC).

***Plasticized PVC in Contact with Liquid.*** When plasticized PVC with an extent of plasticizer ranging from 20 percent to 50 percent is in contact with a liquid, a simultaneous transport of the liquid into, and plasticizer out of the plasticized PVC, is generally observed. In a first attempt, these transports were studied separately, and they were found to be controlled by diffusion. The main conclusions have been drawn from these experiments [26, 27].

- The rate of transport is higher for the liquid than for the plasticizer, especially at the beginning of the process.
- Gradients of concentrations of liquid and plasticizer have been determined.
- The diffusivity for the liquid and for the drug varies during the process.
- In some cases, the amount of liquid located in the polymer passes through a maximum.

Follow-up studies have shown that diffusivity is concentration dependent in a rather complex way, the diffusivity of plasticizer and of liquid being expressed as a function of the concentration of liquid and plasticizer [28, 29]. As a result, the diffusivity for each matter transported varies with time and within the polymer during the process.

Further studies done on the process of evaporation of the liquid which has entered the PVC have illustrated the difficulty of evaporating this liquid because of the presence of gradients of concentration through the polymer and the concentration-dependent diffusivity [30, 31]. From the knowledge of these transports obtained by experiments and calculation, a new method has been elaborated for preparing flexible plasticized PVC with low matter transport [28–31].

The process of transport has been described by using mathematical models based on numerical methods with finite differences. These models take into account all the facts, and especially the simultaneous transport of liquid and plasticizer, the concentration-dependent diffusivities, and the amount of liquid which can be absorbed at equilibrium. Not only the simple case of double transport has been properly predicted [26–29], but also the more complex case of transport when the amount of liquid absorbed by the polymer passes through a maximum [29].

***Elastomers in Contact with Liquid.*** Studies done on the absorption of liquid by elastomers when the elastomer is contacted with the liquid have shown that the process of absorption is rather simple. The liquid enters the elastomer following a diffusional process, with constant or concentration-dependent diffusivity. The desorption process of the liquid is also simple, being controlled by evaporation on the surface and diffusion through the solid. Apparently, absorption and desorption can be followed in succession, without great change in the kinetics, the process being reversible. Perhaps further studies will show that some additives are extracted from the elastomers during absorption, provoking slight but irreversible damage in the solid.

***Polymer-Drug Device in Contact with Gastric Liquid.*** Recent works on polymer-drug devices in which sodium salicylate as a drug is embedded in Eudragit [32] or Carbopol [33] have described the process of transports when these devices are immersed into gastric liquid. Various results have been obtained.

- Two transports simultaneously take place as follows: the liquid penetrates the polymer matrix and dissolves the drug, which then can diffuse through the internal liquid out of the dosage form.
- These transports are controlled by diffusion, with concentration-dependent diffusivities.
- Further studies have shown that the diffusivity of the liquid and drug can be expressed in terms of the concentration of liquid and drug [34].

Mathematical models based on numerical methods with finite differences have been built up in order to describe the whole process [32–35], in the simple case of plane and spherical dosage form, and in other more complex cases with a core and shell.

## 10-2 MODEL FOR MATTER TRANSPORTS BETWEEN PLANE DOSAGE FORMS (SHEETS) AND GASTRIC LIQUID

The following sections are considered in succession [32, 33].

- Experimental, with preparation of dosage forms pressed in thin sheets, and in vitro test description.
- Theoretical, with assumptions, mathematical treatment, and numerical analysis.
- Results, with determination of diffusivities (for the liquid and drug), testing the validity of the model by comparing experimental and theoretical kinetics of matters transported.
- Conclusions.

### 10-2-1 Experimental

#### Materials used for dosage forms

- Drug: Sodium salicylate, in powder form.
- Polymer: Two polymers are used in succession. (1) Eudragit RS (Röhm Pharma), a copolymer of dimethylaminoethylacrylate and ethylmethacrylate with a mean molecular weight of 150,000, in powder form, and (2) Carbopol 934 P (FB Goodrich), polymer of acrylic acid, with a mean molecular weight around 3,000,000, in powder form.

**Preparation of thin sheets.** The drug and polymer (either Eudragit RS or Carbopol 934 P), in powder form, are mixed in a mortar. Sheets are pressed in a steel mold operated by a press at 120°C for 30 s after a 6-minute heating. Several tablets are cut from the sheet. Cohesion after compression is excellent, especially in the case of Eudragit, this polymer being an excellent tablet binder.

**In vitro tests.** Experiments are conducted in closed 250-ml flask, with a controlled rate of stirring. The effect of the rate of stirring on the kinetics of transport is significant, especially at the beginning of the process [26, 36, 37].

**TABLE 10–1** Operational Conditions

| Polymer | Diameter | Thickness | Sample Weight | Liquid |
|---|---|---|---|---|
| Eudragit | 1.5 cm | 0.024 cm | 0.500 g | 100 ml |
| Carbopol | 2.0 | 0.016 | 0.250 g | 100 ml |

The tablet, located in a fiber glass basket, is soaked in synthetic gastric liquid (100 ml) at 37°C. The composition of the liquid is for 1,000 ml of aqueous solution, 80 ml HCl 1 N, 2 g NaCl, pH = 1.2.

Measurements are made in order to obtain the kinetics of matter transports (liquid, drug). At intervals, the tablet is weighed and the concentration of drug released from the device is determined by using a double-beam UV-spectrophotometer (Beckman DB-G) calibrated at 300 nm.

### 10-2-2 Theoretical

**Problem description.** When the dosage form is immersed in the liquid, the liquid penetrates the polymer and dissolves the drug. The drug then diffuses out of the dosage form through the liquid located in the polymer. The problem is studied by considering these two simultaneous transports connected with one another.

**Assumptions.** The following assumptions are made.

1. One-dimensional matter transport is considered through the thin sheet.
2. Both transports are controlled by transient diffusion, with concentration-dependent diffusivity as found from experiments.
3. The concentration at equilibrium is attained for the liquid and drug on both faces of the sheet as soon as the sheet is immersed into the liquid.
4. The preceding concentration at equilibrium on sheet faces is obtained from experiments. This value depends on the ratio of volumes of external liquid and dosage form.
5. The thickness $L$ of the sheet remains constant as diffusion proceeds.

**Mathematical treatment.** The equation of diffusion in one dimension

$$\frac{\partial C}{\partial t} = \frac{\partial}{\partial x}\left(D \cdot \frac{\partial C}{\partial x}\right) \tag{10-3}$$

is applied to the liquid and to the drug, and the diffusivity is a function of concentrations of liquid and drug in each case.

The initial and boundary conditions are:

| | | | | | |
|---|---|---|---|---|---|
| Initial | $t = 0$ | $0 < x < L$ | $C = C_{in}$ | sheet | (10-4) |
| Boundary | $t > 0$ | $x = 0$ | $C = C_{eq}$ | sheet face | (10-5) |
| | | $x = L$ | | | |

Generally, no analytical solution is found. However, for short times

($M_t/M_\infty < 0.1$), the amount of drug transported at time $t$, $M_t$, is small, and the local concentration within the sheet may be assumed to be constant and equal to the original value. The quantity $M_t$ is expressed as a function of the quantity $M_\infty$ transported after infinite time (when equilibrium is reached) by the single equation:

$$\frac{M_t}{M_\infty} = \frac{4}{L}\left(\frac{D \cdot t}{\pi}\right)^{0.5} \tag{10-1}$$

where $D$ is the diffusivity and $L$ the thickness of the sheet.

**Modeling (numerical analysis).** In the cross section of the sheet shown in Figure 10–3, the thickness is divided into $N$ equal finite slices of thickness $\Delta x$ by concentration-reference planes:

$$n \quad \text{for space, with} \quad L = N \cdot \Delta x \tag{10-6}$$

$$i \quad \text{for time, with} \quad t = i \cdot \Delta t \tag{10-7}$$

$n$ and $i$ being integers.

The matter balance written on the plane $n$ enables one to obtain the new concentration at this plane after the elapse of time $\Delta t(t + \Delta t)$, as a function of concentration at the previous time ($t = i \cdot \Delta t$) and at positions $n - 1$, $n$, and $n + 1$.

***Liquid in Polymer Matrix***

$$CN_n^l = \frac{1}{M_n^l}\left[C_{n-l}^1 + (M_n^l - 2) \cdot C_n^l + C_{n+1}^l\right] \tag{10-8}$$

with the dimensionless number at time $t(i \cdot \Delta t)$.

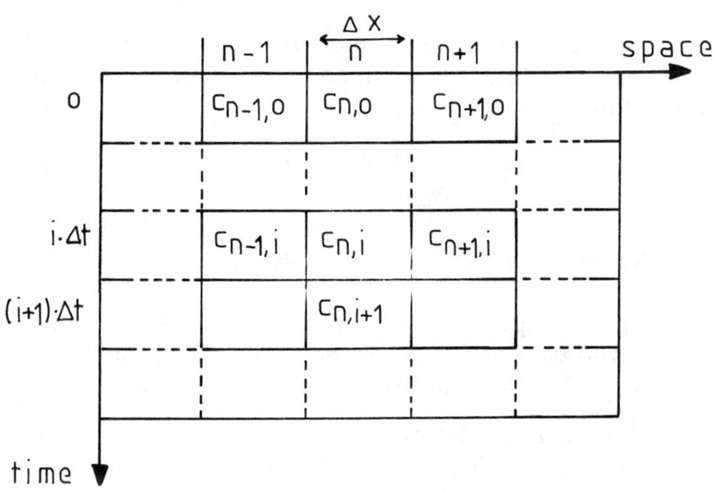


**Figure 10–3** Space-time diagram for calculation of concentrations of drug and liquid.

$$M_n^l = \frac{(\Delta x)^2}{\Delta t} \cdot \frac{1}{D_n^1} \tag{10-9}$$

$D_n^l$ being the diffusivity of the liquid at time $t$ and position $n$.

The diffusivity of the liquid, $D^l$, is expressed in terms of concentrations of liquid and drug, by the following equation, at the time $t$:

$$D_n^l = D_o^l \cdot \exp\left(\frac{-A^l}{C_n^l + \alpha \cdot C_n^d}\right) \tag{10-10}$$

where $A_1$ is a constant, as well as $\alpha$; and $C_n^l$ and $C_n^d$ are concentrations of liquid and drug at place $n$ at the same time $t$. The value of constant $\alpha$ characterizes the difference in contribution of the drug and liquid.

***Drug in Polymer Matrix.*** In the same way, the new concentration of the drug is:

$$CN_n^d = \frac{1}{M_n^d} \cdot [C_{n-1}^d + (M_n^d - 2) \cdot C_n^d + C_{n+1}^d] \tag{10-11}$$

with the dimensionless number at time $t$

$$M_n^d = \frac{(\Delta x)^2}{\Delta t} \cdot \frac{1}{D_n^d} \tag{10-12}$$

The diffusivity of the drug, $D^d$, is also expressed in terms of concentrations of liquid and drug, with the constant $\beta$ characterizing the difference in contribution of the drug and liquid.

$$D_n^d = D_o^d \cdot \exp\left(\frac{-A^d}{C_n^1 + \beta \cdot C_n^d}\right) \tag{10-13}$$

***Amount of Drug and Liquid Transported.*** The amount of each matter located in the sheet at time $t$ is obtained by integrating the preceding concentrations with respect to space.

$$M_t = \frac{L}{N} \cdot \left[C_{o,i} + \sum_{n=1}^{N-1} C_{n,i}\right] \tag{10-14}$$

### 10-2-3 Results with Eudragit RS as Polymer

**Diffusivity and capacity of absorption.** Two results are of interest: the diffusivity and capacity of absorption, as well as the relationship between diffusivity and the concentration of drug or liquid.

***Drug.*** Figure 10–4 shows the amount of sodium salicylate transported into synthetic gastric liquid as a function of the square root of time, for various initial concentrations of drug in devices ranging from 15 to 50 weight percent. The following conclusions can be drawn.

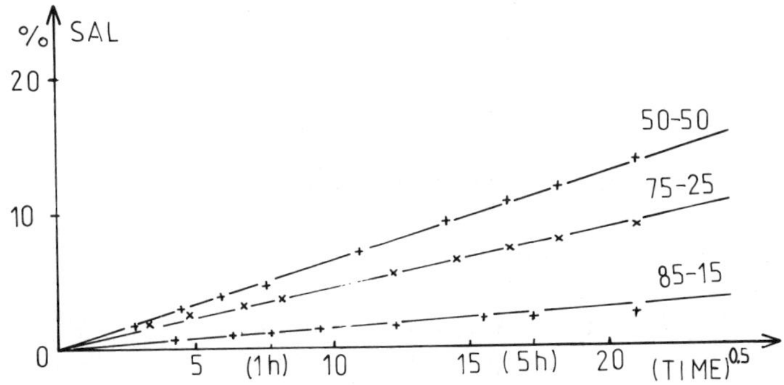


**Figure 10–4** Amount of drug transported (weight % of initial sample) versus square root of time. Initial concentrations of drug ranging from 15 to 50 (weight %) Eudragit RS. In vitro tests. pH = 1.2 at 37°C. (Reprinted from *Int. J. Pharm.*, 27, (1985), 233–243, A. Droin, C. Chaumat, M. Rollet, J. L. Taverdet, and J. M. Vergnaud, ref. 32, by courtesy of Elsevier Sci. Biomedical Div.)

1. Straight lines are obtained, the amount of drug transported being proportional to (time)$^{0.5}$.
2. The higher the initial concentration of drug, the higher the rate of transport of the drug in liquid.
3. The amount of drug transported at equilibrium (obtained for times of about 15 –20 h) increases with the initial amount of drug in the device (Figure 10–5).
4. Diffusivity of the drug, obtained from the slope of straight lines in Figure 10–4 by using equation (10–4), is about constant and does not

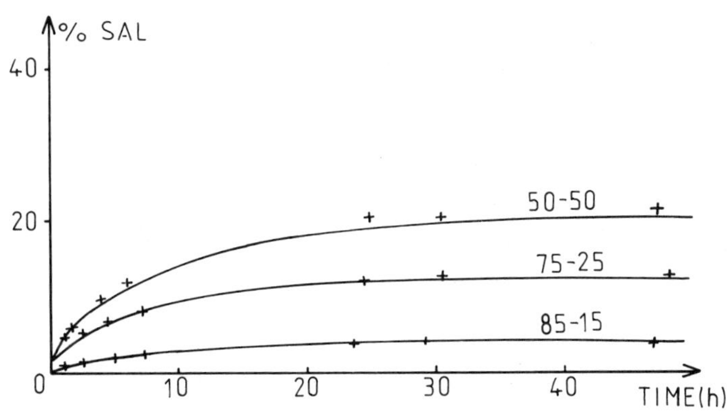


**Figure 10–5** Kinetics of release of drug, for various initial concentrations of drug in the device. Eudragit RS. In vitro test. pH = 1.2 at 37°C. (Reprinted from *Int. J. Pharm.*, 27 (1985), 233–243, A. Droin, C. Chaumat, M. Rollet, J. L. Taverdet, and J. M. Vergnaud, ref. 32, by courtesy of Elsevier Sci. Biomedical Div.)

depend on the drug concentration. This fact results from the second and third conclusions.

***Liquid.*** The amount of liquid transported into the plane device is expressed as a function of the square root of time (Figure 10–6), for various initial concentrations of the drug in devices. Straight lines are also obtained.

The amount of liquid absorbed by the device at equilibrium (after a time around 10–15 h) is shown in Figure 10–7, as well as in Table 10–2. The diffusivity in the logarithm form for the liquid is proved to be inversely proportional to the initial drug concentration devices (Figure 10–8). The value of diffusivity and capacity of absorption at equilibrium are cited in Table 10–2.

The diffusivity of the liquid is expressed in terms of the initial concentration of drug by:

$$D^l = \exp\left(-\frac{40.4}{C^1 + C^s} - 13.3\right) \quad (\text{cm}^2/\text{s}) \tag{10-15}$$

while the diffusivity of the drug is constant.

$$D^d = 16.5 \times 10^{-8} \quad (\text{cm}^2/\text{s}). \tag{10-16}$$

**Validity of the model.** Following the fifth assumption, the sheet for calculation is considered as constant when diffusion proceeds. However,

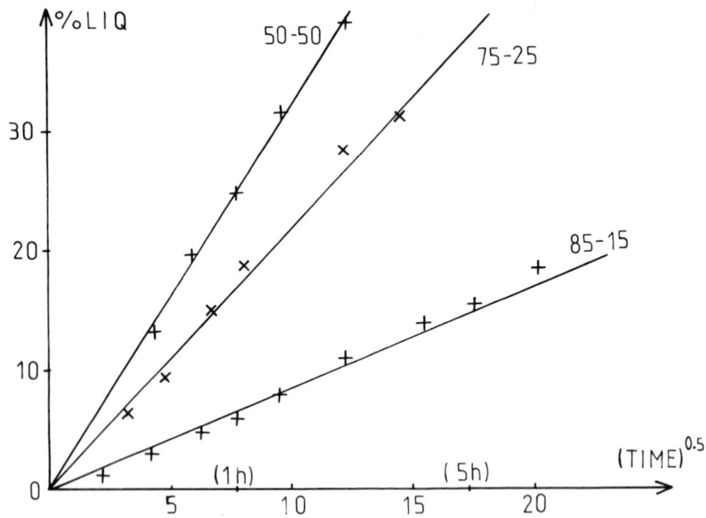


**Figure 10–6** Amount of liquid transported (weight % of initial sample) versus square root of time. Initial drug concentration ranging from 15 to 50. Eudragit RS. pH = 1.2 at 37°C. (Reprinted from *Int. J. Pharm.*, 27, (1985), 233–243, A. Droin, C. Chaumat, M. Rollet, J. L. Taverdet, and J. M. Vergnaud, ref. 32, by courtesy of Elsevier Sci. Biomedical Div.)

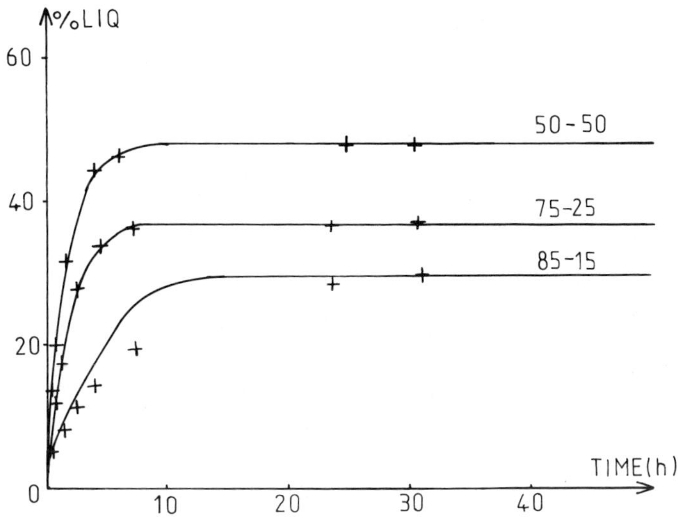


**Figure 10–7** Kinetics of absorption of liquid by various devices, for various initial drug concentrations. Eudragit RS. pH = 1.2 at 37°C. (Reprinted from *Int. J. Pharm.*, 27, (1985), 233–243, A. Droin, C. Chaumat, M. Rollet, J. L. Taverdet, and J. M. Vergnaud, ref. 32, by courtesy of Elsevier Sci. Biomedical Div.)

a not large but significant swelling can be observed, as the rate of liquid transport is higher than that of the drug. A frame of reference fixed with respect to the initial sheet is considered. Then the thickness of the sheet is constant and equal to the thickness of the original unswollen sheet.

Figures 10–5 and 10–7 illustrate the validity of the model for the sodium salicylate and liquid transported, respectively (continuous line = theoretical results; plots = experimental data). A good agreement is obtained not only for the transport of drug, but also for the transport of liquid.

In fact, these results test the validity of the model, and also the experimental data obtained, especially the diffusivity and its concentration dependency.

Profiles of concentration are calculated in Figure 10–9 for the drug and liquid, as they are developed through the sheet thickness for various times (initial concentration of drug: 50 percent; thickness = 0.022 cm).

**TABLE 10–2** Diffusivity and Capacity of Absorption

| Eudragit-Drug | $M_\infty$ drug | $M_\infty$ liquid | $D \times 10^8$ drug | $D \times 10^8$ liquid |
|---|---|---|---|---|
| 50–50 | 20.5 | 48 | 16.5 | 70.5 |
| 75–25 | 12.5 | 37 | 16.5 | 37.5 |
| 85–25 | 4 | 29.5 | 16.5 | 11 |

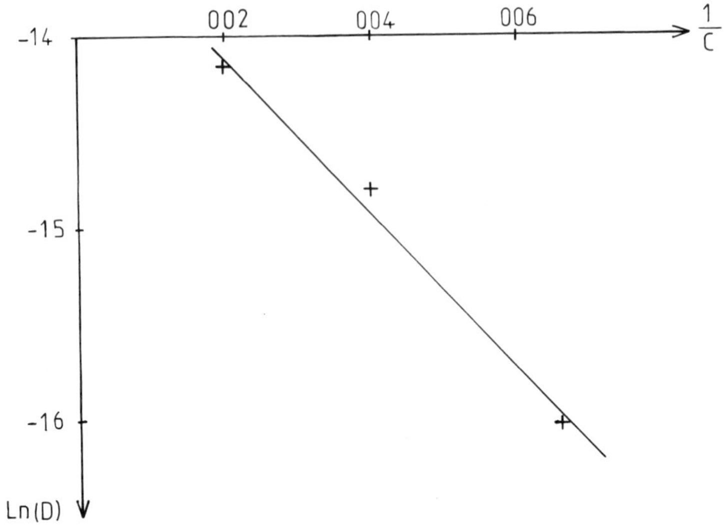


**Figure 10–8** Log diffusivity of liquid as a function of (initial drug concentration)$^{-1}$. Eudragit RS. pH = 1.2 at 37°C. (Reprinted from *Int. J. Pharm.*, 27, (1985), 233–243, A. Droin, C. Chaumat, M. Rollet, J. L. Taverdet, and J. M. Vergnaud, ref. 32, by courtesy of Elsevier Sci. Biomedical Div.)

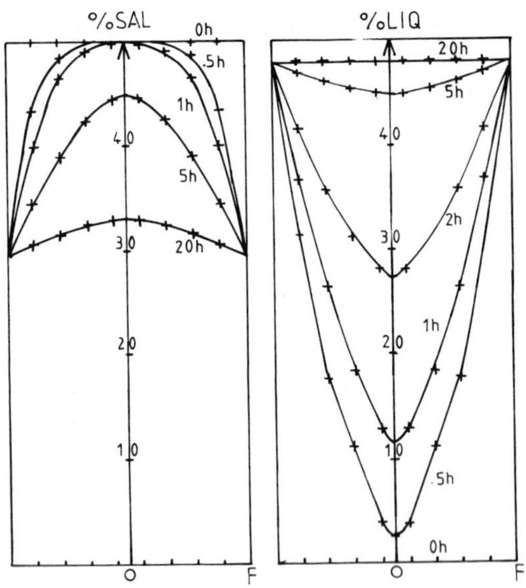


**Figure 10–9** Profiles of concentration of sodium salicylate and liquid through the sheet thickness. Initial drug concentration = 50%. Eudragit RS. Thickness = 0.022 cm at 37°C. pH = 1.2. (Reprinted from *Int. J. Pharm.*, 27 (1985), 233–243, A. Droin, C. Chaumat, M. Rollet, J. L. Taverdet, and J. M. Vergnaud, ref. 32, by courtesy of Elsevier Sci. Biomedical Div.)

### 10-2-4 Results with Carbopol as Polymer

**Drug diffusivity and capacity of absorption.** The amount of drug transported into gastric liquid varies about proportionally with the square root of time, with an increase in the slope at about 2 h, and that for initial drug concentrations ranging from 15 to 20 weight percent (Figure 10–10).

The diffusivity of a drug is not constant, but concentration dependent. It is expressed in terms of the concentration of drug and liquid, as shown in equation (10-13), with β equal to 1, showing that drug and liquid play the same role for calculation (equation 10-16).

The amount of drug transported into gastric liquid, $M_\infty$, increases with the initial drug concentration in sheets. However, a large amount of drug remains in dosage form.

**Liquid diffusivity and capacity of absorption.** The square root of time dependence with the amount of liquid which enters the polymer is illustrated in Figure 10–11.

The value of diffusivity varies with the initial drug concentrations. The capacity of absorption of liquid by the sheets is very large, and its values increase with the amount of Carbopol in the sheet.

The capacity of absorption in Table 10–3 is expressed in percent of

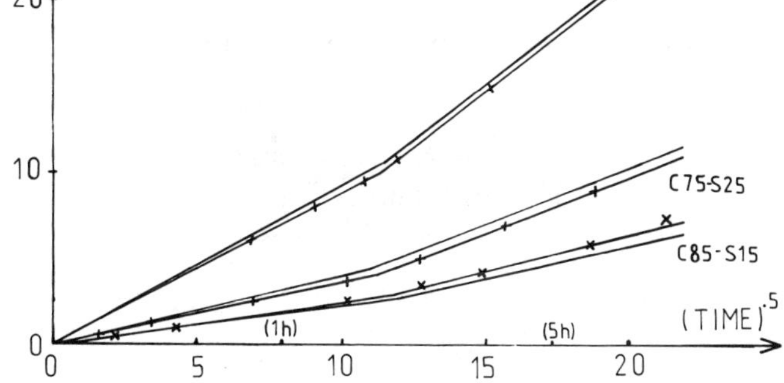


**Figure 10–10** Amount of drug (sodium salicylate) transported versus square root of time, for various compositions. Carbopol. pH = 1.2 at 37°C. (Reprinted from *Drug Dev. Ind. Pharm.*, 13 (1987), 67–81, I. Malley, J. Bardon, M. Rollet, J. L. Taverdet, and J. M. Vergnaud, ref. 33, by courtesy of Marcel Dekker Inc.)

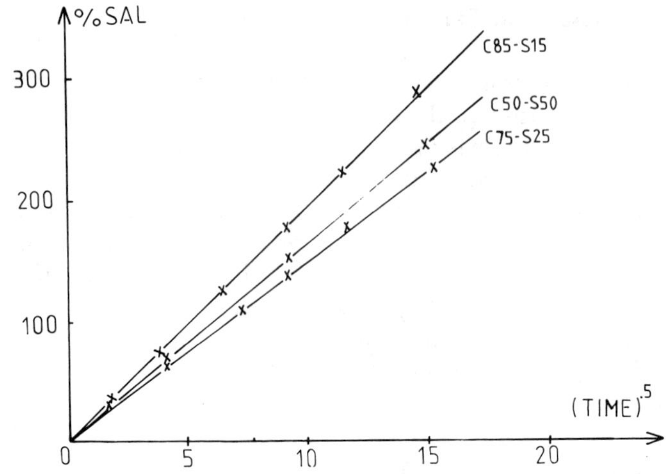


**Figure 10–11** Amount of liquid transported versus square root of time, for various compositions. Carbopol. pH = 1.2 at 37°C. (Reprinted from *Drug Dev. Ind. Pharm.*, 13, (1987), 67–81, I. Malley, J. Bardon, M. Rollet, J. L. Taverdet, and J. M. Vergnaud, ref. 33, by courtesy of Marcel Dekker Inc.)

the weight of the initial sheet, in the same way as the composition of initial sheets.

The diffusivity for the liquid and the drug are expressed in terms of concentration and liquid by the equations:

$$\text{Drug} \qquad D^d = D_o \cdot \left(1 + \frac{C^1}{C^1 + C^d}\right) \tag{10-17}$$

$$\text{Liquid} \qquad D^1 = \exp\left(\frac{-13.8}{C^1 + C^d} - 14.7\right) \tag{10-18}$$

$C^l$ and $C^d$ being the concentrations, or volume fraction of liquid and drug in the sheet.

**Validity of the model.** The validity of mathematical simulations undertaken by using the model and data concerned with diffusivity and ca-

**TABLE 10–3** Diffusivity and Capacity of Absorption

| Carbopol-Drug | $D^1 \times 10^8$ (cm$^2$/s) | $D^d \times 10^8$ (cm$^2$/s) | $M^1_\infty$ (%) | $M^d_\infty$ (%) |
|---|---|---|---|---|
| 50–50 | 16 | 3.3 | 235 | 38 |
| 75–25 | 23 | 3.6 | 315 | 19 |
| 85–15 | 30 | 6.7 | 500 | 7.5 |

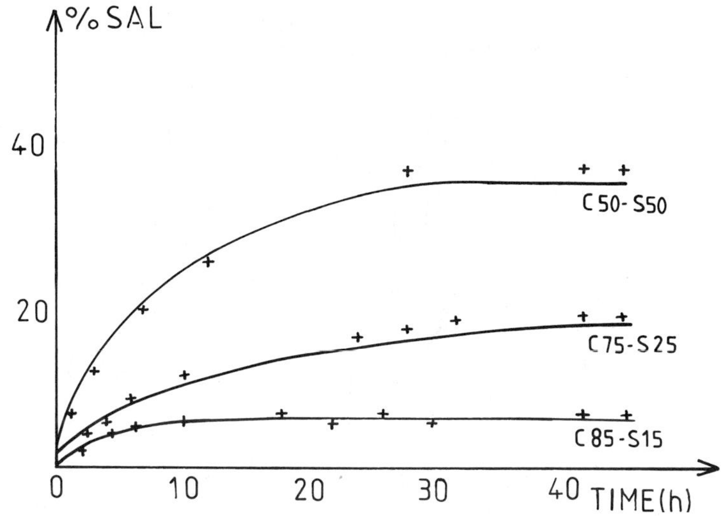


**Figure 10–12** Kinetics of release of drug, for various initial concentrations of drug. Carbopol. pH = 1.2 at 37°C. (Reprinted from *Drug Dev. Ind. Pharm.*, 13, (1987), 67–81, I. Malley, J. Bardon, M. Rollet, J. L. Taverdet, and J. M. Vergnaud, ref. 33, by courtesy of Marcel Dekker Inc.)

pacity of absorption is tested in Figures 10–12 and 10–13, by comparing the theoretical kinetics (full line) to experimental values (plots).

Profiles of concentration of drug and liquid developed through the thickness of the sheet are drawn in Figure 10–14, when the initial drug concentration is 25 percent.

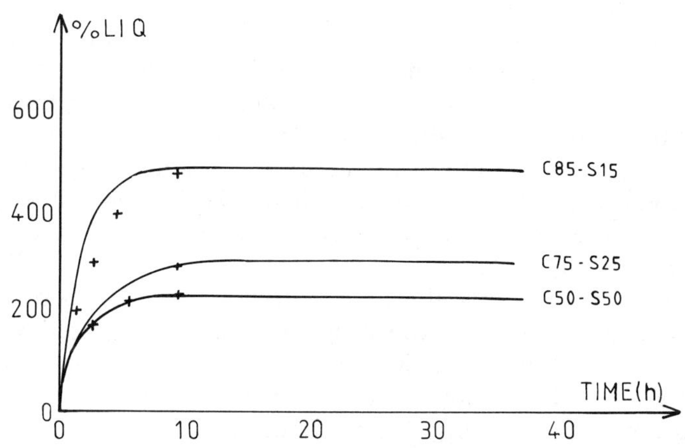


**Figure 10–13** Kinetics of absorption of liquid, for various initial concentrations of drug. Carbopol. pH = 1.2 at 37°C. (Reprinted from *Drug Dev. Ind. Pharm.*, 13, (1987), 67–81, I. Malley, J. Bardon, M. Rollet, J. L. Taverdet, and J. M. Vergnaud, ref. 33, by courtesy of Marcel Dekker Inc.)

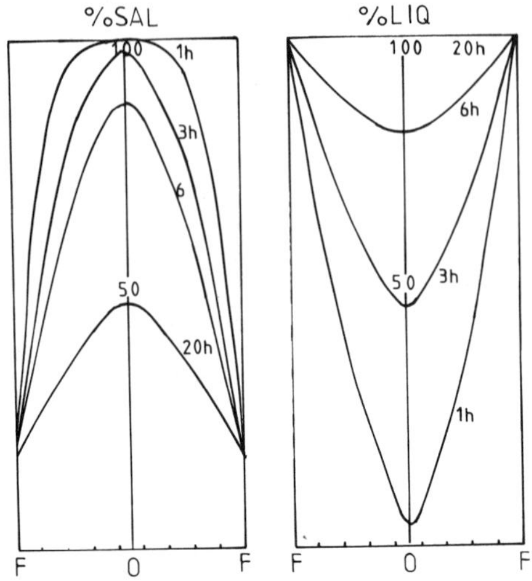


**Figure 10–14** Profiles of concentration developed through the thickness of sheet. Carbopol. pH = 1.2 at 37°C. Initial drug concentration = 25%. (Reprinted from *Drug Dev. Ind. Pharm.*, 13, (1987), 67–81. I. Malley, J. Bardon, M. Rollet, J. L. Taverdet, and J. M. Vergnaud, ref. 33, by courtesy of Marcel Dekker Inc.)

### 10-2-5 Conclusions in the Case of Sheet

The case of sheet is very simple because there is a one-dimensional transport, and the contribution of each slice to the transport of material is the same (see the case of spherical bead).

However, a double transport takes place; the liquid enters the polymer, dissolves the drug, and then enables the dissolved drug to leave the sheet through the liquid located in the sheet. Both these transports are controlled by diffusion, and the diffusivities are concentration dependent. Moreover, the concentration of the liquid in the sheet plays an important role, and the diffusivity of the drug depends largely on this value. Thus, the transports of drug and liquid are connected with each other, the liquid allowing the drug to diffuse.

The mathematical model is able to describe these facts properly. The kinetics calculated either for the drug or liquid transported are in good agreement with experiments. Moreover, interesting information can be obtained by using the model, such as the profiles of concentration of drug and liquid developed through the thickness of the sheet.

A part of the drug is transported from the sheet into the external liquid. This fact is due to the relatively low solubility of sodium salicylate under the conditions (pH = 1.2). Experiments done with the same sheet

(0.5 g drug-Eudragit, and 0.25 g with drug-Carbopol) and larger amount of liquid find higher extractions of the drug.

## 10-3 MODEL OF TRANSPORT PROCESS WHEN pH OF GASTRIC LIQUID IS VARIED

The pH of gastric liquid in contact with the dosage form may vary for several reasons depending on time, on position in the stomach, and food, among other parameters. The main purpose of this chapter is to show that transports of gastric liquid and drug can be studied for various values of the pH, by using experiments and mathematical models [38].

### 10-3-1 Experimental

#### Materials used for dosage forms

- Drug: Sodium salicylate, in powder form.
- Polymer: Eudragit RS is the polymer (copolymer of dimethylamino-ethylacrylate and ethylmethacrylate) in powder form.

**Preparation of thin sheets.** Thin tablets are prepared by pressing the mixture of drug and Eudragit (25–75 weight %) as shown in the previous section on drug delivery.

**Measurements of matter transports.** A tablet of about 0.250 g, previously inserted in a fiberglass basket, is soaked into liquid (100 ml) kept at constant temperature (37°C) at a controlled rate of stirring. Samples of liquid are taken at regular times for analysis and the tablet is weighed. Various compositions are prepared for the liquid in order to have several pH values.

### 10-2-2 Theoretical

The model described in the last section is used.

### 10-2-3 Results

Special attention is given to the variation of diffusivity and the capacity of absorption with the pH of the liquid. The value chosen for the pH of the liquid ranges from 1.9 to 7.4.

**Amount of matter transported at equilibrium.** The amount of drug transported into the liquid increases regularly with time, for the various values chosen for the pH of the liquid. The higher the pH, the larger the amount of drug transported (Figure 10–15).

The amount of liquid transported into the dosage form is plotted against time for various pHs within the 1.9–7.4 range (Figure 10–16). As shown in these curves, the behavior of the liquid in the dosage form is quite different according to the pH value. For instance, the amount of liquid increases regularly with time for the lower values of pH (1.9–2.25). But for higher values of pH, the amount of liquid increases with time, passes through a maximum, and then decreases. Moreover, the maximum value of the liquid transported decreases with the pH.

**Diffusivity.** Diffusivity is obtained from the curves expressing the amount of liquid (and of drug) as a function of the square root of time. Straight lines are obtained for various pHs of the liquid [13].

The variation of the diffusivity with the pH of the liquid is drawn for the drug (Figure 10–17) and for the liquid (Figure 10–18). From these straight lines, the following equations are obtained.

Drug $$D^d = 2.5 \times 10^{-7} \times \exp\left(-\frac{0.3}{\text{pH}}\right) \quad (\text{cm}^2/\text{s}) \tag{10-19}$$

$$D^l = 1.5 \times 10^{-6} \times \exp(0.13 \times \text{pH}) \times \exp\left(\frac{-40.1}{C^d + C^1}\right) \tag{10-20}$$

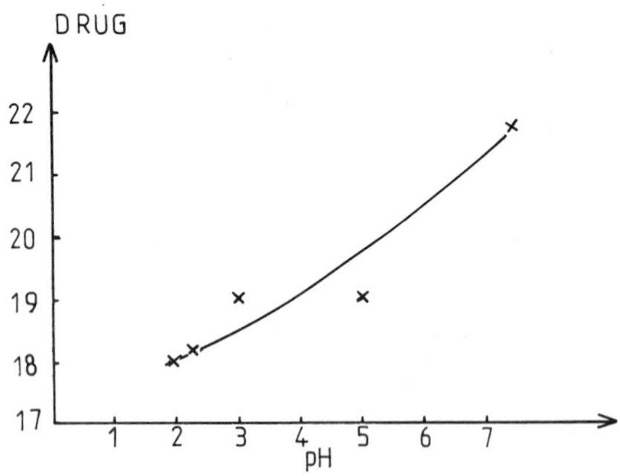


**Figure 10–15** Amount of drug transported at equilibrium versus pH of liquid. Sodium salicylate-Eudragit (25–75 wt %). (Reprinted from *Int. J. Pharm.*, 32, (1986), 143–150, A. Eddine, A. Droin, J. L. Taverdet, and J. M. Vergnaud, ref. 38, by courtesy of Elsevier Sci. Biomedical Div.)

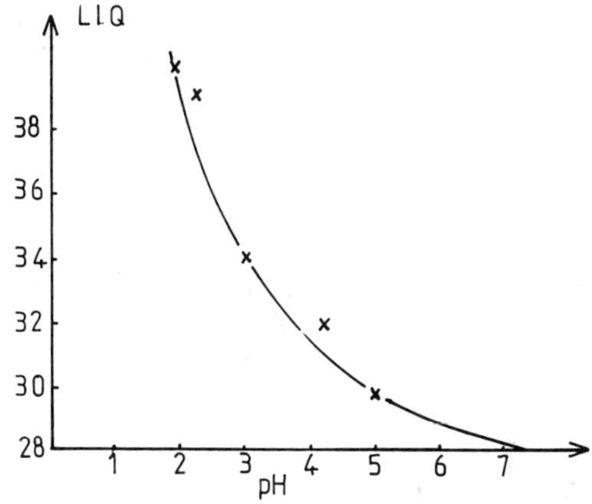


**Figure 10–16** Amount of liquid transported at equilibrium versus pH of liquid. Sodium salicylate-Eudragit (25–75 wt %). (Reprinted from *Int. J. Pharm.*, 32, (1986), 143–150, A. Eddine, A. Droin, J. L. Taverdet, and J. M. Vergnaud, ref. 38, by courtesy of Elsevier Sci. Biomedical Div.)

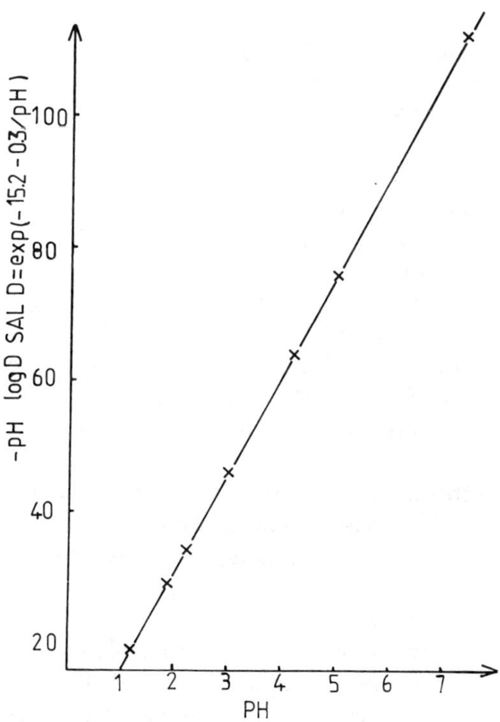


**Figure 10–17** Variation of diffusivity of drug with pH of liquid. Sodium salicylate-Eudragit (25–75 wt %). (Reprinted from *Int. J. Pharm.*, 32, (1986), 143–150, A. Eddine, A. Droin, J. L. Taverdet, and J. M. Vergnaud, ref. 38, by courtesy of Elsevier Sci. Biomedical Div.)

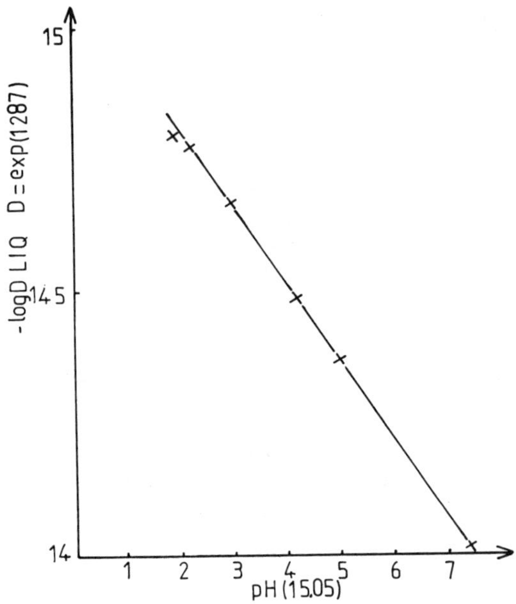


**Figure 10–18** Variation of diffusivity of liquid with pH of liquid. Sodium salicylate-Eudragit (20–75 wt %). (Reprinted from *Int. J. Pharm.*, 32, (1986), 143–150, A. Eddine, A. Droin, J. L. Taverdet, and J. M. Vergnaud, ref. 38, by courtesy of Elsevier Sci. Biomedical Div.)

Drug diffusivity increases slowly with the pH of liquid, while diffusivity of liquid increases highly.

There is no special problem for measuring the amount of drug at equilibrium, and thus for calculating diffusivity by using equation (10-1).

The problem is more complicated for determining the amount of liquid at equilibrium, because of the maximum value. Two different values can be chosen: (1) the value obtained for a long time, by considering the decrease in the amount of liquid with time, and (2) the maximum value of the liquid. The maximum value of the liquid is chosen in the present study.

The values of the amount of matter transported at equilibrium defined in this way are pH dependent. The amount of drug increases with pH (Figure 10–19), while the amount of liquid decreases (Figure 10-20).

**Modeling of kinetics of transport.** Modeling the process is obtained by using the preceding numerical analysis for a sheet. The kinetics of transport are calculated either for the drug (Figure 10–19) or for the liquid (Figures 10–20 and 10–21), by considering the various values of the pH used in experiments. As shown in these figures, a good agreement between theoretical and experimental values is obtained for the drug. In the case

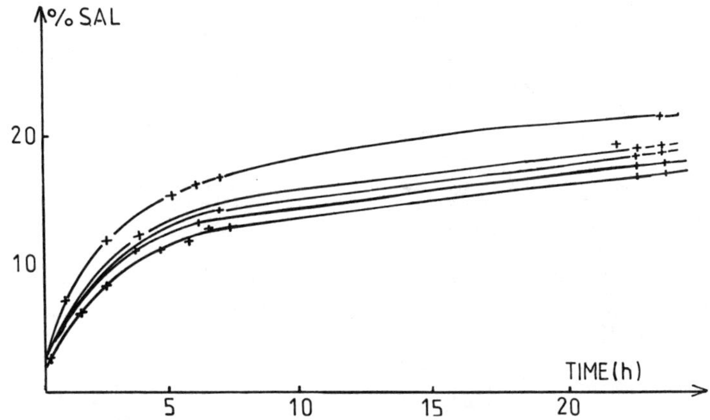


**Figure 10–19** Kinetics of drug transported in the liquid, for various pHs. Sodium salicylate-Eudragit (25–75 wt %). (Reprinted from *Int. J. Pharm.*, 32, (1986), 143–150, A. Eddine, A. Droin, J. L. Taverdet, and J. M. Vergnaud, ref. 38, by courtesy of Elsevier Sci. Biomedical Div.)

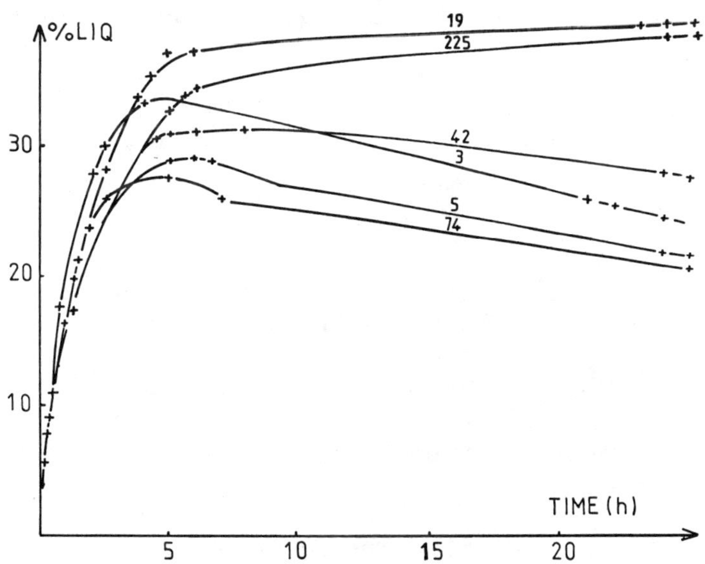


**Figure 10–20** Kinetics of liquid transported in the dosage form, for various pHs. Sodium salicylate-Eudragit (25–75 wt %). (Reprinted from *Int. J. Pharm.*, 32, (1986), 143–150, A. Eddine, A. Droin, J. L. Taverdet, and J. M. Vergnaud, ref. 38, by courtesy of Elsevier Sci. Biomedical Div.)

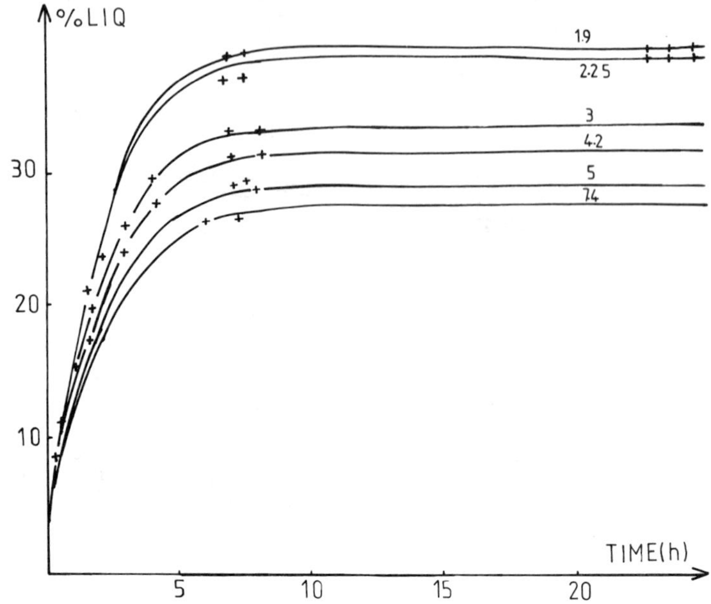


**Figure 10–21** Kinetics (calculated) of liquid transported in the dosage form, for various pHs. Sodium salicylate-Eudragit (25–75 wt %). (Reprinted from *Int. J. Pharm.*, 32, (1986), 143–150, A. Eddine, A. Droin, J. L. Taverdet, and J. M. Vergnaud, ref. 38, by courtesy of Elsevier Sci. Biomedical Div.)

of liquid, as the value at the maximum is selected for the value at equilibrium, the kinetic curves are different in experiments and calculation for long times.

### 10-3-4 Conclusions

The model is able to describe the process of matter transport for the liquid and drug, for various pHs of the liquid. Calculations are made in the case of a constant value of the pH.

However, the model is able to follow the process when the pH is varied during the experiment. The variation of pH with time has to be introduced in the model, as the diffusivity and amount of matter transported are pH dependent.

## 10-4 MODEL FOR MATTER TRANSPORTS BETWEEN SPHERICAL DOSAGE FORMS AND GASTRIC LIQUID

The following sections are presented.

- Experimental, with preparation of spheric forms, and in vitro test description.

- Theoretical with mathematical treatment, assumptions, and numerical analysis.
- Results, with determination of diffusivities, testing the validity of the model.
- Conclusions [34].

### 10-4-1 Experimental

**Materials used for dosage form**

- Drug: Sodium salicylate, in powder form.
- Polymer: Eudragit RS (Röhm Pharma), copolymer of dimethylaminoethylacrylate and ethylmethacrylate with a mean molecular weight of 150,000, in powder form.

**Preparation of spheric dosage forms.** Both these materials are mixed in a mortar and then transformed into a homogeneous thick paste, after pulverization with a small amount of ethanol (which is a solvent for Eudragit). Spheric dosage forms are obained by pressing the paste in a mold. Beads are then dried at room temperature until complete evaporation of ethanol (>2 days).

Various spheres are prepared and tested:

- with same radius (0.43 cm) – (0.4–0.5 g) and various compositions, the drug percent ranging from 25 to 75 wt %.
- with same composition (50 wt % drug) and various sizes (0.36–0.43–0.62–0.87 cm).

**In vitro tests.** The spheric dosage form (0.4–0.5 g), inserted in a fiberglass basket, is immersed in synthetic gastric liquid (100 ml) at 37°C. The composition of the liquid is for pH = 1.2 (80 ml HCL 1N, 2 g Na Cl for 1000 ml of aqueous solution).

Experiments are carried out in a closed flask with a rotating bar magnet. At intervals, the solid is weighed and a sample of liquid is taken for analysis by UV-spectrometry (Bechman DB-G calibrated at 300 nm).

### 10-4-2 Theoretical

**Description of the process in the drug-polymer device.** When the dosage form is in contact with gastric liquid, two matter transports take place successively (and simultaneously). The liquid enters the polymer and dissolves the drug, thus enabling the drug to leave the dosage form through the liquid located in the polymer. These two matter transports are considered simultaneously.

**Assumptions.** Several assumptions are made:

1. The dosage form is spherical in shape, and the drug is intimately dispersed in the polymer matrix.
2. Radial transports are considered.
3. Two matter transports take place simultaneously: the liquid enters the polymer, while the drug leaves the dosage form.
4. Both these transports are controlled by transient diffusion, and they are connected with each other.
5. Diffusivities of these transports depend on the concentrations of drug and liquid.
6. Concentrations of matter transported (drug, liquid) on the surface attain the value at equilibrium as soon as the process starts.
7. In spite of the swelling of the polymer, a frame of reference is fixed with respect to the initial dosage form for calculation.

**Mathematical treatment.** The equation of radial diffusion follows Fick's equation with concentration-dependent diffusivity.

$$\frac{\partial C}{\partial t} = \frac{1}{r^2} \cdot \frac{\partial}{\partial r}\left[D \cdot r^2 \cdot \frac{\partial C}{\partial r}\right] \tag{10-21}$$

while Fick's first law is also considered:

$$F = -D \cdot \frac{\partial C}{\partial r} \qquad F = \text{flux (matter per unit area and unit time)} \tag{10-22}$$

The initial and boundary conditions follow the first and sixth assumptions, respectively.

| | | | | | |
|---|---|---|---|---|---|
| Initial | $t = 0$ | $r < R$ | $C = C_{in}$ | sample | (10-23) |
| Boundary | $t > 0$ | $r = R$ | $C = C_{eq}$ | surface | (10-24) |

No analytical solution is found for the problem, essentially because of the double transport and concentration-dependent diffusivities.

For very short tests, and very small amounts of matter transported (drug or liquid), $M_t$, the actual concentration in the sheet may be considered as constant and equal to the initial value at the beginning of the test. The amount $M_t$ is then expressed in terms of $M_\infty$, the amount transported after infinite time when equilibrium is attained.

$$\frac{M_t}{M_\infty} = \frac{6}{R} \cdot \left(\frac{D \cdot t}{\pi}\right)^{0.5} \tag{10-25}$$

where $R$ is the radius of the sphere and $D$ is diffusivity.

**Modeling of the process (numerical analysis).** This complex problem of double transport is resolved by using an explicit numerical method with finite differences. The dosage form is divided into concentric spheres with the following radius, $r + \Delta r$, $r$, $r - \Delta r$. The number of concentric spheres is related to the thickness $\Delta r$ of each spherical part and to the radius $R$ of dosage form.

$$R = N \cdot \Delta R \tag{10-26}$$

Each position is defined by an integer $n$:

$$r = n \cdot \Delta r \cdot \tag{10-27}$$

Concentrations of drug and liquid are constant within each spherical part of thickness $\Delta r$ located between two adjacent spheres, at the same time $t$.

The matter balance is calculated during the short elapse of time $\Delta t$ within the spherical part of thickness $\Delta r$ and centered on the position $r$ defined by the integer $n$ (Figure 10–22), by using Fick's first law.

***Within the Dosage Form.*** The new concentration at position $r$ (integer $n$) after the elapse of time $\Delta t$ is then expressed in terms of concen-

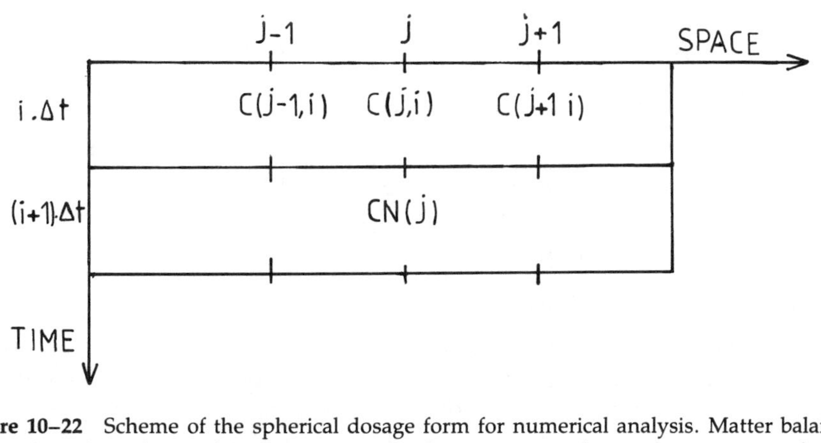


**Figure 10–22** Scheme of the spherical dosage form for numerical analysis. Matter balance within the spherical part of thickness $\Delta r$ centered on position $r$, during the increment of time $\Delta t$. (Reprinted from *Int. J. Pharm.*, 40, (1987), 33–41, J. Y. Armand, F. Magnard, J. Bouzon, M. Rollet, J. L. Taverdet, and J. M. Vergnaud, ref. 34, by courtesy of Elsevier Sci. Biomedical Div.)

tration at the previous time in the same place and in the two adjacent places.

$$CN_n = C_n + \frac{\Delta t}{n^2(\Delta r)^2} \cdot [J_{n-1/2} - J_{n+1/2}] \tag{10-28}$$

$J$ being function of diffusivity and concentration:

$$J_{n-1/2} = \left(n - \frac{1}{2}\right)^2 \cdot D_{n-1/2} \cdot (C_{n-1} - C_n) \tag{10-29}$$

The diffusivity at the intermediate position ($n - \frac{1}{2}$) can be obtained from diffusivities at the positions $n$ and $n + 1$ by either way, using one of the following equations:

$$D_n = D_o \cdot \exp(Cte \cdot C_n) \tag{10-30}$$

$$D_n = D_o \cdot \exp Cte(C - C_n) \tag{10-31'}$$

$$D_n + D_o \cdot \exp \frac{Cte}{C_n} \tag{10-30''}$$

***Middle of the Dosage Form.*** The preceding equations obtained for the sphere cannot be used for the middle of the sphere ($r = o$, $n = o$). The matter balance is then calculated for the sphere of radius $\Delta r/2$ located at the middle of the dosage form. The new concentration after elapse of time $\Delta t$ is:

$$CN_o = C_o + 24 \cdot \frac{\Delta t}{(\Delta r)^2} J_{1/2} \tag{10-31}$$

where $J$, as written in the preceding equation, becomes:

$$J_{1/2} = \left(\frac{1}{2}\right)^2 \cdot D_{1/2} \cdot (C_o - C_1) \tag{10-32}$$

***Surface of Dosage Form.*** The concentration of the drug and liquid on the surface of dosage forms is the concentration at equilibrium, following the fourth assumption.

$$C_o = C_{eq} \qquad \text{liquid, drug} \tag{10-33}$$

***Amount of Matter Located in Dosage Form.*** The amount of drug remaining (or of liquid entered) in dosage form at time $t$ is obtained by integrating the concentration of drug at this time with respect to space.

$$M_t = 4\pi \cdot \int_o^R r^2 \cdot C_{r,t} \cdot dr \tag{10-34}$$

This expression is rewritten by using the finite differences:

$$M_t = 4\pi(\Delta r)^3 \cdot \left[\frac{C_o}{24} + \sum_{n=1}^{n-2} n^2 C_n + \frac{9}{8}(n - 1)^2 C_{n-1} + \frac{3}{8} n^2 \cdot C_n\right] \quad (10\text{-}35)$$

The preceding equations can be used for calculating the concentration at any place and any time, either for the liquid or drug, if diffusivities are known.

### 10-4-3 Results with Spherical Dosage Form

**Diffusivity and capacity of absorption.** The diffusivity expressed in terms of concentration, as well as the amount of matter transported after infinite time, are determined. The variation of drug released and of liquid absorbed, as a function of the square root of time, is illustrated in Figures 10–23 and 10–24 for various initial concentrations of drug in dosage forms. The weight of drug located in each dosage form (drug and polymer) is between 0.4 and 0.5 g, while the volume of gastric liquid is 100 ml. These values are of interest not only for the amount of matter transported after infinite time but also for diffusivity, because of the rather low solubility of the drug in acid liquid.

Two conclusions can be drawn:

1. Straight lines are obtained for the liquid (Figure 10–24), passing through the axes origin.

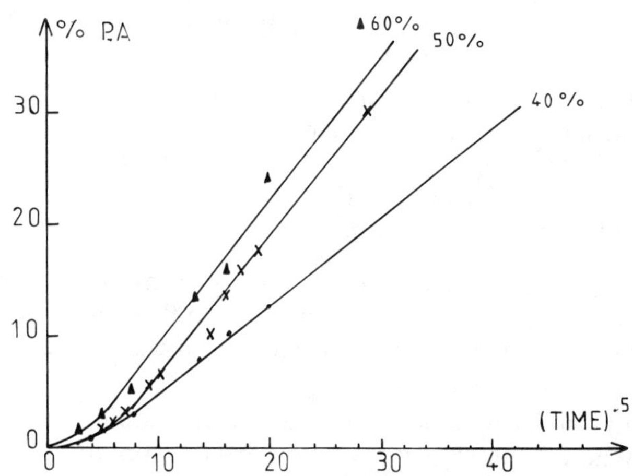


**Figure 10–23** Drug (sodium salicylate) transported as a function of square root of time, for various initial drug concentrations in spheric dosage forms. pH = 1.2 at 37°C. Radius = 0.435 cm. (Reprinted from *Int. J. Pharm.*, 40, (1987), 33–41, J. Y. Armand, F. Magnard, J. Bouzon, M. Rollet, J. L. Taverdet, and J. M. Vergnaud, ref. 34, by courtesy of Elsevier Sci. Biomedical Div.)

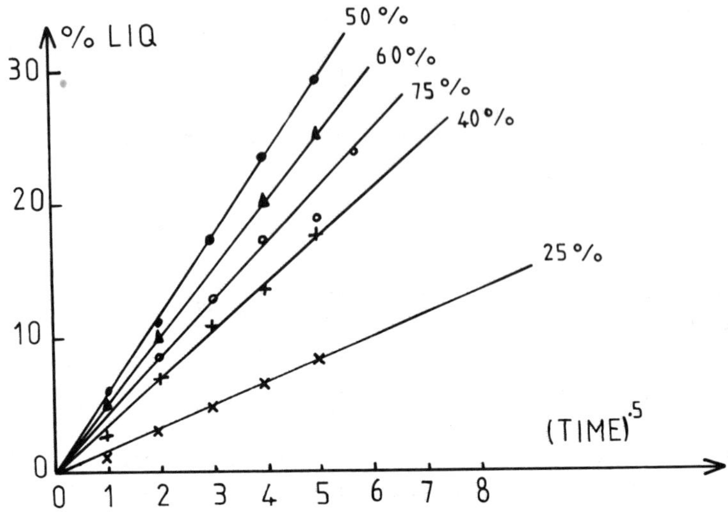


**Figure 10–24** Gastric liquid absorbed by spheric dosage forms as a function of square root of time, for various initial drug concentrations. pH = 1.2 at 37°C. Radius = 0.435 cm. (Reprinted from *Int. J. Pharm.*, 40, (1987), 33–41, J. Y. Armand, F. Magnard, J. Bouzon, M. Rollet, J. L. Taverdet, and J. M. Vergnaud, ref. 34, by courtesy of Elsevier Sci. Biomedical Div.)

2. Contrary to the liquid, the slopes of the curves expressing the matter released versus the square root of time, are not constant at the beginning of the process. Starting with a very low value when the dosage form is immersed in the liquid, the slopes increase regularly with time up to a constant value.

***Amount of Matter Transported at Equilibrium.*** The amount of matter transported at equilibrium (after a long time) is obtained from the kinetic curves shown in Figures 10–25, 10–26, and 10–27 for various initial concentrations of drug in dosage form.

Two results are worth noting:

1. The transport of liquid is controlled by transient diffusion, with a vertical tangent for the kinetic curve when the dosage form is in contact with liquid. This fact corresponds with the constant value of diffusivity shown previously.
2. The drug transport is more complex, with an oblic tangent for kinetic curves when the dosage form is immersed. From this result, the process can be described as follows: The liquid enters the polymer, dissolves the drug, and then enables the drug to leave the dosage form through the liquid located in the polymer. The diffusivity for the drug is very low at the beginning, and then increases with the concentration of liquid up to a constant value attained for a high concentration of liquid.

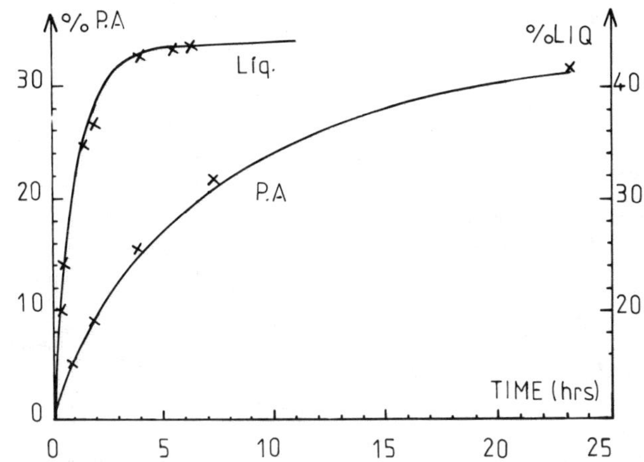


**Figure 10–25** Kinetics of the matter transported (drug, liquid). Drug-polymer 50–50 wt %. Radius = 0.435 cm. pH = 1.2 at 37°C. +:experiments. –theoretical. (Reprinted from *Int. J. Pharm.*, 40, (1987), 33–41, J. Y. Armand, F. Magnard, J. Bouzon, M. Rollet, J. L. Taverdet, and J. M. Vergnaud, ref. 34, by courtesy of Elsevier Sci. Biomedical Div.)

The data obtained for diffusivity and amount of matter transported at infinite time (at equilibrium) are shown in Table 10–4. $M_\infty$ is expressed in weight percent of the initial drug for the drug, and in weight percent of the polymer for the liquid.

Some results are of interest:

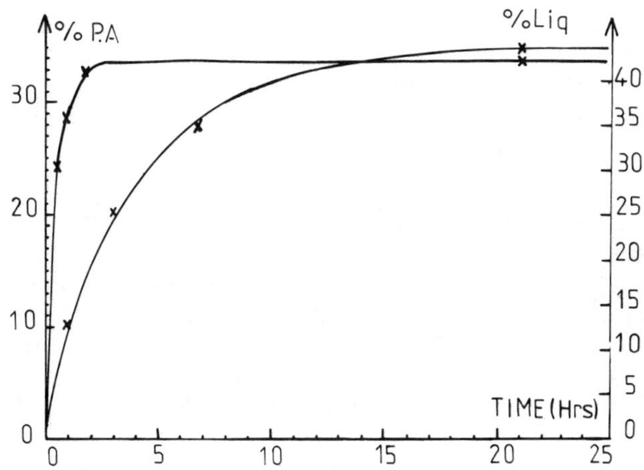


**Figure 10–26** Kinetics of the matter transported (drug, liquid). Drug-polymer 50–50 wt %. Radius 0.31 cm. pH = 1.2 at 37°C. +:experiments. –theoretical. (Reprinted from *Int. J. Pharm.*, 40, (1987), 33–41, J. Y. Armand, F. Magnard, J. Bouzon, M. Rollet, J. L. Taverdet, and J. M. Vergnaud, ref. 34, by courtesy of Elsevier Sci. Biomedical Div.)

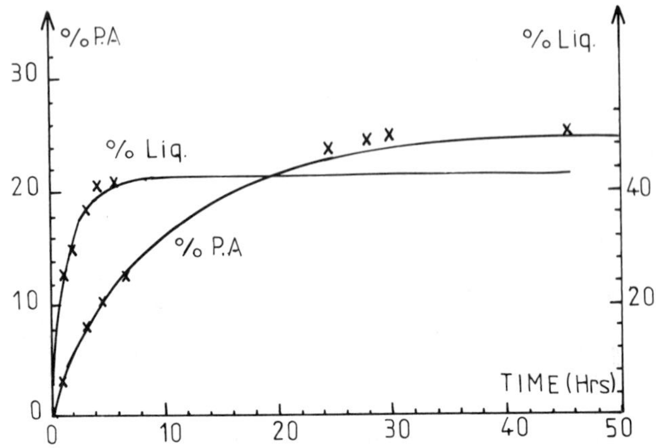


**Figure 10–27** Kinetics of the matter transported (drug, liquid). Drug-polymer 40–60 wt %. Radius = 0.435 cm. pH = 1.2 at 37°C. +:experiments. –theoretical. (Reprinted from *Int. J. Pharm.*, 40, (1987), 33–41, J. Y. Armand, F. Magnard, J. Bouzon, M. Rollet, J. L. Taverdet, and J. M. Vergnaud, ref. 34, by courtesy of Elsevier Sci. Biomedical Div.)

1. The amount of liquid transported after infinite time expressed in terms of the polymer weight is about constant, ranging from 76 to 90 wt %.
2. The amount of drug transported at equilibrium as a fraction of initial weight of drug is about constant, ranging from 62 to 70 percent.
3. The diffusivity for the liquid is constant.

$$D_1 = 4.3 \times 10^{-6} \quad (\text{cm}^2/\text{s}). \tag{10-36}$$

4. The diffusivity for the drug is expressed in terms of the concentration of liquid in the dosage form by

$$D_d = 6 \times 10^{-3} \times \exp - \frac{400}{C_1} \quad (\text{cm}^2/\text{s}). \tag{10-37}$$

whatever the initial concentration of drug.

**Validity of the model.** The experimental and theoretical kinetics of transport for the drug and liquid drawn in Figures 10–25 to 10–27 illustrate

**TABLE 10–4** Diffusivity and Amount of Matter Transported at Equilibrium

| Drug-Polymer | Radius | $D_0^d$ | $D_l$ | $M_\infty^d(\%)$ | $M_\infty^1(\%)$ |
|---|---|---|---|---|---|
| 50–50 | 0.435 | $6\times10^{-3}$ | $4.3 \times 10^{-6}$ | 66 | 88 |
| 50–50 | 0.31 | $6\times10^{-3}$ | $4.3 \times 10^{-6}$ | 70 | 90 |
| 40–60 | 0.435 | $6\times10^{-3}$ | $4.3 \times 10^{-6}$ | 62 | 76 |

the validity of the mathematical model for both matter transports. A good agreement is obtained not only for the drug but also for the liquid, for various drug concentrations and sizes of dosage forms.

Calculations are done by using a simple way to obtain the diffusivity in position $n - \frac{1}{2}$. The mean diffusivity is calculated between $D_n$ and $D_{n-1}$.

$$D_{n-1/2} = \frac{1}{2}(D_n + D_{n-1}) \tag{10-38}$$

Moreover, the profiles of concentration of drug and liquid developed through the dosage form during the process are shown in Figure 10–28 as they are obtained by calculation. Some results can be drawn from these curves:

1. Steep gradients of concentration of liquid and drug are developed in the dosage form, especially next to the surface.
2. The dosage form swells, as the rate of transport is higher for the liquid than for the drug.
3. A constant value is reached on the surface, as soon as the process starts, following the fourth assumption.
4. After infinite time, a part of the drug remains in the dosage form. This fact is due to the low solubility of the drug, and to the ratio of the volume of liquid and drug.

### 10-4-4 Conclusions in the Case of Spherical Form

The case of a sphere is of interest because many dosage forms are spherical in shape. There is a one-dimensional transport, but the contribution of each spherical slice to the transport is not the same, the volume

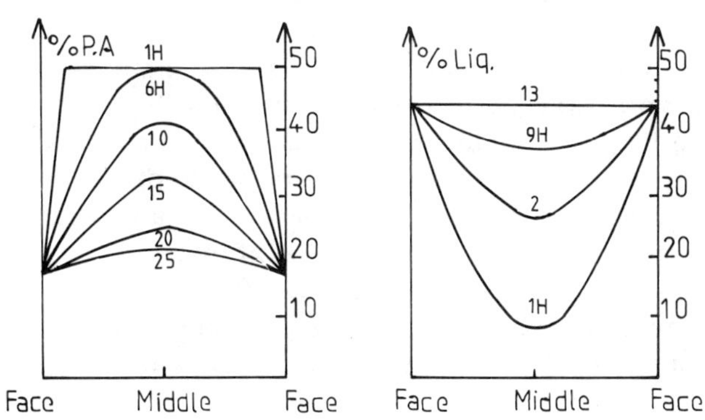


**Figure 10–28** Profiles of concentration developed through the dosage form. – Left:drug. – Right:liquid. Drug-polymer: 50–50 wt %. Radius = 0.435 cm. pH = 1.2 at 37°C. (Reprinted from *Int. J. Pharm.*, 40, (1987), 33–41, J. Y. Armand, F. Magnard, J. Bouzon, M. Rollet, J. L. Taverdet, and J. M. Vergnaud, ref. 34, by courtesy of Elsevier Sci. Biomedical Div.)

of each slice being proportional to radius squared. Under these conditions, the behavior of the external slice is more effective than that of any internal slice with a lower radius.

The diffusivity is calculated in a simple way, by considering the mean value of diffusivities in adjacent places. Of course, other ways can be used with various expressions of diffusivity in terms of concentration. Moreover, the concentration in position $n - \frac{1}{2}$ can be the mean value of those in adjacent places $n$ and $n + 1$.

As shown in the case of sheet, a double transport is observed. The liquid enters the polymer, dissolves the drug, and enables the drug to leave the dosage form through the liquid located in the form. Both these transports are controlled by diffusion, with a concentration dependence for the diffusivity of the drug. The diffusivity of the drug can be expressed in terms of the actual concentration of liquid.

The mathematical model describes this process properly, the kinetics calculated for the drug and liquid being in good agreement with experiments. Moreover, the profile of concentration of drug and liquid developed through the dosage form can be calculated by using the model.

Only a part of the drug initially located in dosage forms is released in gastric liquid. This fact has already been explained by considering the low solubility of the drug under these conditions. An increase in the ratio of the volume of gastric liquid and drug in dosage form is responsible for a higher amount of drug released at equilibrium.

## 10-5 MODEL FOR DOSAGE FORMS MADE OF A CORE AND SHELL

### 10-5-1 Introduction: Purpose with this Dosage Form

With dosage forms prepared by dispersing the drug in a polymer matrix and by pressing the mixture into various shapes such as sheet [32, 33] or sphere [34], the matter transports are controlled by transient diffusion. As a result, the kinetics of delivery of the drug exhibit a very high rate at the beginning of the process (illustrated by a vertical tangent at time zero), which then decreases regularly with time.

Some simple devices able to deliver the drug with a constant rate have been prepared by surrounding the previously described dosage form (made by dispersing the drug in a polymer) by an erodible polymer such as Gelucire [39, 40].

Another way to obtain a constant rate of delivery consists of preparing devices able to deliver the drug with kinetics illustrated by curve $b$ in Figure 10–29, and mixing these devices with the previously described devices

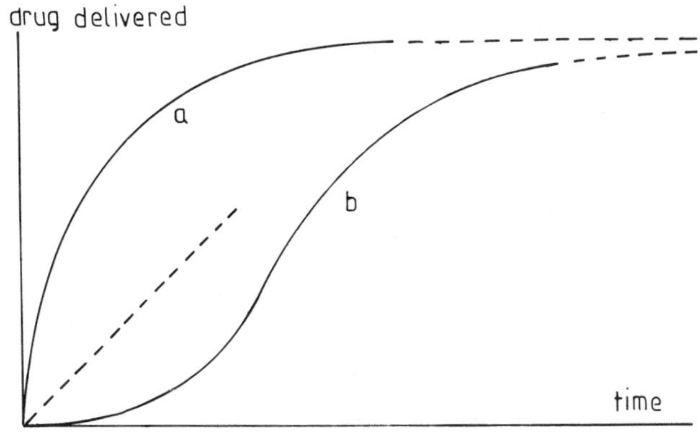


**Figure 10–29** Scheme of kinetics of drug delivery. *a:* diffusional curve with classical device; *b:* with the device studied in the chapter.

exhibiting kinetics of delivery following the shape of curve *a* in Figure 10–29.

These new devices, with kinetics of delivery following curve *b* in Figure 10–30, consist of a core and shell. The core is obtained by dispersing the drug in a polymer such as Eudragit, and the shell is made of pure Eudragit [41].

### 10-5-2 Experimental

#### Materials used

- Drug: Sodium salicylate, in powder form.
- Polymer: Eudragit RS (Röhm Pharma), copolymer of dimethylaminoethylacrylate and ethylmethacrylate with a mean molecular weight of 150,000 in powder form.

**Preparation of spherical dosage forms.** The device consists of a core and shell. The core is obtained by dispersing the drug in Eudragit and pressing the mixture into spherical beads as described earlier. The core is obtained with the mixture of drug and Eudragit 50-50 in weight.

The core is surrounded by a shell made of pure Eudragit, by using a paste of Eudragit and ethanol. The final device is then dried at room temperature. The thickness of the shell is determined by the weight, and it can be controlled by conditions of soaking [41].

**Conditions of testing (in vitro).** In vitro experiments are carried out in a closed flask with high stirring. The device (400–500 mg), inserted in

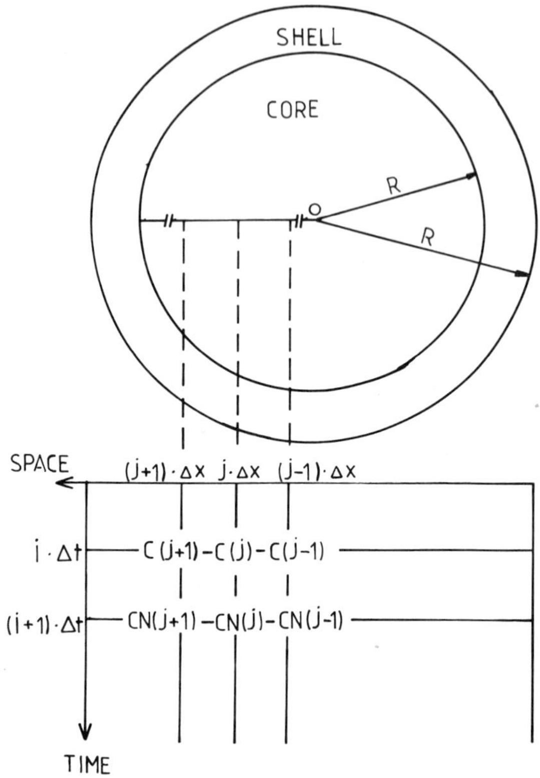


**Figure 10–30** Dosage form with core and shell, and scheme for numerical analysis. (Reprinted from *Int. J. Pharm.*, 45, (1988), 217–227, H. Liu, P. Magron, J. Bouzon, and J. M. Vergnaud, ref. 41, by courtesy of Elsevier Sci. Biomedical Div.)

a fiberglass basket, is immersed into synthetic gastric liquid (100 ml at 37°C and pH 1.2).

The device is weighed and a sample of liquid is taken for analysis at intervals. The amount of drug released is determined by using UV-spectrometry (Beckman DB-G calibrated at 300 nm).

### 10-5-3 Theoretical

**Description of the process.** As two matter transports take place, one concerned with the liquid entering the polymer and dissolving the drug and the other with the drug leaving the device, these two transports are studied simultaneously.

**Assumptions.** Certain assumptions are made for the core and shell.

1. Dosage forms are made of two concentric spheres. In the core, the drug is properly dispersed in Eudragit. The shell is pure Eudragit with constant thickness.
2. Two transports takes place simultaneously, as shown earlier.
3. These two transports are controlled by transient diffusion and are connected with each other.
4. Diffusivities for these transports are concentration dependent.
5. The concentration of liquid on the external surface of the dosage form is attained as soon as the process starts.
6. A partition coefficient exists for the drug between the external surface of the dosage form and the liquid of finite volume in which the form is immersed.
7. Another partition coefficients exists for the liquid on both surfaces at the interface between the core and shell.
8. A frame of reference is fixed with respect to the original dosage form, in spite of the swelling.

**Mathematical treatment.** Fick's laws are written in case of radial transport through the sphere, with concentration-dependent diffusivity.

$$F = -D \cdot \frac{\partial C}{\partial x} \tag{10-22}$$

$$\frac{\partial C}{\partial t} = \frac{1}{r^2} \times \frac{\partial}{\partial r}\left[D \cdot r^2 \cdot \frac{\partial C}{\partial r}\right] \tag{10-21}$$

Initial and boundary conditions are:

| | | | | | |
|---|---|---|---|---|---|
| Initial | $t = 0$ | $0 \le r < R_1$ | $C = C_{in}$ | drug | |
| | | $R_1 < r \le R_2$ | $C = 0$ | drug | (10-39) |
| Boundary | | $0 \le r < R_2$ | $C = 0$ | liquid | |
| | $t > 0$ | $r = R_2$ | $C = C_{eq}$ | liquid | (10-40) |

In the case of a short test, when the amount of matter transported (drug, liquid), $M_t$, is low, the diffusivity can be calculated from the following equation expressing its variation with the square root of time.

$$\frac{M_t}{M_\infty} = \frac{6}{R}\left(\frac{D \cdot t}{\pi}\right)^{0.5} \tag{10-25}$$

**Modeling of the process (numerical analysis).** As no analytical solution can be found for this complex problem, a numerical method with finite differences is used. The dosage form with the two concentric spheres (core and shell) is shown in Figure 10–30). The core ($R_1$) and shell ($R_2$) are divided into concentric spheres with constant thickness between them ($\Delta r_1$ and $\Delta r_2$, respectively).

$$\text{Core } \Delta r_1 \text{ with } r = n \cdot \Delta r_1 \qquad 0 \leq n \leq N_1 \tag{10-41}$$

$$\text{and } R_1 = N_1 \cdot \Delta r_1$$

$$\text{Shell } \Delta r_2 \text{ with } r = R_1 + (n - N_1) \cdot \Delta r_2 \qquad N_1 \leq n \leq N \tag{10-42}$$

$$\text{and } R_2 = R_1 + (N - N_1) \cdot \Delta r_2$$

The matter balance is determined in various places. *Middle of the sample* $n = 0$ (for liquid and drug).

$$CN_0 = C_0 + 24 \frac{\Delta t}{(\Delta r_1)^2} J_{1/2} \tag{10-43}$$

with

$$J_{1/2} = \left(\frac{1}{2}\right)^2 \cdot D_{1/2} \cdot (C_0 - C_1) \tag{10-44}$$

The diffusivity $D_{\frac{1}{2}}$, at position $\frac{1}{2}$, is expressed in terms of the concentration at positions 0 and 1 by using one of the following equations:

$$D_n = D_0 \cdot \exp(Cte \cdot C) \tag{10-30}$$

$$D_n = D_0 \cdot \exp Cte(C - C_n) \tag{10-30'}$$

$$D_n = D_0 \cdot \exp \frac{Cte}{C_n} \tag{10-30''}$$

The matter balance is determined *within the core* $1 \leq n \leq N_1 - 1$ (for the liquid and drug).

$$CN_n = C_n + \frac{\Delta t}{n^2 \cdot (\Delta r)^2} \cdot [J_{n-1/2} + J_{n+1/2}] \tag{10-45}$$

with

$$J_{n-1/2} = \left(n - \frac{1}{2}\right)^2 \cdot D_{n-1/2} \cdot (C_{n-1} - C_n) \tag{10-46}$$

The matter balance is determined *within the shell* $N_1 + 1 \leq n \leq N - 1$ (for liquid and drug).

$$CN_n = C_n + \frac{\Delta t}{n^2 (\Delta r_2)^2} \cdot [J_{n-1/2} + J_{n+1/2}] \tag{10-47}$$

***Interface between Core and Shell,* n = N₁.** Special attention is given to this interface, in order to account for the discontinuity of the concentration of matter (drug and liquid). For the liquid,

$$CN^l_{N_1} = C^l_{N_1} - \frac{\Delta x}{3 \cdot B}\left(1 - \frac{1}{N_1}\right)^2 \cdot [CN^l_{N_1-1} - C^l_{N_1-1}] - \frac{\Delta x_2}{3 \cdot B}\left(1 + \frac{\Delta x_2}{R_1}\right)^2 \cdot [CN^l_{N_1+1} - C^l_{N_1+1}] + \frac{8 \cdot \Delta t \cdot \Delta x_2}{3 \cdot B \cdot R_1^2} \cdot G^l_{N_1+1} - \frac{4 \cdot \Delta t \cdot \Delta x}{3 \cdot B \cdot R_1^2} G^l_{N_1} \tag{10-48}$$

The concentration of liquid varies discontinuously through the interface as follows:

$$C^{l_2}_{N_1} = C^{l_1}_{N_1} \cdot \frac{C^{l_2}_{eq}}{C^{l_1}_{eq}} \tag{10-49}$$

where $C^{l_2}_{N_1}$ is the concentration at the interface on the side in contact with the shell; $C^{l_1}_{N_1}$ is the concentration at the interface on the side in contact with the core; and $C^{l_1}_{eq}$ and $C^{l_2}_{eq}$ are the concentrations of liquid at equilibrium in the core and in the shell. For the drug,

$$CN^d_{N_1} = C^d_{N_1} - \frac{\Delta x}{3(\Delta x + \Delta x_2)}\left(2 - \frac{1}{N_1}\right)^2 \cdot (CN^d_{N_1-1} - CN^d_{N_1}) - \frac{\Delta x_2}{3(\Delta x + \Delta x_2)}\left(1 + \frac{\Delta x_2}{R_1}\right)^2 \cdot (CN^d_{N_1+1} - C^d_{N_1+1}) + \frac{4 \cdot \Delta t \cdot \Delta x_2}{(\Delta x + \Delta x_2)R_1^2} \cdot G^d_{N_1+1} - \frac{4 \cdot \Delta t \cdot \Delta x}{(\Delta x + \Delta x_2) \cdot R_1^2} \cdot G^d_{N_1} \tag{10-50}$$

by writing

$$B = \Delta x + \Delta x_2 \cdot \frac{C^{l_2}_{eq}}{C^{l_1}_{eq}} \tag{10-51}$$

and

$$G^{l_2}_{N_1+1} = \left(\frac{R_1}{\Delta x_2} + \frac{1}{2}\right)^2 \cdot (C^{l_2}_{N_1+1} - C^{l_2}_{N_2}) \tag{10-52}$$

$$G^d_{N_1+1} = \left(\frac{R_1}{\Delta x_2} + \frac{1}{2}\right)^2 \cdot (C^d_{N_1+1} - C^d_{N_1}) \cdot (D^d_{N_1+1} - D^d_{N_1}) \tag{10-53}$$

***External Surface of Dosage Form,* n = N**

$$C_N^1 = C_{eq}^1 \qquad \text{liquid} \tag{10-54}$$

$$CN_N^d = \frac{K \cdot (V_1 \cdot C_{core\ in}^d - M_s^d)}{V + \dfrac{2}{3}\pi R_2^2 \cdot \Delta r_2 \cdot K} \tag{10-55}$$

where the partition coefficient $K$ is

$$K = \frac{M_s^d - M_\infty}{M_\infty} \tag{10-56}$$

with $M_s^d$ the amount of drug in the dosage form at time $t$, excluding the amount located in the half slide next to the external surface. $V_1$ and $V$ are the volumes of the core (radius $R_1$) and dosage form (radius $R_2$).

***Amount of Drug and Liquid Remaining in Dosage Form.*** The amount of drug and liquid located in the dosage form can be calculated at any time, by integrating the concentration of drug and liquid in the dosage form with respect to space.

### 10-5-4 Results with Spherical Dosage Form: Core and Shell

**Diffusivity and matter transported at equilibrium.** Short tests are done in order to determine the amount of matter transported (drug, liquid) in short times. Diffusivities are calculated from the straight lines obtained by plotting these amounts as a function of the square root of time [34, 41].

The diffusivity is expressed in terms of concentration as follows:

$$\text{Drug} \qquad D^d = 2.9 \times 10^{-6} \times \exp\left(\frac{-125}{C^l}\right) \quad \text{core and shell} \tag{10-57}$$

$$\text{Liquid} \qquad D^l = 4.5 \times 10^{-4} \times \exp\left(\frac{-320}{C^l + C^d}\right) \quad \text{core} \tag{10-58}$$

$$\text{Liquid} \qquad D^l = 1.4 \times 10^{-7} \quad \text{shell} \tag{10-59}$$

The values of matter transported at equilibrium are obtained from long experiments, by extrapolating for infinite time. They are calculated as a function of the initial weight of the core and shell.

Certain conclusions are worth noting:

**TABLE 10–5** Amount of Matter Transported at Equilibrium

| Drug (core, shell) | Liquid (core) | Liquid (shell) |
|---|---|---|
| 38% | 55% | 29% |

1. Diffusivity of liquid is constant in pure Eudragit (shell). It does not depend on the initial concentration of drug in the core.
2. Drug diffusivity depends largely on the concentration of liquid.
3. The amount of liquid transported at equilibrium is not the same in the core and shell; it is higher in the core, where the drug is initially located.

**Validity of the model.** Experimental and theoretical kinetics of matter transported are shown in Figures 10–31 and 10–32, for about the same radius of the core, and various thicknesses of the shell. In all cases, good agreement between theoretical and experimental results is observed, proving the validity of the model.

The profiles of concentration developed through the sample are also determined (Figure 10–33) [41]. The effect of the thickness of the shell on the kinetics of drug release can be determined precisely with the help of the model. Certain results can be drawn:

1. For very thin shells, the kinetics of drug release follow curve *a* in Figure 10–29.
2. For thicker shells, the kinetics of drug release follow curve *b* in Figure 10–29.

The diffusivity at position $n - 1/2$ can also be dertermined by using the following equation, but only when the gradient of concentration is not very steep.

$$D_{n-1/2} = \frac{1}{2}(D_n + D_{n-1}) \tag{10-60}$$

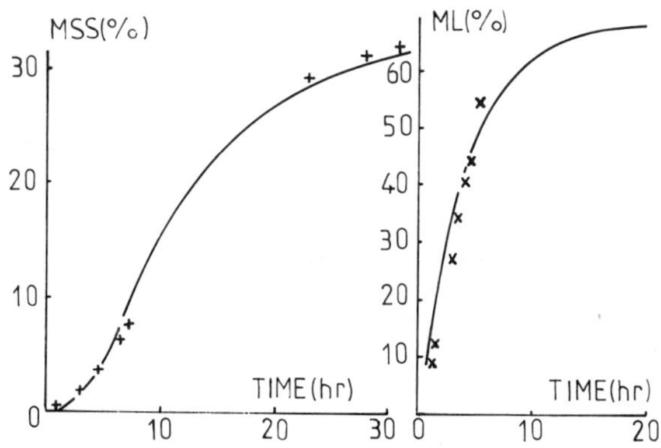


**Figure 10–31** Kinetics of release of drug (left) and liquid (right) with dosage form made of core and shell. Radius of core = 0.425 cm. Thickness of shell = 0.0055 cm. (Reprinted from *Int. J. Pharm.*, 45, (1988), 217–227, H. Liu, P. Magron, J. Bouzon, and J. M. Vergnaud, ref. 41, by courtesy of Elsevier Sci. Biomedical Div.)

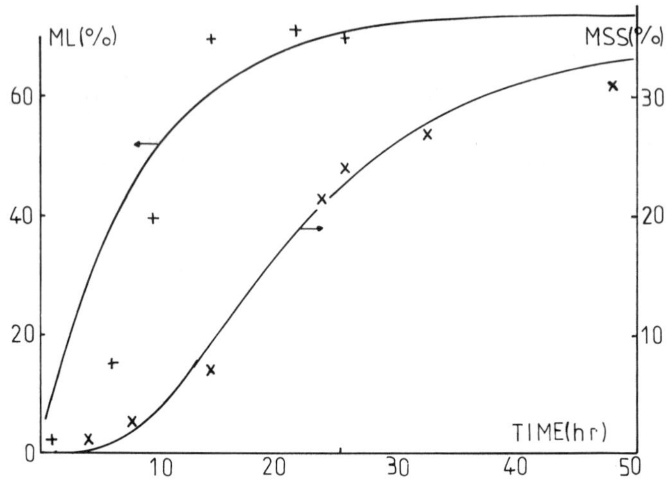


**Figure 10–32** Kinetics of release of drug and liquid with dosage form made of core and shell. Radius of core = 0.441 cm. Thickness of shell = 0.0105 cm. (Reprinted from *Int. J. Pharm.*, 45, (1988), 217–227, H. Liu, P. Magron, J. Bouzon, and J. M. Vergnaud, ref. 41, by courtesy of Elsevier Sci. Biomedical Div.)

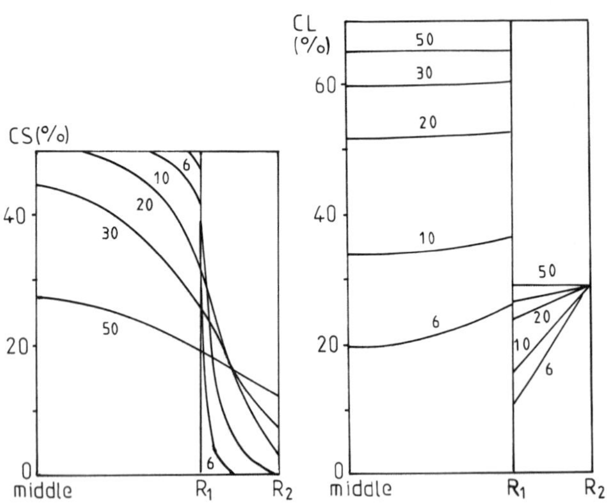


**Figure 10–33** Profiles of concentration of the drug (left) and liquid (right) developed through the dosage form. Radius of core = 0.435 cm. Thickness of shell = 0.02 cm. (Reprinted from *Int. J. Pharm.*, 45, (1988), 217–227, H. Liu, P. Magron, J. Bouzon, and J. M. Vergnaud, ref. 41, by courtesy of Elsevier Sci. Biomedical Div.)

**Conclusions.** Mathematical models are able to describe the process of drug delivery in the complex case of dosage form made of a core and shell, by considering the transport of liquid and drug through the shell and core.

These forms with core and shell, with a convenient thickness of the shell, deliver the drug with a kinetics described by curve *b* in Figure 10–29.

The mean diffusivity is found from diffusivity obtained at adjacent places, but other ways of calculation can be used. For instance, the concentration can be calculated in any place from the values of concentrations in adjacent places.

## REFERENCES

1. A. Zaffaroni, "New approaches to drug administration," 31st Intern. Congress of Pharmac. Sci., Washington, DC (1971), 19–20.
2. J. Godbillon, J. Richard, A. Gerardin, J. Moppert, D. Leory, and F. Theeuwes, "Plasma concentration profiles of metoprolol achieved after administration of oral osmotic controlled-release systems (OROS)," in eds. J. M. Aiache and J. Hirtz, 1st Europ Cong. of Biopharmac and Pharmacokinet. *Proceed, V.1 Technique et Documentation, Biopharmaceuties* (1981), 437–466.
3. Ibid., p. 163.
4. T. Higuchi, Osmotic dispenser with collapsible supply container, U.S. Pat, 3,760,805, September 15, 1973.
5. M. J. Kendall, D. B. Jack, K. L. Woods, S. J. Laugher, C. P. Quarterman, and V. A. John, "Comparison of the pharmacodynamic and pharmacokinetic profiles of single and multiple doses of a commercial slow-release Metoprolol formulation with a new OROS Delivery System," *Br. J. Clin. Pharmacol.*, 13, (1982), 393–398.
6. K. Heilman, in *Therapeutic Systems: Rate-controlled drug-delivery; Concept and Development*. New York: Thième Stratton Inc. (1984), 1–34.
7. J. Heller, "Biodegradable polymers in controlled drug delivery," *CRC Critical Reviews in Therap. Drug Carrier Systems*, 1, (1984), 39–90.
8. J. Feijen, "Controlled drug delivery based on the use of synthetic polymers," Roven: XIV Meet. of French Polymer Group, November 1984.
9. M. Vergnaud-Trouin and J. M. Vergnaud, "Modélisation de la libération du principe actif à partir d'une matrice polymérique," *STP Pharma*, (Avril–Mai 1986), 410–415.
10. B. Focher, A. Marzetti, V. Sarto, P. L. Baltrane, and P. Carmitti, "Cellulosic materials: structure and enzymatic hydrolysis relationships," *J. Appl. Polym. Sci.*, 29, (1984), 3329–3338.
11. H. Fessi, J. P. Marty, F. Puisieux, and J. T. Carstensen, "Square root of time dependence of matrix formulations with low drug content," *J Pharma. Sci.*, 71, (1982), 749–752.
12. E. Touitou and M. Donbrow, "Drug release from non-disintegrating hydro-

philic matrices sodium salicylate as model drug," *Int. J. Pharm.*, 11, (1982), 355–364.

13. M. Nicklasson, A. Brodin, and L. O. Sundelof, "Studies of some characteristics of molecular dissolution kinetics from rotating discs," *Int. J. Pharm.*, 23 (1985), 97–108.
14. E. Teillaud et Y. Pourcelat-Roubeau, "Validation d'un nouveau modèle de libération in-vitro," *Labo Pharma Probl. Tech.*, 32, (1984), 279–283.
15. N. A. Peppas, R. Gurny, E. Doekler, and P. Buri, "Modelling of drug diffusion through swellable polymeric systems," *J. Membrane Sci.*, 7, (1980), 241–253.
16. C. Brossard, D. Lefort des Ylouses, D. Duchene, F. Puisieux, and J. T. Carstensen, "Dissolution of a soluble drug substance from vinyl polymer matrices," *J. Pharm. Sci.*, 72, (1983), 162–169.
17. N. A. Peppas and N. M. Franson, "The swelling interface number as a criterion for prediction of diffusional solute release mechanisms in swellable polymers," *J. Polym. Sci. Phys. Ed.*, 21, (1983), 983–997.
18. D. Wouessidjewe, C. Brossard, J. J. Goujon, F. Puisieux, and J. P. Devissaguet, "Preparation de microgranules de trinitrine à libération prolongée par pelliculage en milieu aqueux," *Labo Pharma Probl. Tech.*, 32, (1984), 772–777.
19. G. S. Kumar, V. Kalpagam, and U. S. Nandi, "Biodegradation of gelatin-polyethylacrylate copolymers," *J. Appl. Polym. Sci.*, 29, (1984), 3075–3085.
20. T. Higuchi, "Rate of release of medicaments from ointment bases containing drugs in suspension," *J. Pharm. Sci.*, 50 (1961), 874–875.
21. W.-I. Higuchi, N. F. A. Ho, and H. P. Merkle, "Design of oral drug delivery systems: past, present and future," *Drug, Dev. Ind. Pharm.*, 9, (1983), 1227–1239.
22. P. I. Lee, "Diffusional release of a solute from a polymeric matrix. Approximate analytical solutions," *J. Membr. Sci.*, 7, (1980), 255–275.
23. H. B. Hopfenberg and K. C. Hsu, "Swelling-controlled constant rate delivery systems," *Polym. Engng. Sci.*, 18, (1978) 1186–1191.
24. N. A. Peppas, "Release of bioactive agents from swellable polymers: theory and experiments," in *Recent Advances in Drug Delivery Systesm*," eds. J. M. Anderson and S. W. Kim (NY: Plenum Publishing Corp., (1984), pp. 279–289.
25. N. A. Peppas and S. Segot-Chicq, "Les dispositifs à libération controlée pour la délivrance des principes actifs médicamentaux. Modélisation des mécanismes diffusionels," *STP Pharm*, 1, (1985), 208–216.
26. D. Messadi and J. M. Vergnaud, "Simultaneous diffusion of benzyl alcohol into plasticized PVC and of plasticizer from polymer into liquid," *J. Appl. Polym. Sci.*, 26, (1981), 2315–2324.
27. D. Messadi, J. L. Taverdet, and J. M. Vergnaud, "Plasticizer migration from plasticized PVC into liquids. Effect of several parameters on the transfer," *I and EC Prod. Res. Develop.*, 22, (1983), 142–146.
28. J. L. Taverdet and J. M. Vergnaud, "Study of transfer process of liquid into and plasticizer out of plasticized PVC in using short test," *J. Appl. Polym. Sci.*, 29, (1984), 3391–3400.
29. J. L. Taverdet and J. M. Vergnaud, "Modelization of matter transfers between plasticized PVC and liquids in case of a maximum for liquid-time curves," *J. Appl. Polym. Sci.*, 31, (1986), 111–122.
30. A. Aboutaybi, J. Bouzon, and J. M. Vergnaud, "Modelling the process of drying

of plasticized PVC previously immersed in n-hexane," *Europ. Polym. J.* 25, (1989), 1013–1018.

31. A. Aboutaybi, H. David, B. Touchard, J. Bouzon, and J. M. Vergnaud, "Drying of plasticized PVC previously immersed into liquids," in *Drying of Solids*, ed. Mujumdar, Sakri, (1989).
32. A. Droin, C. Chaumat, M. Rollet, J. L. Taverdet, and J. M. Vergnaud, "Model of matter transfers between sodium salicylate-Eudragit matrix and gastric liquid," *Int. J. Pharmac.*, 27, (1985), 233–243.
33. I. Malley, J. Bardon, M. Rollet, J. L. Taverdet, and J. M. Vergnaud, "Modelling of controlled drug release in case of Carbopol-sodium salicylate matrix in gastric liquid," *Drug Develop. Ind. Pharm.*, 13, (1987), 67–81.
34. J. Y. Armand, F. Magnard, J. Bouzon, M. Rollet, J. L. Taverdet, and J. M. Vergnaud, "Modelling of the release of drug in gastric liquid from spheric galenic forms with Eudragit matrix," *Int. J. Pharm.*, 40, (1987), 33–41.
35. N. Farah, J. Bouzon, M. Rollet, J. L. Taverdet, and J. M. Vergnaud, "Dry emulsion: a sustained release form. Modelling of drug transfers in liquid," *Int. J. Pharm.*, 36, (1987), 81–88.
36. D. Messadi and J. M. Vergnaud, "A new approach to the study of plasticizer migration form PVC into methanol," *J. Appl. Polym. Sci.*, 26 (1981), 667–677.
37. D. Messadi and J. M. Vergnaud, "Plasticizer transfer from plasticized PVC into Ethanol-water mixtures," *J. Appl. Polym. Sci.*, 27, (1982), 3945–3955.
38. A. Eddine, A. Droin, J. L. Taverdet, and J. M. Vergnaud, "Effect of pH on drug release between Sodium salicylate-Eudragit compound and gastric liquid: modelling of the process," *Int. J. Pharm.*, 32, (1986), 143–150.
39. P. Magron, M. Rollet, J. L. Taverdet, and J. J. Vergnaud, "Spherical oral polymer-drug device with two polymers for constant drug delivery," *Int. J. Pharm.*, 38, (1987), 91–97.
40. N. Laghoueg, J. Paulet, J. L. Taverdet, and J. M. Vergnaud, "Oral polymer-drug devices with a core and an erodible shell for constant drug delivery," *Int. J. Pharm.*, 50, (1989), 133–139.
41. H. Liu, P. Magron, J. Bouzon, and J. M. Vergnaud, "Spherical dosage form with a core and shell. Experiments and modelling." *Int. J. Pharm.*, 45, (1988), 217–227.

Chapter Eleven

# Release of Active Agent in Agriculture

## 11-1 INTRODUCTION

A general problem exists in agriculture and in pharmaceutical therapy: how to control the release of active agent in the first case, and the delivery of drug in the second one.

Active agents (herbicides, pesticides, chemical fertilizer, and so on) are often used in agriculture, and the main problem is to prolong the duration of their action over a longer period of time. In fact, these agents can be swept along by water during rain, resulting in a high concentration of active agent in ground water provoking pollution, and a low activity of active agent on soil surface.

A simple device is considered in this chapter. The active agent (ethoprophos as pesticide, and anilin as simulating agent) is dispersed in a polymer (EVAc) available in small beads. So the matter transports through the polymer play a leading role.

The following problems are studied:

1. Test conditions simulating the release.
2. Agent-polymer device.
3. Modeling the process of release.
4. Study of various parameters of interest.

**Test conditions simulating the release.** It is difficult to do real experiments by measuring the propagation of the active agent through the soil medium. Moreover, there is no unique soil, but various soils with different properties.

A test is described, in order to simulate the process of release of active agent out of the system. This is a hard test for the agent-polymer system because the system is completely immersed in water, with a strong stirring. The amount of water as well as the weight of the system are determined, since the ratio of the volume of active agent and water has been found to be of high concern for the process of release [1–3].

**Active agent-polymer device.** Among the qualities that the device must have, some of them are worth noting:

- The polymer must be available in large quantities with a low price.
- Small beads of polymer are of interest in order to disperse the agent.
- The polymer device contains a large amount of active agent.

For all these reasons, EVAc (Atochem), a copolymer of ethylene-vinyl acetate, is selected for the polymer. The beads of polymer have a radius of about 2 mm and an average weight of 0.03 g.

Two agents are selected:

- Ethoprophos (Rhone-Poulenc) as active agent.
- Anilin, as simulating agent.

The device is prepared by immersing polymer beads into the pure active agent, and provoking absorption of this liquid by the polymer. The saturated polymer is then called the agent-polymer device.

**Modeling the process of release.** According to the general purpose of this book, the process of release (desorption) of active agent in water is studied as well as the process of absorption of this agent by the polymer. Mathematical models are built in order to describe the process of absorption and desorption under the test conditions.

**Parameters of interest.** As the polymer plays the leading role in the process, some parameters characterizing this polymer and the process of release are especially studied.

- the vinyl acetate percentage (VAc wt %) in EVAc copolymers.
- the melt index, a physical property related to the viscosity of molten polymer, which increases as the mean molecular weight of polymer decreases.

## 11-2 MODEL FOR MATTER TRANSPORTS BETWEEN EVAc AND LIQUID (ACTIVE AGENT, WATER)

### 11-2-1 Experimental

**Materials used**

- Polymer, beads of EVAc (Atochem), copolymers of ethylene vinyl acetate, with weight % VAc ranging from 0 to 40. The beads have a radius of about 0.2 cm and an average weight of 0.03 g.
- Liquid, the active agent is Ethoprophos (Rhone-Poulenc Agroch). It has a slight solubility in water.
- Anilin is used for simulating the active agent [4].

**Apparatus for testing.** A closed 500 ml flask is used, with controlled temperature and controlled stirring in order to get uniform concentration in the solution during desorption. Absorption experiments are carried out by soaking the pure beads in pure active agent (Ethoprophos, or anilin). Desorption is obtained by immersing in pure water the presaturated bead. The amount of water is of 200 ml for one bead of about 0.03 g.

**Analytical measurements.** Absorption—The bead is weighed at intervals, so as to determine the kinetics of absorption of liquid.

Desorption—At intervals, the bead is weighed and a 1-ml sample of aqueous solution is taken. The concentration of Ethoprophos is measured by gas chromatography equipped with a photometric detector sensitive to phosphorus, after dilution in toluene. The concentration of anilin in water is measured by UV-spectrometry (Beckman DB-G calibrated at 196 nm).

### 11-2-2 Theoretical

**Description of the process.** From the matter balance calculations, it is shown that the process of desorption is very simple with only the transport of active agent, as water does not enter EVAc.

A swelling of the bead is observed during absorption, especially for high VAc percentage and for low value of the melting index. A shrinkage occurs during desorption.

**Assumptions.** The following assumptions are made:

1. Beads are spherical in shape.
2. Absorption is controlled by transient diffusion with constant diffusivity.
3. The concentration of active agent is uniform through the bead at the

end of absorption, after infinite time, as well as at the beginning of absorption.

4. The desorption process consists of only the transport of active agent out of the bead. Water does not enter the polymer.
5. Desorption is controlled by transient diffusion with constant diffusivity.
6. During desorption, the volume of water with respect to that of the bead is very large, and the concentration of active agent is very low in water.
7. During absorption and desorption, the concentration of active agent attains the equilibrium value on the surface of the bead as soon as the sample is immersed into the liquid (active agent or water).
8. A frame of reference with the polymer is taken in spite of the change in volume during absorption and desorption, and the radius of the bead is constant and equal to the radius of the original empty bead.

**Mathematical treatment.** The change of concentration of active agent in the spherical bead is described by Fick's laws.

$$F = -D \cdot \frac{\partial C}{\partial r} \tag{11-1}$$

$$\frac{\partial C}{\partial t} = D \cdot \left[ \frac{\partial^2 C}{\partial r^2} + \frac{2}{r} \frac{\partial C}{\partial r} \right] \tag{11-2}$$

with constant diffusivity.

Initial and boundary conditions are:

| | | | | | |
|---|---|---|---|---|---|
| Initial | $t = o$ | $o \leq r < R$ | $C = C_{in}$ | polymer | (11-3) |
| Boundary | $t > o$ | $r = R$ | $C = C_{eq}$ | surface | (11-4) |

where $R$ is the radius of the bead, and $C_{in}$ and $C_{eq}$ are the concentrations at the beginning and at the end of the process (absorption, desorption), respectively.

In the case of constant diffusivity, the kinetics of matter transported can be described by the series:

$$\frac{M_\infty - M_t}{M_\infty} = \frac{6}{\pi^2} \cdot \sum_{n=1}^{\infty} \frac{1}{n^2} \cdot \exp\left( - \frac{n^2 \pi^2}{R^2} Dt \right) \tag{11-5}$$

where $M_t$ and $M_\infty$ are the amount of liquid transported at time $t$ and after infinite time, respectively, and $n$ is an integer.

In the case of short times, when the amount of matter transported is low:

$$\frac{M_t}{M_\infty} < 0.2 \tag{11-6}$$

This amount of matter transported, $M_t$, is proportional to the square root of time.

$$\frac{M_t}{M_\infty} = \frac{6}{R}\left[\frac{D \cdot t}{\pi}\right]^{0.5} \tag{11-7}$$

### 11-2-3 Results with Anilin as Liquid

**Diffusivity and capacity of absorption.** Diffusivity is determined either for absorption of the liquid or for desorption in water. The amount of liquid previously absorbed by the EVAc bead is completely desorbed in water.

***Absorption.*** As shown in Figure 11–1, the amount of anilin absorbed is proportional to the square root of time, for short times when $M_t/M_\infty < 0.2$.

The diffusivity is calculated by using equation (11-7). Various values are obtained for different weight % VAc in polymer beads (Table 11–1).

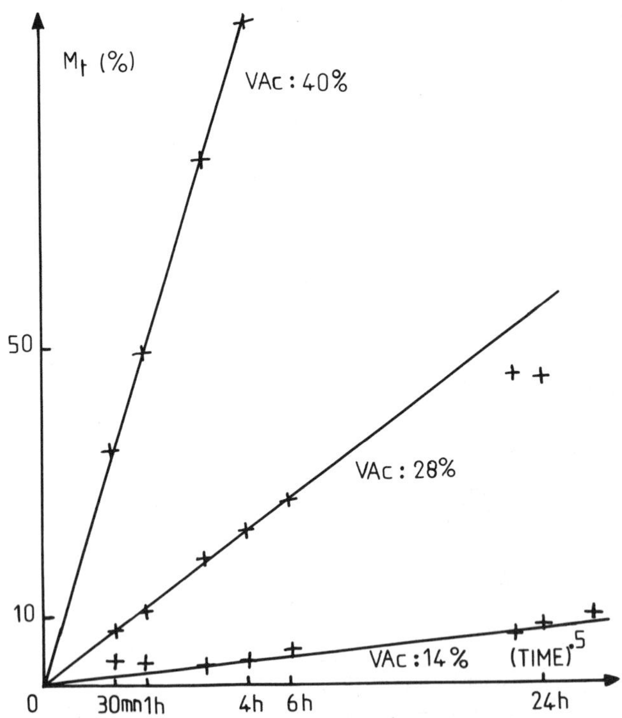


**Figure 11–1** Amount of anilin absorbed as a function of the square root of time, by 100 g EVAc, with various % VAc, at 25°C. (Reprinted from *J. Control. Release*, 8, (1988), 151–156, H. David, J. Bouzon, and J. M. Vergnaud, ref. 4, by courtesy of Elsevier Science Publ.)

**TABLE 11–1** Diffusivity and $M_\infty$, for Absorption of Anilin

| Diffusivity and $M_\infty$ | | | | Temperature (°C) |
|---|---|---|---|---|
| VAc (wt %) | 14 | 28 | 40 | |
| $D \times 10^8$ (cm²/s) | 1.5 | 2.7 | 3.5 | 17 |
| $M_\infty$ (%) | 11 | 45 | 160 | |
| $D \times 10^8$ (cm²/s) | 2 | 5.5 | 4.5 | 25 |
| $M_\infty$ (%) | 12 | 50 | 200 | |
| $D \times 10^8$ (cm²/s) | 4.5 | 6 | 5 | 35 |
| $M_\infty$ (%) | 18 | 70 | 190 | |

The capacity of absorption, or the amount of liquid absorbed after infinite time, varies also with % VAc percentage in polymer (Table 11–1).

***Desorption.*** The values of diffusivity in the case of desorption are shown for various VAc percentages in polymer and various temperatures (Table 11–2) and their variation is illustrated in Figure 11–2.

From the values of the amount of anilin desorbed (Table 11–2), it is clear that almost the whole quantity of anilin previously absorbed is desorbed in water. The variation of the capacity of absorption with the VAc percentage in polymer is considered for various temperatures (Figure 11–3).

**Validity of the model.** The validity of the model is determined by comparing the kinetics of absorption and desorption, as they are obtained either by experiments or by calculation, using the model with equation (11-5).

As shown in Figure 11–4 for absorption and in Figure 11–5 for desorption, a good agreement exists between these values, proving the validity of the model.

**Effect of VAc percentage and temperature.** The effect of these parameters is relevant either for absorption or desorption. The following results are drawn:

**TABLE 11–2** Diffusivity and $M_\infty$, for Desorption of Anilin

| Diffusivity and $M_\infty$ | | | | Temperature (°C) |
|---|---|---|---|---|
| VAc (wt %) | 14 | 28 | 40 | |
| $D \times 10^8$ (cm²/s) | 6 | 7 | 7 | 12 |
| $M_\infty$ (%) | 8 | 43 | 158 | |
| $D \times 10^8$ (cm²/s) | 5 | 7.5 | 20 | 25 |
| $M_\infty$ (%) | 12 | 48 | 195 | |
| $D \times 10^8$ (cm²/s) | 6 | 18 | 30 | 35 |
| $M_\infty$ (%) | 18 | 70 | 190 | |

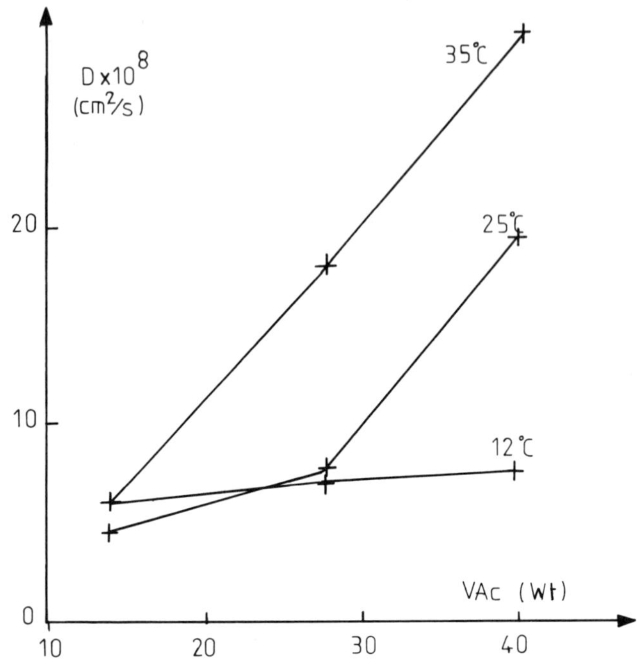


**Figure 11–2** Diffusivity for desorption of anilin as a function of VAc percentage, for various temperatures. (Reprinted from *J. Control. Release*, 8, (1988), 151–156, H. David, J. Bouzon, and J. M. Vergnaud, ref. 4, by courtesy of Elsevier Science Publ.)

1. The higher the VAc percentage in the polymer, the higher the desorption capacity which parallels the results found for absorption (Figures 11–4 and 11–5).
2. The effect of temperature on absorption and desorption is also of importance, a higher temperature being responsible for a higher rate and higher capacity (Figures 11–2 and 11–3).

### 11-2-4 Results with Ethoprophos as Liquid

**Diffusivity and capacity of absorption.** The amount of matter transported varies linearly with the square root of time, as in the case of anilin, for short times when $M_t/M_\infty < 0.2$.

In the same way, all the quantity of Ethoprophos previously absorbed is desorbed in water, and the capacity of absorption characterizes the polymer. The diffusivity and especially the absorption varies with the VAc percentage in polymer (Table 11–3).

**Validity of the model.** The validity of the model is tested by comparing experimental and theoretical kinetics of absorption and desorption.

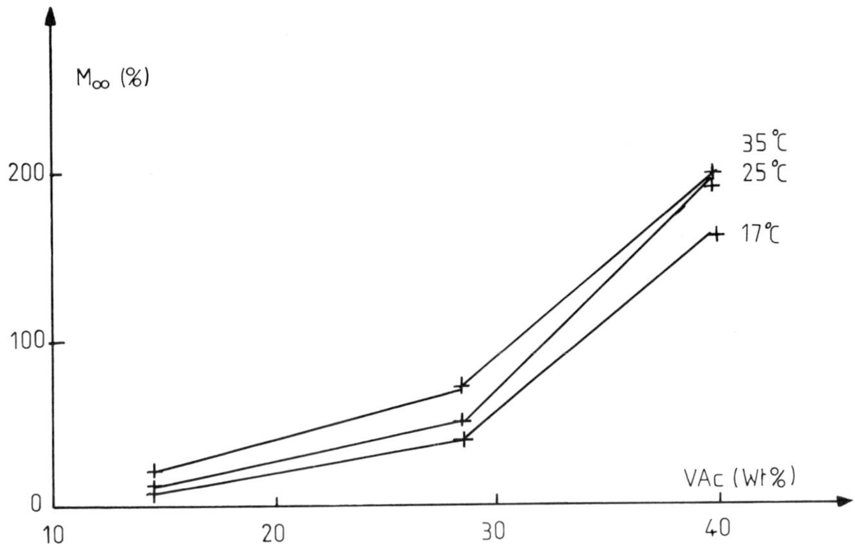


**Figure 11–3** Capacity of absorption (and desorption of anilin by EVAc (100 g) as a function of VAc percentage at various temperatures. (Reprinted from *J. Control. Release*, 8, (1988), 151–156, H. David, J. Bouzon, and J. M. Vergnaud, ref. 4, by courtesy of Elsevier Science Publ.)

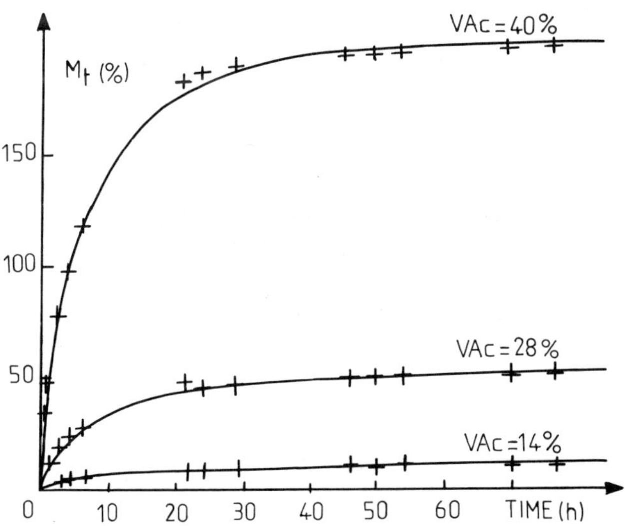


**Figure 11–4** Validity of the model, for absorption of anilin by EVAc with various VAc percentages at 25°C ($M_t$: amount of anilin absorbed by 100 g EVAc). (Reprinted from *J. Control. Release*, 8, (1988), 151–156, H. David, J. Bouzon, and J. M. Vergnaud, ref. 4, by courtesy of Elsevier Science Publ.)

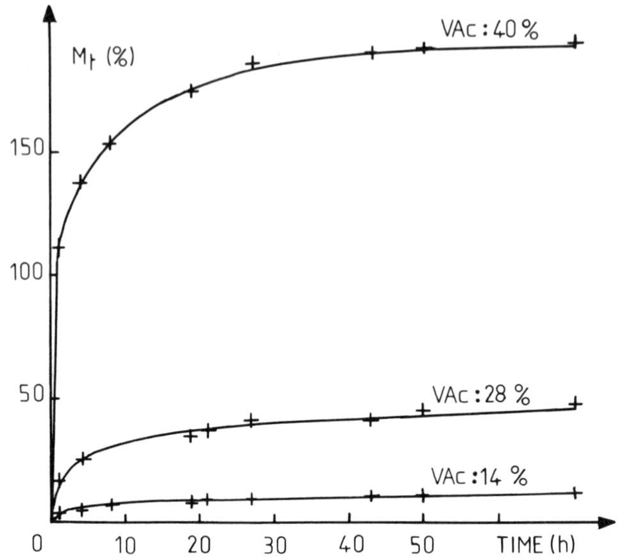


**Figure 11–5** Validity of the model, for desorption of anilin from EVAc with various VAc percentages at 25°C. ($M_t$: amount of anilin desorbed by 100 g EVAc). (Reprinted from *J. Control. Release*, 8, (1988), 151–156, H. David, J. Bouzon, and J. M. Vergnaud, ref. 4, by courtesy of Elsevier Science Publ.)

As shown in Figure 11–6 in case of absorption and in Figures 11–7 and 11–8 for the desorption, the model describes the process with accuracy, and for various values of VAc percentages ranging from 0 to 40.

**Effect of VAc and melt index.** Two parameters characterize the polymer, the VAc percentage and the melt index, the second parameter being connected with the mean molecular weight.

The effect of the VAc percentage on the kinetics of liquid absorbed and desorbed is illustrated in Figures 11–6, 11–7, and 11–8. In fact, the diffusivity varies slightly with VAc percentage, while the capacity of absorption is highly increased with the VAc percentage, as shown in Figure 11–9.

**TABLE 11–3** Diffusivity ($cm^2/s$) and $M_\infty$, for Various VAc Percentages

| % VAc | 0 | 5 | 14 | 24 | 28 | 33 | 40 |
|---|---|---|---|---|---|---|---|
| $M_\infty$ (%) | 2 | 2.5 | 8 | 30 | 50 | 110 | 260 |
| $D \times 10^8$ absorption | 0.2 | 1.2 | 4 | 1 | 0.7 | 0.8 | 1 |
| $D \times 10^8$ desorption | 10 | 1 | 2 | 2 | 1 | 4 | 6 |

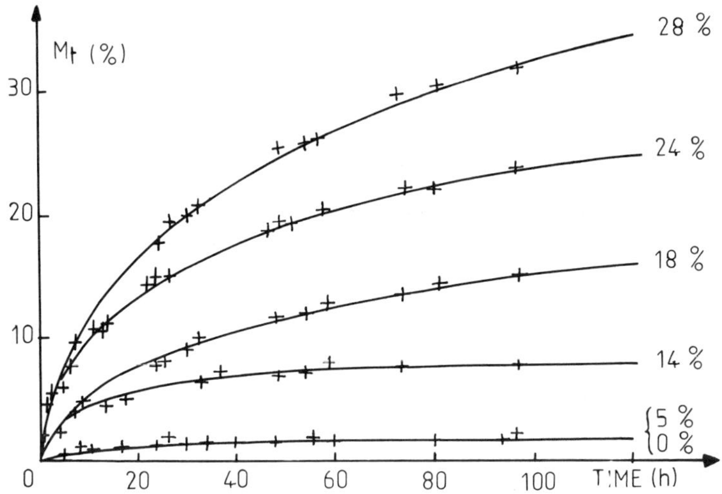


**Figure 11–6** Validity of the model, for absorption of Ethoprophos by EVAc with various VAc percentages at 25°C ($M_t$: amount of Ethoprophos absorbed by 100 g EVAc). (Reprinted from *Europ. Polym. J.*, 25, (1989), 1007–1011, H. David, J. Bouzon, and J. M. Vergnaud, ref. 5, by courtesy of Pergamon Press.)

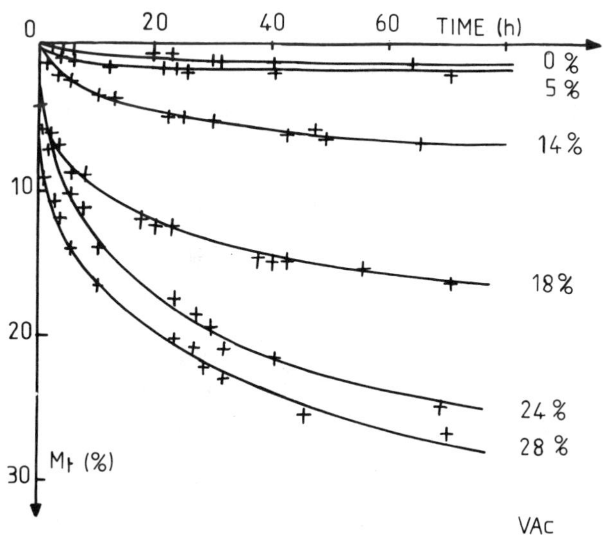


**Figure 11–7** Validity of the model, for desorption of Ethoprophos by EVAc, with various VAc percentages at 25°C ($M_t$: amount absorbed by 100 g EVAc). (Reprinted from *Europ. Polym. J.*, 25, (1989), 1007–1011, H. David, J. Bouzon, and J. M. Vergnaud, ref. 5, by courtesy of Pergamon Press.)

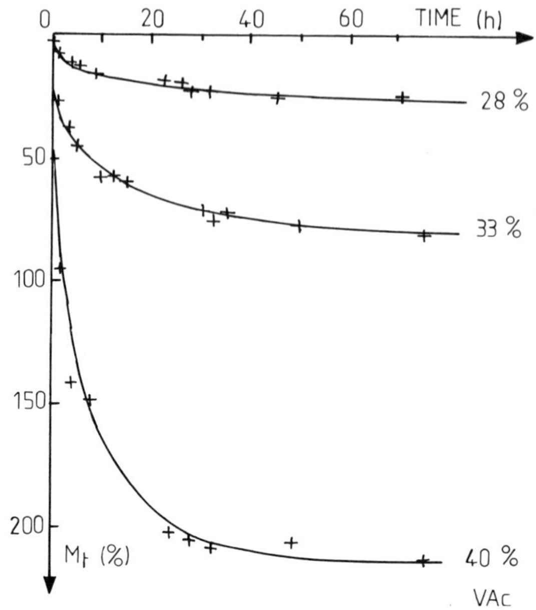


**Figure 11–8** Validity of the model, for desorption of Ethoprophos by EVAc with various % VAc at 25°C ($M_t$: amount absorbed by 100 g EVAc). (Reprinted from *Europ. Polym. J.*, 25 (1989), 1007–1011, H. David, J. Bouzon, and J. M. Vergnaud, ref. 5, by courtesy of Pergamon Press.)

The value of the melt index of polymer is also of importance to the capacity of absorption, as indicated in Figure 11–10. The effect of this parameter on the diffusivity is very low.

### 11-2-5 Conclusions

In the case of absorption and desorption of liquids as different as anilin and Ethoprophos by EVAc polymer, some conclusions are drawn:

1. The process is simple, either for absorption or desorption. The whole quantity of liquid previously absorbed is completely desorbed. Moreover, as water does not enter the polymer, the transport of active agent only takes place during desorption.
2. The process is controlled by transient diffusion with constant diffusivity, and modeling is obtained by using an analytical solution of Fick's equations.
3. Three parameters are of concern for the process, and especially for the capacity of absorption: the temperature, the VAc percentage, and the value of the melt index.

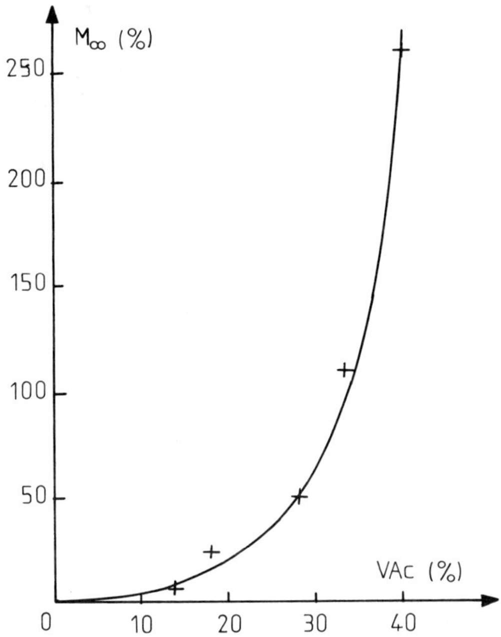


**Figure 11–9** Capacity of absorption of EVAc for Ethoprophos as a function of % VAc, at 25°C. (Reprinted from *Europ. Polym. J.*, 25, (1989), 1007–1011, H. David, J. Bouzon, and J. M. Vergnaud, ref. 5, by courtesy of Pergamon Press.)

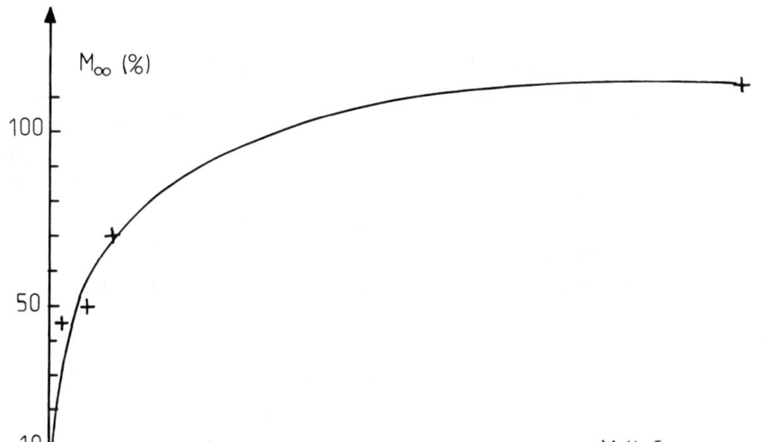


**Figure 11–10** Capacity of absorption of EVAc for Ethoprophos as a function of the melt index, at 25°C and % VAc = 28. (Reprinted from *Europ. Polym. J.*, 25, (1989), 1007–1011, H. David, J. Bouzon, and J. M. Vergnaud, ref. 5, by courtesy of Pergamon Press.)

## 11-3 MODELING THE TRANSPORTS WITH CORE AND SHELL SYSTEMS MADE OF EVAc AND OF EVAc AND SILICONE

In the previous chapter, the process of absorption and desorption was studied by considering a simple spherical bead. The release of preabsorbed active agent is controlled by transient diffusion. This means that the rate of transport is high at the beginning (with a vertical tangent for kinetic curves at the origin of time) and then decreases regularly with time.

In order to control the release of active agent, more sophisticated devices are prepared and studied. They consist of a core and shell, and in this chapter the core and shell are made of EVAc with various VAc contents.

The core and shell must exhibit specific properties. The core must have a high capacity of absorption for active agent and a rather low diffusivity in the case of desorption. The shell must have a good flexibility, to be able to follow the swelling or shrinkage of the core and a low rate of transport of active agent, which can be obtained by a low diffusivity and/or a low capacity of absorption.

Two examples are given when:

- the core is made of EVAc with 28% VAc and the shell with EVAc with 14% VAc [6, 7].
- the core is made of EVAc with 40% VAc and the shell with silicone elastomer [8].

### 11-3-1 Experimental

**Preparation of samples.** EVAc samples composed of a core and shell are prepared as follows: EVAc beads (core) with a high capacity of absorption (28% VAc by wt) are weighed and immersed in a gel of EVAc (with 14% VAc by wt) in trichlorethylene at 40°C. The sample is then extracted and dried at room temperature. The thickness of the shell surrounding the core is determined by selecting the proper trichlorethylene percent as well as the temperature of the gel.

Samples with a core made of EVAc and a shell made of silicone are prepared as follows: The EVAc bead used as the core (40% VAc) is immersed in a mixture of silicone (RTV 501, Rhone-Poulenc) and catalyst (XY70, 0.5 wt %). After 10 min at room temperature, a layer of polymerized silicone surrounds the core forming a shell. The thickness is a function of the degree of polymerization of the silicone elastomer when the core is immersed.

**Test for absorption and desorption of active agent.** For absorption, the beads are soaked in pure active agent (Ethoprophos, anilin) in a flask at 25°C with gentle stirring. The beads are weighed at intervals, for determining the kinetics of absorption.

To test desorption, the beads previously saturated in active agent are immersed in water (200 ml) in a flask at 25°C, with controlled stirring. Measurements are made at intervals. The beads are weighed, and a small amount (1 ml) of the aqueous solution is extracted for analysis.

The concentration of anilin is determined by UV spectrometry (Beckman DB-G calibrated at 196 nm). The concentration of Ethoprophos is measured by using gas chromatography equipped with a photometric detector sensitive to phosphorus, after dilution in toluene.

### 11-3-2 Theoretical

**Description of the process.** As shown in the case of the core alone, the process of matter transport is very simple either for absorption or desorption.

- During desorption of the active agent, only this matter transport is observed.
- The process is controlled by transient diffusion.
- A swelling and shrinkage of the sample take place during absorption and desorption.

**Assumptions.** The following assumptions can be drawn from the knowledge of the process.

1. The device consists of two concentric spheres of radius $R_1$ and $R_2$. The core is made of highly permeable EVAc (28 wt % VAc) and the shell of low permeability EVAc (14 wt % VAc).
2. In the case of desorption in water, only the transport of active agent is observed.
3. The transport of active agent throughout the device is controlled by transient diffusion either for absorption or in desorption, with constant diffusivities.
4. The concentration of active agent on the external surface of the device reaches the equilibrium value as soon as the device is in contact with the liquid (pure liquid for absorption, water for desorption).
5. A discontinuity exists in the transport properties at the core-shell interface (diffusivity, capacity of absorption).
6. For calculation, a frame of reference is taken for the device. The dimensions of the bead are those of the empty bead both in absorption and in desorption.

**Mathematical treatment.** The change of concentration of liquid in the device is described by Fick's law:

$$\frac{\partial C}{\partial t} = \frac{1}{r^2} \cdot \frac{\partial}{\partial r} \left[ D \cdot r^2 \cdot \frac{\partial C}{\partial r} \right] \tag{11-8}$$

where $D$ is the diffusivity, $C$ is the concentration of liquid, and $r$ the radius of the spherical surface considered.

Initial and boundary conditions are:

| | | | | | |
|---|---|---|---|---|---|
| Absorption | $t = 0$ | $0 \leq r < R_2$ | $C = 0$ | sample | (11-9) |
| | $t > 0$ | $r = R_2$ | $C = C_{eq}$ | external surface | (11-10) |
| Desorption | $t = 0$ | $0 \leq r < R_1$ | $C = C_{eq}$ | core | (11-11) |
| | | $R_1 < r < R_2$ | $C = C_{eq}$ | shell | |
| | $t > 0$ | $r = R_2$ | $C = 0$ | external surface | (11-12) |

For short times, when the amount of matter transported is low, the square root of time equation is considered:

$$\frac{M_t}{M_\infty} = \frac{6}{R} \left[ \frac{D \cdot t}{\pi} \right]^{0.5} \quad \text{for} \quad \frac{M_t}{M_\infty} < 0.1 \tag{11-7}$$

The kinetics of absorption and desorption for each EVAc polymer used for the core and shell can be described by the well-known equation, when these polymers are spherical in shape, and when the diffusivity is constant.

$$\frac{M_\infty - M_t}{M_\infty} = \frac{6}{\pi^2} \cdot \sum_{n=1}^{\infty} \frac{1}{n^2} \cdot \exp \left[ - \frac{n^2 \pi^2}{R^2} Dt \right] \tag{11-5}$$

where $M_t$ and $M_\infty$ are the amount of matter transported at time $t$, and after infinite time; $R$ is the radius of sphere, $n$ is an integer; and $D$ is the constant diffusivity.

***Modeling: Numerical Analysis.*** In the general case with concentration-dependent diffusivities in the core and shell, no analytical solution can be found, and the problem is then solved by using a numerical method with finite differences. The device is divided into $N$ spheres with a thickness between them of

$$\Delta r_1 = \frac{R_1}{N_1} \qquad \text{for the core} \tag{11-13}$$

$$\Delta r_2 = \frac{R_2 - R_1}{N_2} \qquad \text{for the shell} \tag{11-14}$$

Each concentric sphere is associated with the integer $n$ (Figure 11–11) with $N = N_1 + N_2$.

$$r = n \cdot \Delta r_1 \qquad 0 \le n \le N_1 \qquad \text{core} \tag{11-15}$$

$$r = R_1 + (n - N_1) \cdot \Delta r_2 \qquad N_1 \le n \le N \qquad \text{shell} \tag{11-16}$$

The balance of matter is calculated for each concentric spherical layer located between the radii $r - \Delta r/2$ and $r + \Delta r/2$ . From these calculations, the new concentrations after the elapse of time $\Delta t$ is obtained in terms of the concentrations at the same place and adjacent places.

*Core,* with $1 \le n \le N_1 - 1$

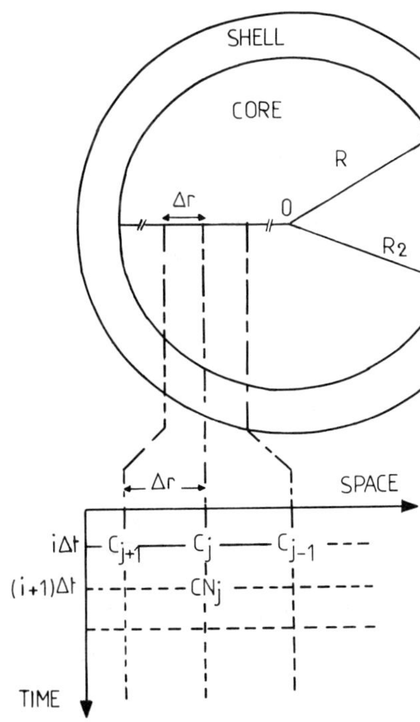


**Figure 11–11** Scheme of spherical sample with core and shell.

The new concentration $CN_n$ is given as a function of the preceding concentrations $C_n$:

$$CN_n = C_n + \frac{\Delta t}{\left(n^2 + \frac{1}{12}\right)\cdot(\Delta r_1^2)} [J_{n-1/2} - J_{n+1/2}] \tag{11-17}$$

with

$$J_{n-1/2} = \left(n - \frac{1}{2}\right)^2 \cdot D_{n-1/2} \cdot (C_{n-1} - C_n) \tag{11-18}$$

while the diffusivity at position $n - \frac{1}{2}$ can be obtained in terms of concentration by one of the equations shown earlier:

$$D_n = D_o \cdot \exp \cdot Cte \cdot C_n \tag{11-19}$$

$$D_n = D_o \cdot \exp\ Cte(C - C_n) \tag{11-19'}$$

$$D_n = D_o \cdot \exp \frac{Cte}{C_n} \tag{11-19''}$$

*Middle of core,* with $n = 0$

The matter balance is calculated within the sphere of radius $\Delta r_1/2$:

$$CN_o = C_o + 24 \frac{\Delta t}{(\Delta r_1)^2} \cdot J_{1/2}$$

*Shell,* with $N_1 + 1 \leq n \leq N - 1$

### 11-3-3 Results

Three points are of interest:

1. The study of the process of matter transports for each pure material (EVAc with various compositions, silicone elastomers), and measurements of diffusivities and capacity of absorption.
2. The validity of the model, by considering the kinetics of absorption and desorption.
3. The study of parameters concerned with the shell: thickness, diffusivity, capacity of absorption.

## Diffusivity and capacity of absorption

***Diffusivity.*** Straight lines are obtained by plotting the amount of liquid absorbed and desorbed as a function of the square root of time, by considering the polymers: EVAc with various VAc percentages and silicone elastomer, and various liquids such as anilin and Ethoprophos. Some of these lines are shown in the case of anilin and EVAc with 28 % VAc and 14 % VAc (Figure 11–12) [6].

The values of diffusivity are given in various cases as well as the capacity of absorption in the case of anilin with the various polymers (Table 11–4).

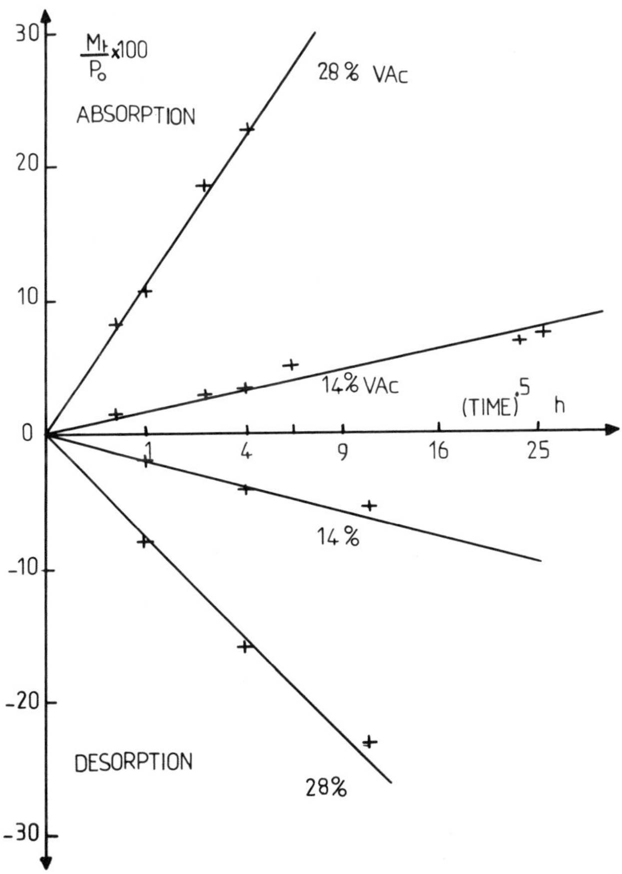


**Figure 11–12** Amount of anilin absorbed and desorbed as a function of the square root of time, by spherical samples of EVAc with 14% VAc and 28% VAc ($P_o$ is the weight of empty bead). (Reprinted from *Europ. Polym. J.*, 25, (1989), 89–94, H. David, J. Bouzon, and J. M. Vergnaud, ref. 6, by courtesy of Pergamon Press.)

**TABLE 14–4** Diffusivity ($D \times 10^7$ cm²/s) and $M_\infty$ for Anilin

| Materials | $D$ abs $\times 10^7$ | $D$ des $\times 10^7$ | $M_\infty$ abs | $M_\infty$ des |
|---|---|---|---|---|
| EVAc (40% VAc) | 1 | 2 | 120 | 120 |
| EVAc (28% VAc) | 1.2 | 3 | 45 | 45 |
| EVAc (14% VAc) | 0.6 | 0.4 | 15 | 15 |
| Silicone elastomers | 25 | 130 | 1.5 | 1.5 |

***Capacity of Absorption.*** The amount of liquid absorbed is completely desorbed in water, in the case of anilin and Ethoprophos with the various polymers, as shown in Table 11–4 for anilin. Two facts are worth noting:

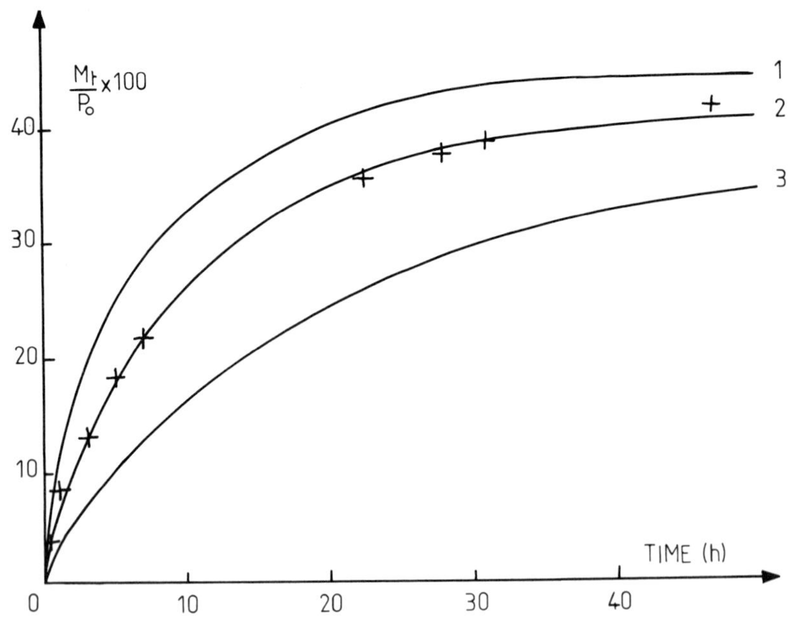


**Figure 11–13** Kinetics of absorption of anilin by various samples: 1: pure EVAc bead with 28% VAc (R = 0.191 cm); 2: device with core (EVAc with 28% VAc, $R$ = 0.191 cm) and shell (EVAc with 14% VAc and $\Delta R$ = 0.007 cm); 3: device with core (EVAc with 28% VAc, $R$ = 0.191 cm) and shell (EVAc with 14% VAc and $\Delta R$ = 0.018 cm); $M_t$: weight of anilin absorbed; $P_o$: weight of empty device. (Reprinted from *Europ. Polym. J.*, 25, (1989), 89–94, H. David, J. Bouzon, and J. M. Vergnaud, ref. 6, by courtesy of Pergamon Press.)

1. The capacity of absorption increases largely with the VAc percentage in EVAc copolymers.
2. In the case of silicone elastomer, the diffusivity is very high, while the capacity of absorption is low.

**Validity of the model.** The validity of the model is tested by comparing the experimental and theoretical kinetics for absorption of the liquid and for the desorption of liquid in water. As shown in Figure 11–13 for absorption, and in Figure 11–14 for desorption of anilin, a good agreement is obtained between these results. Three curves are drawn:

1. Curve 1 obtained with pure EVAc with 28% VAc.
2. Curve 2 obtained with the device consisting of a core made of EVAc

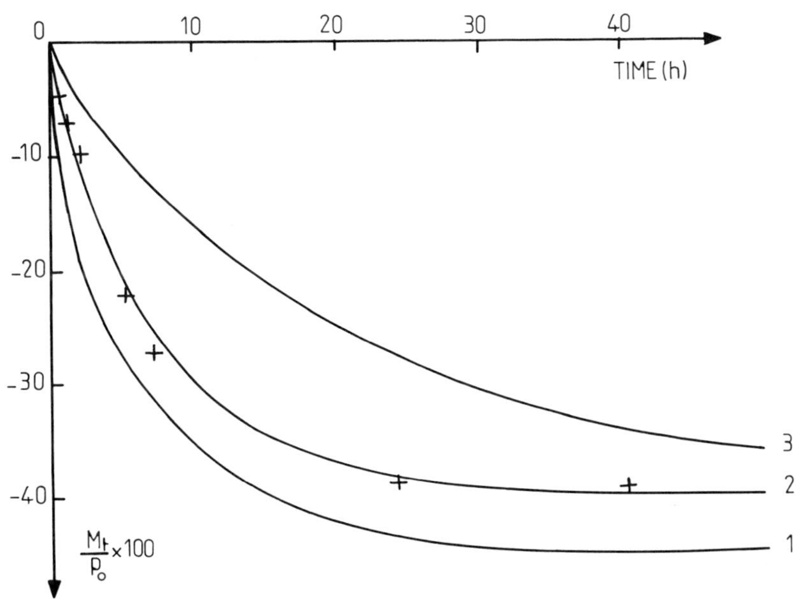


**Figure 11–14** Kinetics of desorption of anilin by various samples: 1: pure EVAc bead with 28% VAc ($R$ = 0.191 cm); 2: device with core (EVAc with 28% VAc, $R$ = 0.191 cm) and shell (EVAc with 28% VAc, $\Delta R$ = 0.007 cm); 3: device with core (EVAc with 28% VAc, $R$ = 0.191 cm) and shell (EVAc with 14% VAc, $\Delta R$ = 0.018 cm); $M_t$: weight of anilin absorbed, $P_o$: weight of empty device. (Reprinted from *Europ. Polym. J.*, 25, (1989), 89–94, H. David, J. Bouzon, and J. M. Vergnaud, ref. 6, by courtesy of Pergamon press.)

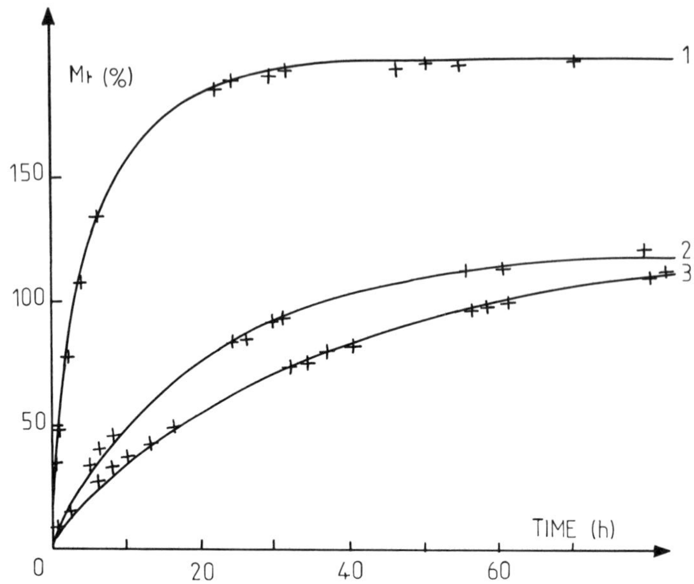


**Figure 11–15** Kinetics of absorption of anilin by various samples: 1: pure EVAc bead with 40% VAc ($R$ = 0.17 cm); 2: device with core (EVAc with 40% VAc, $R$ = 0.17 cm) and shell (silicone elastomer RTV 501, $\Delta R$ = 0.07 cm); 3: device with core (EVAc with 40% VAc, $R$ = 0.17 cm) and shell (silicone elastomer RTV 501, $\Delta R$ = 0.122 cm). (Reprinted from *Europ. Polym. J.*, 25, (1989), 939–945, H. David, J. Bouzon, and J. M. Vergnaud, ref. 8, by courtesy of Pergamon Press.)

with 28% VAc ($R_1$ = 0.191 cm) and a shell with 14% VAc ($\Delta R$ = 0.007 cm).

3. Curve 3 obtained with the device consisting of a core made of EVAc with 28% VAc ($R_1$ = 0.191 cm) and a shell with 14% VAc ($\Delta R$ = 0.018 cm).

The validity of the model is also tested in the case of anilin and a device consisting of pure EVAc with 40% VAc for the core and of silicone elastomer (RTV 501) for the shell. Three curves are obtained for absorption (Figure 11–15) and for desorption (Figure 11–16) [8].

1. Curve 1 with pure EVAc with 40% VAc ($R$ = 0.17 cm).
2. Curve 2 with the device consisting of a core made of EVAc with 40% VAc ($R_1$ = 0.17 cm) and a shell with silicone elastomers: RTV 501 ($\Delta R$ = 0.07 cm).

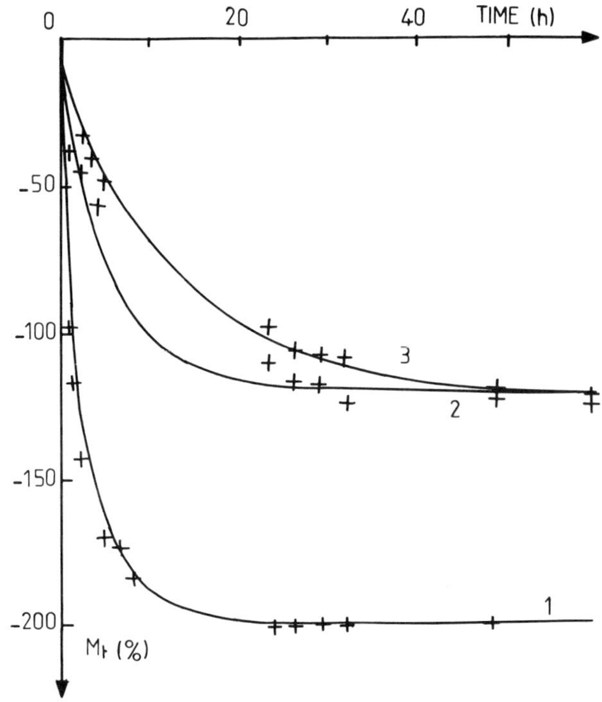


**Figure 11–16** Kinetics of desorption of anilin by various samples: 1: pure EVAc bead with 40% VAc ($R$ = 0.17 cm); 2: device with core (EVAc with 40% VAc, $R$ = 0.17 cm) and shell (silicone elastomer RTV 501, $\Delta R$ = 0.07 cm); 3: device with core (EVAc with 40% VAc, $R$ = 0.17 cm) and shell (silicone elastomer RTV 501, $\Delta r$ = 0.122 cm). (Reprinted from *Europ. Polym. J.*, 25, (1989), 939–945, H. David, J. Bouzon, and J. M. Vergnaud, ref. 8, by courtesy of Pergamon Press.)

3. Curve 3 with the device consisting of a core made of EVAc with 40% VAc ($R_1$ = 0.17 cm) and of shell made of silicone elastomer RTV 501 ($\Delta R$ = 0.122 cm).

The profiles of concentration are drawn during desorption as they are obtained from calculation with the model, either in the case of a shell made of EVAc (Figure 11–17) and in the case of a shell made of silicone elastomer (Figure 11–18). The device used in Figure 11–17 consists of a core (EVAc with 28% VAc and $R_1$ = 0.191 cm) and of a shell (EVAc with 14% VAc and $\Delta r$ = 0.007 cm). The device used in Figure 11–18 consists of a core (EVAc with 40% VAc and $R_1$ = 0.175 cm) and of a shell (silicone elastomer RTV 501 and $\Delta R$ = 0.122 cm).

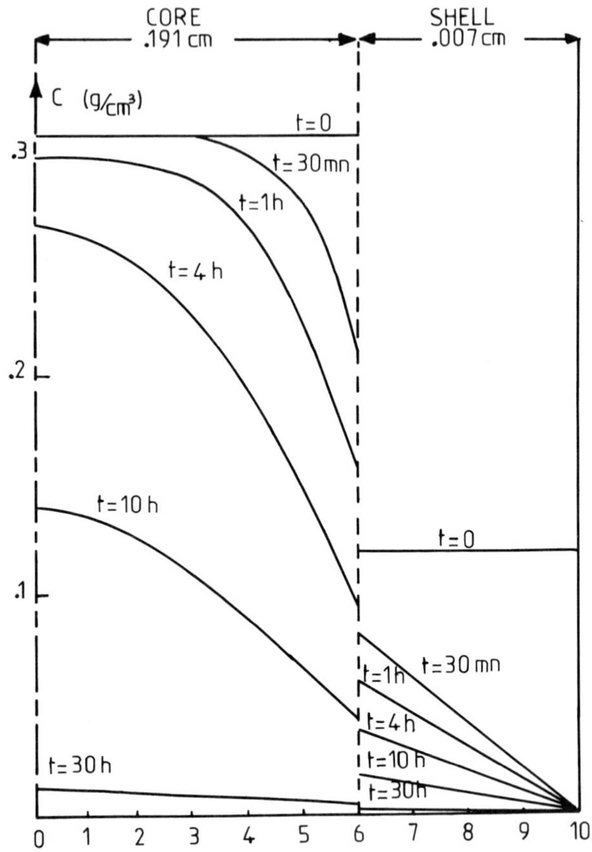


**Figure 11–17** Profiles of concentration of anilin during desorption, at various times. Device with core and shell made of EVAc. Core with 28% VAc and $R_1 = 0.191$ cm. Shell with 14% VAc and $\Delta r = 0.007$ cm. (Reprinted from *Europ. Polym. J.*, 25, (1989), 89–94, H. David, J. Bouzon, and J. M. Vergnaud, ref. 6, by courtesy of Pergamon Press.)

**Effect of parameters on the kinetics of desorption.** Three parameters characterizing the shell are of interest:

- the thickness of the shell
- the capacity of absorption, $M_\infty$
- the diffusivity of liquid

The effect of these parameters on the kinetics of desorption of anilin is determined by using two different cores.

***Core Made of EVAc with 28% VAc and $R_1 = 0.191$ cm.*** The following three figures are obtained with this core and various thicknesses of the shell: 0.007 cm (Figure 11–19); 0.018 cm (Figure 11–20); 0.05 cm (Figure 11–21) and various characteristics for the polymeric shell:

- Curve 1: no shell
- Curve 2: shell with $D = 4 \times 10^{-8}$ (cm²/s) and $M_\infty = 15\%$
- Curve 3: shell with $D = 10^{-8}$ and $M_\infty = 15\%$
- Curve 4: shell with $D = 2 \times 10^{-9}$ and $M_\infty = 15\%$

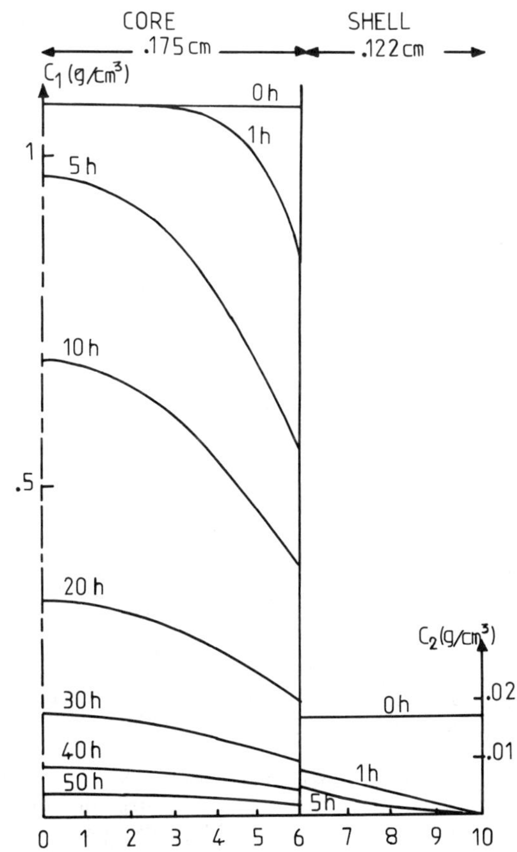


**Figure 11–18** Profiles of concentration of anilin during desorption, at various times. Device with core (made of EVAc with 40% VAc and $R_1 = 0.175$ cm) and shell (made of silicone elastomer with $\Delta R = 0.122$ cm). (Reprinted from *Europ. Polym. J.*, 25, (1989), 939–945, H. David, J. Bouzon, and J. M. Vergnaud, ref. 8, by courtesy of Pergamon Press.)

- Curve 5: shell with $D = 2 \times 10^{-9}$ and $M_\infty = 9.4\%$
- Curve 6: shell with $D = 2 \times 10^{-9}$ and $M_\infty = 5\%$

Of course, the thickness of the shell plays a significant role for controlling the release of anilin. However, the values of the diffusivity and capacity of absorption of the shell are certainly the relevant factors.

***Core Made of EVAc with 40% VAc and $R_1$ = 0.17 cm.*** The following six curves drawn in Figure 11–22 are obtained for various values of the

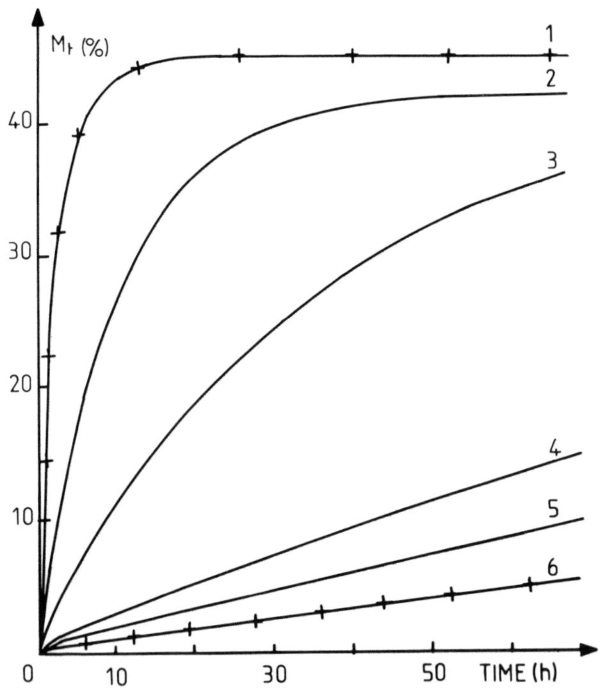


**Figure 11–19** Effect of parameters of shell on kinetics of desorption of anilin in water. Core made of EVAc with 28% VAc and $R_1 = 0.191$ cm and shell with a thickness of 0.007 cm. Curve 1: no shell. Curve 2: shell with $D = 4 \times 10^{-8}$ ($cm^2/s$) and $M_\infty = 15\%$. Curve 3: shell with $D = 10^{-8}$ and $M_\infty = 15\%$. Curve 4: shell with $D = 2 \times 10^{-9}$ and $M_\infty = 15\%$. Curve 5: shell with $D = 2 \times 10^{-9}$ and $M_\infty = 9.4\%$. Curve 6: shell with $D = 2 \times 10^{-9}$ and $M_\infty = 5\%$. (Reprinted from *Plast. Rubber Process Appl.*, 11, (1989), 9–16, H. David J. Bouzon, and J. M. Vergnaud, ref. 7, by courtesy of Elsevier Applied Sci. Publ.)

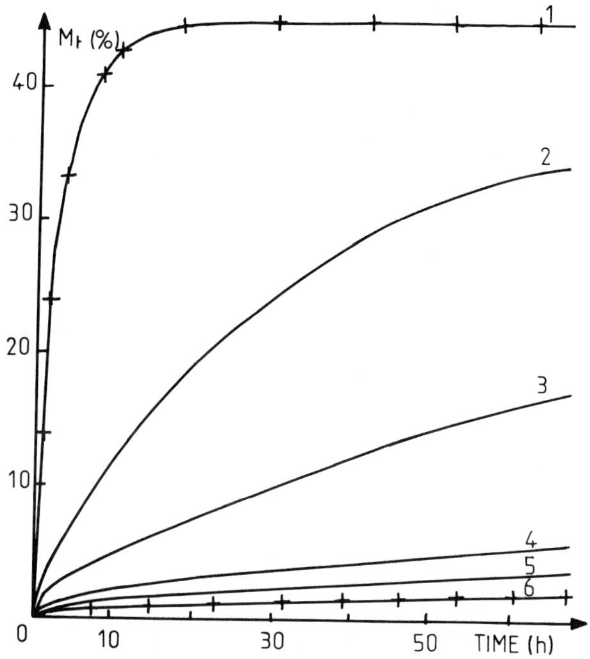


**Figure 11–20** Effect of parameters of shell on kinetics of desorption of anilin in water. Core made of EVAc with 28% VAc and $R_1$ = 0.191 cm and shell with a thickness of 0.018 cm. Curve 1: no shell. Curve 2: shell with $D = 4 \times 10^{-8}$ (cm²/s) and $M_\infty$ = 15%. Curve 3: shell with $D = 10^{-8}$ and $M_\infty$ = 15%. Curve 4: shell with $D = 2 \times 10^{-9}$ and $M_\infty$ = 15%. Curve 5: shell with $D = 2 \times 10^{-9}$ and $M_\infty$ = 9.4%. Curve 6: shell with $D = 2 \times 10^{-9}$ and $M_\infty$ = 5%. (Reprinted from *Plast. Rubber Process Appl.*, 11, (1989), 9–16, H. David, J. Bouzon, and J. M. Vergnaud, ref. 7, by courtesy of Elsevier Applied Sci. Publ.)

capacity of absorption, while the other characteristics are constant: thickness = 0.07 cm, $D = 13 \times 10^{-6}$ (cm²/s). The effect of the capacity of absorption given to the polymer used for the shell on the kinetics of desorption is described in this figure.

- Curve 1: $M_\infty$ = 0.1%
- Curve 2: $M_\infty$ = 0.5%
- Curve 3: $M_\infty$ = 1%
- Curve 4: $M_\infty$ = 1.5% (silicone)
- Curve 5: $M_\infty$ = 20%
- Curve 6: no shell

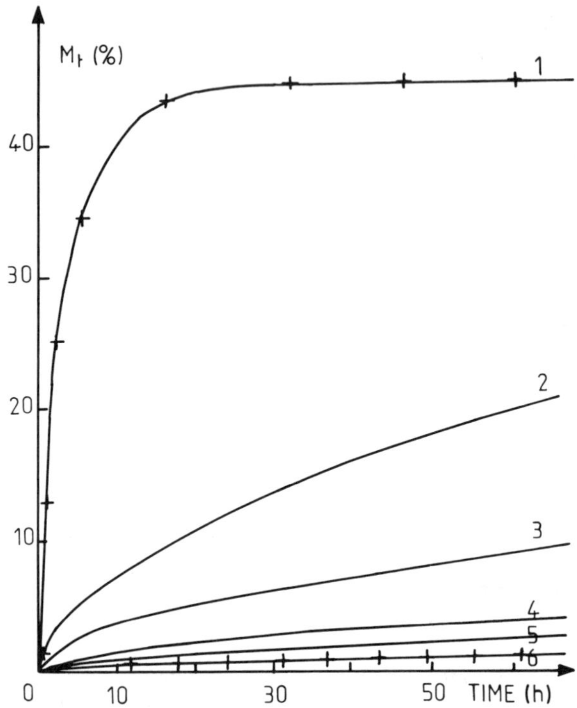


**Figure 11–21** Effect of parameters of shell on kinetics of desorption of anilin in water. Core made of EVAc with 28% VAc and $R_1 = 0.191$ cm and shell with a thickness of 0.05 cm. Curve 1: no shell. Curve 2: shell with $D = 4 \times 10^{-8}$ (cm²/s) and $M_\infty = 15\%$. Curve 3: shell with $D = 10^{-8}$ and $M_\infty = 15\%$. Curve 4: shell with $D = 2 \times 10^{-9}$ and $M_\infty = 15\%$. Curve 5: shell with $D = 2 \times 10^{-9}$ and $M_\infty = 9.4\%$. Curve 6: shell with $D = 2 \times 10^{-9}$ and $M_\infty = 5\%$. (Reprinted from *Plast. Rubber Process Appl.*, 11, (1989), 9–16, H. David, J. Bouzon, and J. M. Vergnaud, ref. 7, by courtesy of Elsevier Applied Sci. Publ.)

***Conclusions.*** The first interesting conclusion concerns the model. It is able not only to describe the process of absorption and desorption in the case of complex devices with core and shell, but it is also able to determine the kinetics of release of liquid obtained with special characteristics given to the polymeric shell. It is thus possible to predict the properties necessary for the polymer used in a shell, and then it is of help for screening research of polymers.

The other important conclusion concerns the shell which plays the leading role for controlled release. Two parameters characterize the nature of the polymer, as diffusivity and capacity of absorption, while the thickness of the shell can be chosen by the user.

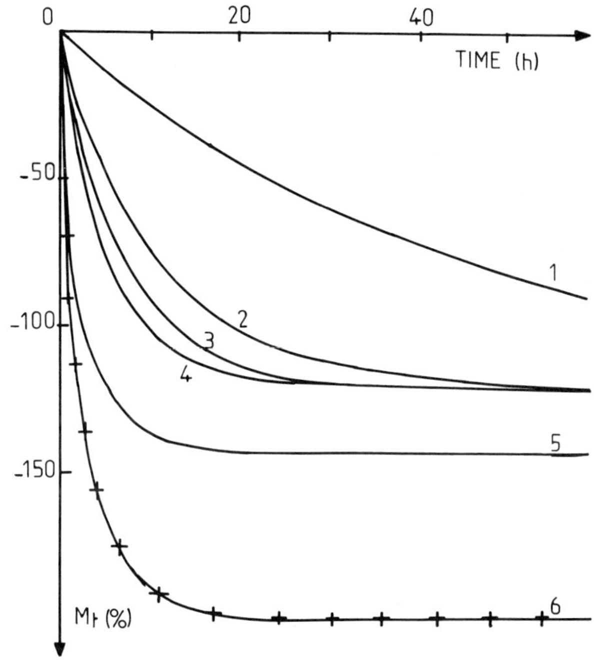


**Figure 11–22** Effect of the capacity of absorption of anilin by the shell on the kinetics of desorption at 25°C. Core of EVAc with 40% VAc and $R_1 = 0.17$ cm. Shell with $\Delta r = 0.07$ cm and $D = 13 \times 10^{-6}$ (cm²/s). 1: $M_\infty = 0.1\%$. 2: $M_\infty = 0.5\%$. 3: $M_\infty = 1\%$. 4: $M_\infty = 1.5\%$ (silicone). 5: $M_\infty = 20\%$. 6: no shell. (Reprinted from *Europ. Polym. J.*, 25, (1989), 939–945, H. David, J. Bouzon, and J. M. Vergnaud, ref. 8, by courtesy of Pergamon Press.)

## REFERENCES


1. D. Messadi, M. Hivert, and J. M. Vergnaud, "A new approach to the study of plasticizer migration from PVC into methanol," *J. Appl. Polym. Sci.*, 26, (1981), 667–677.
2. D. Messadi and J. M. Vergnaud, "Plasticizer transfer from plasticized PVC into ethanol-water mixtures," *J. Appl. Polym. Sci.*, 27, (1982), 3945–3955.
3. D. Messadi and J. M. Vergnaud, "Simultaneous diffusion of benzyl alcohol into plasticized PVC and of plasticizer from polymer into liquid," *J. Appl. Polym. Sci.*, 26, (1981), 2315–2324.
4. H. David, J. Bouzon, and J. M. Vergnaud, "Absorption of anilin by EVAc copolymers and desorption into water," *J. Control. Release*, 8, (1988), 151–156.
5. H. David, J. Bouzon, and J. M. Vergnaud, "Controlled absorption and release of an active agent by using EVAc beads. Effect of various parameters," *Europ. Polym. J.*, 25, (1989), 1007–1011.
6. H. David, J. Bouzon, and J. M. Vergnaud, "Modelling of absorption and de-

sorption of a liquid by a polymer device made of a core and shell," *Europ. Polym. J.*, 25, (1989), 89–94.

7. H. David, J. Bouzon, and J. M. Vergnaud, "Modelling of desorption of liquid from an EVA polymer device composed of a core and shell," *Plast. Rubber Proces. Applic.*, 11, (1989), 9–16.
8. H. David, J. Bouzon, and J. M. Vergnaud, "Modelling of matter transfer with a polymer device made of a core and shell. Effect of the capacity of absorption by the silicone shell," *Europ. Polym. J.*, 25, (1989), 939–945.

# Chapter Twelve

# Drying of Coating

## 12-1 INTRODUCTION

Many factors have been found to affect the life of a paint film, and some of these factors are interdependent. Some factors are completely outside the influence of the paint formulator, and for instance, the durability of the paint may be solely dependent on the vagaries of the weather and the diligence of the painter. Other parameters, such as the concentration of the solvent, the temperature when it is deposited, and the time of drying of the first layer before depositing the second one, may be of importance in paint preparation.

The purpose of this part devoted to the drying of coatings is to attain a further insight into the nature of the process involved in the drying of paints. The general problem is rather complicated because of the following facts.

1. The process of drying combines two different factors, that is, evaporation of the solvent from the surface, and also diffusion of the solvent through the polymer and solid ingredients.
2. Following the loss of solvent, a shrinkage of the coating may be significant especially when the coating is flexible. During some drying stages, the coating solidifies, becoming hard and stiff, and sometimes

being constrained by adhesion to the supporting substrate. This shrinkage has also been observed in the desorption of liquid from rubber [1].

3. The rate of evaporation depends on various factors such as the nature of the solvent and its volatility, temperature which affects not only the volatility but also the rate of diffusion, and air motion.
4. The rate of transport of the solvent through the paint is controlled by diffusion, and the diffusivity may be concentration dependent. This concentration dependence of the diffusivity has also been found in the case of plasticized PVC in contact with a liquid [2–4]

A second objective is to develop a model capable of describing the whole process of drying by considering all the facts, that is, the diffusion of the solvent through the coating and the evaporation from the surface. The model is based on a numerical method with finite differences. Simulations of the process enable us to predict the time necessary to dry a coating of a given thickness. The difficult case of paints made of several superimposed layers is also investigated.

In the case at hand, the solvent is not absorbed by the additives of the paint, and no reaction takes place. The problems which appear in the latter case are quite different from the problems of diffusion-evaporation of the solvent. With epoxy resin, for instance, the process is essentially controlled by the cure of the reaction, and the temperature plays an important role with the obvious statement that the higher the temperature, the shorter the time of cure. As the cure reaction can be expressed in terms of an activation energy and a pre-exponential factor by an Arrhenius equation, it is possible to predict the time of cure at any temperature [5, 6].

## 12-2 DRYING OF ONE-LAYER COATING, EFFECT OF THE THICKNESS, TEMPERATURE, AND CONCENTRATION OF SOLVENT: MODELING OF THE PROCESS

The process of drying of a paint film consists of the evaporation of the solvent. In fact, the process is more complex, and the desorption of the solvent is controlled either by diffusion of the solvent through the paint or by evaporation from the surface. Moreover, the diffusivity is often concentration dependent, and a shrinkage of the paint may be observed. The rate of evaporation is proportional to the actual concentration of the solvent on the surface of the paint.

Because of these complex facts, no analytical solution can generally be found to describe the process of evaporation, and the problem is resolved by using a numerical method with finite differences. The following parameters are of interest:

- the thickness of the paint
- the temperature of the paint
- the concentration of solvent in the paint

### 12-2-1 Experimental

**Materials.** The coating used in fire retardancy (PI 248 from JIVAL) has a density of 1.25 with a dry extract of 72.4 wt %. The coating has the following components:

- Pliolite resin: 10.8 wt %
- $T_i\ O_2$ pigment: 6.4 wt %
- Fire-retarding agents: 55.2 wt %
- Solvent (Xylene + Benzene F): 27.6 wt %

This coating dried tack-free within 8 h, and a second layer can be applied after 24 h.

**Coating preparation and measure of the loss in weight.** Coatings of various thicknesses are applied onto glass plates. They are then exposed to motionless air at a constant temperature selected within the range 20°C to 46°C, and they are weighed at intervals.

The influence of the solvent concentration on the drying process is determined by using paints with various solvents concentrations ranging from 22 to 34.5 wt %. The rate of evaporation of the pure solvent is determined at various temperatures by weighing the amount of solvent remaining on a plate at various time intervals.

### 12-2-2 Theoretical

**Assumptions.** The following assumptions are made in order to describe all the facts known regarding the process.

1. The process is controlled by evaporation of the solvent from the surface and by diffusion through the paint.
2. The solvent concentration is uniform through the paint at the start of the drying process.
3. One-dimensional diffusion through the paint thickness is considered.

4. The rate of evaporation of solvent from the surface of the paint is taken to be either constant or proportional to the concentration of solvent on the surface.
5. The transport of solvent through the paint is controlled by transient diffusion, with concentration-dependent diffusivity.
6. Despite shrinkage of the coating during drying, a frame of reference is fixed with respect to the paint thickness.

**Mathematical treatment.** The equation of one-dimensional diffusion is:

$$\frac{\partial C}{\partial t} = \frac{\partial}{\partial x}\left(D \cdot \frac{\partial C}{\partial x}\right) \tag{12-1}$$

where $C$ is the concentration of solvent in the coating at abscissa $x$ and time $t$ and $D$ is the concentration-dependent diffusivity.

The initial and boundary conditions are:

| | | | | | |
|---|---|---|---|---|---|
| Initial | $t = 0$ | $0 < x < L$ | $C = C(n,o) = C_{in}$ | | (12-2) |
| Boundary | $t > 0$ | $0 < x < L$ | $C = C(n,i)$ | | |
| | | $x = L$ | $C = C(o,i)$ | surface | (12-3) |

The rate of solvent evaporation on the surface of the paint is proportional to the concentration of the solvent (7):

$$F = \frac{F_o \cdot C(o,i)}{C(o,o)} \tag{12-4}$$

where $F_o$ is the initial rate of evaporation, $C(o,o)$ is the concentration of solvent at the coating surface at time $t = o$, and $C(o,i)$ is the concentration at the same place and time $i \cdot \Delta t$.

**Numerical analysis.** As no analytical solution can be obtained for equations (12-1) and (12-4) because of the concentration dependence of the diffusivity, a numerical method with finite differences is used to solve the problem.

The sheet thickness $L$ is divided into $N$ constant increments of space $\Delta x$, and the space-time region is covered by means of a grid of rectangles of sides $\Delta x$ and $\Delta t$, $\Delta t$ being the increment of time (Figure 12–1).

The coordinates are expressed by using the integers $i$ and $n$ defined by:

$$t = i \cdot \Delta t \qquad x = n \cdot \Delta x \quad \text{and} \quad L = N \cdot \Delta x \tag{12-5}$$

From the matter balance calculated in the slice centered on the position $n$, the new concentration of solvent in this slice after the elapse of

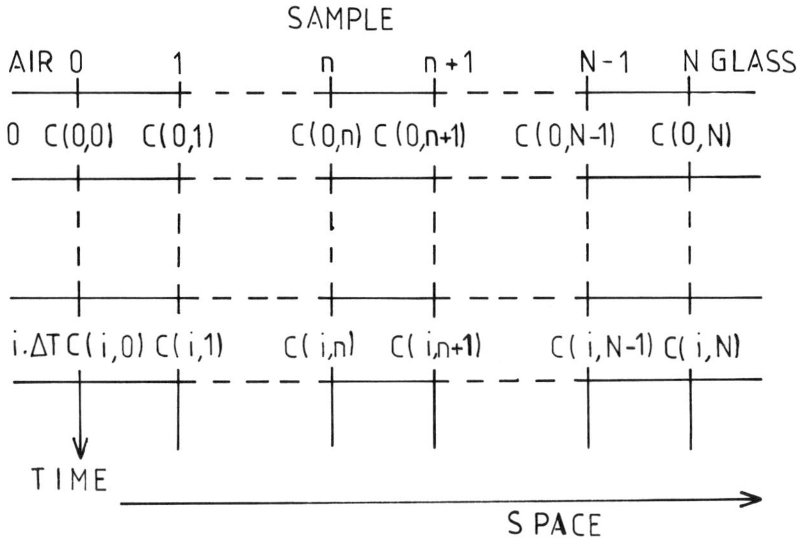


**Figure 12–1** Space-time diagram for numerical analysis.

time $\Delta t$ is expressed in terms of the concentration at the previous time and at the same position $n$, and at the adjacent two slices $n - 1$ and $n + 1$:

$$CN(n) = \frac{1}{M_n}[C(n + 1) + (M_n - 2)\cdot C(n) + C(n - 1)] \tag{12-6}$$

$$C(n,i + 1) = \frac{1}{M_{n,i}}[C(n + 1,i) + (M_{n,i} - 2)\cdot C(n,i) + C(n - 1,i)] \tag{12-6'}$$

where $CN(n)$ or $C(n,i + 1)$ is the concentration of solvent at place $n$ and time $(i + 1)\Delta t$ and $C(n - 1,i)$ is the concentration at place $(n - 1)$ and time $i\cdot\Delta t$.

The dimensionless number is:

$$M(n,i) = \frac{(\Delta x)^2}{\Delta t}\cdot\frac{1}{D(n,i)} \tag{12-7}$$

where the diffusivity $D$ is expressed in terms of position and time (being concentration dependent).

For the coating plane adjacent to the glass plate, the thickness $\Delta x/2$ has been chosen [7]. The new concentration of the coating in contact with the glass plate is:

$$CN(N) = \frac{1}{M(N)}[2C(N - 1) + (M - 2)\cdot C(N)] \tag{12-8}$$

For the surface of the paint, the solvent diffusion is considered through a slice of thickness $\Delta x/2$, as well as the rate of evaporation on the surface. The new concentration on the surface is then:

$$CN(o) = C(o) + \frac{1}{3}[C(1) - CN(1)] + \frac{8}{3 \cdot M(1)} \cdot \left[C(1) - C(o) - \frac{8}{3}F \cdot \frac{\Delta t}{\Delta x}\right] \tag{12-9}$$

The amount of solvent remaining in the coating at time $i\Delta t$ is obtained by integrating the concentration with respect to space.

$$M_t = \left[\frac{3}{8}C(o) + \frac{9}{8} \cdot C(1) + \sum_{2}^{N-1} C(n) + \frac{1}{2}C(N)\right] \cdot \Delta x \tag{12-10}$$

### 12-2-3 Results

Before testing the validity of the model by comparing experimental and calculated results, it is necessary to determine the diffusivity for various solvent concentrations in the paint at various temperatures in order to establish a relationship between diffusivity and these parameters.

#### Determination of diffusivity and rate of evaporation

***Diffusivity.*** The kinetics of drying are experimented with for paints having different solvent concentrations at various temperatures. The short-test technique [4] is used for determining the diffusivity in terms of the solvent concentration, since the solvent concentrations vary little and can be assumed to be constant over a short period of time.

By plotting the amount of solvent evaporated as a function of the square root of time, the diffusivity is obtained from the slope of the resulting straight line if the total amount of solvent $M_\infty$ is known.

$$\frac{M_t}{M_\infty} = \frac{2}{L}\left[\frac{D \cdot t}{\pi}\right]^2 \tag{12-11}$$

where $M_t$ and $M_\infty$ are the amounts of solvent evaporated at time $t$ and infinite time, respectively.

The values of diffusivity obtained for various concentrations of solvent are expressed as a function of these solvent concentrations (Figure 12-2). The straight lines obtained conform to the following relationship:

$$D = D_o \cdot \exp\left(-\frac{Cte}{C}\right) \tag{12-12}$$

The values of coefficient $D_o$ in equation (12-12) are plotted against the reciprocal of the temperature (Figure 12–3).

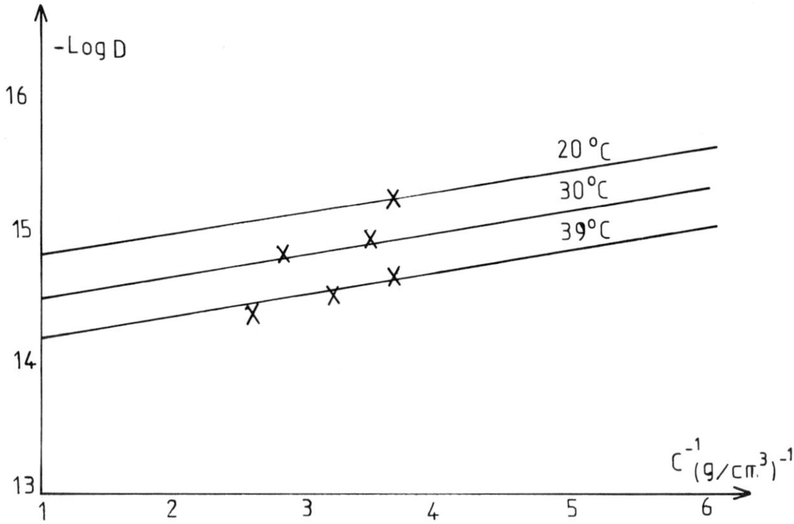


**Figure 12–2** Concentration dependence of diffusivity at various temperatures. (Reprinted from *Prog. Organ. Coatings*, 15, (1987), 163–172, H. P. Blandin, J. C. David, and J. M. Vergnaud, ref. 7, by courtesy of Elsevier Sequoia Publ.)

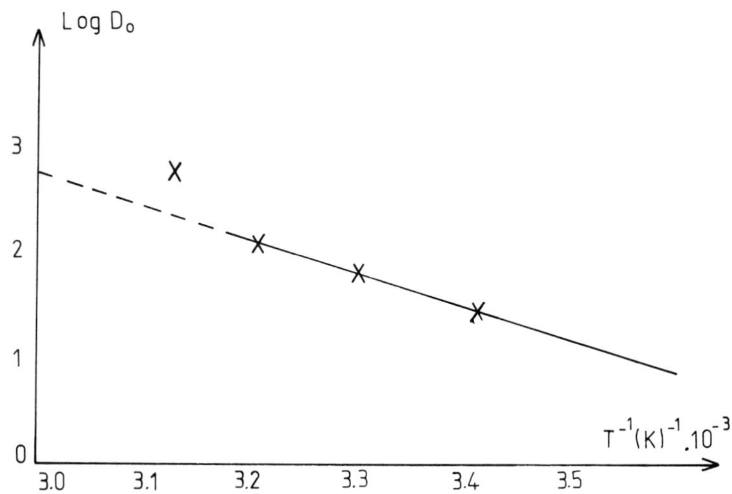


**Figure 12–3** Variation of diffusivity ($D_o$) with temperature. (Reprinted from *Prog. Organ. Coatings*, 15, (1987), 163–172, H. P. Blandin, J. C. David, and J. M. Vergnaud, ref. 7, by courtesy of Elsevier Sequoia Publ.)

From the straight line shown in Figure 12–3, the diffusivity can be expressed in terms of the solvent concentration and the temperature by the relationship:

$$D = 9.3 \times 10^{-3} \times \exp\left(-\frac{2921}{T}\right) \times \exp\left(-\frac{0.163}{C}\right) \quad (12\text{-}13)$$

where the temperature $T$ is expressed in $K$ and the solvent concentration in g/cm³.

***Rate of Evaporation.*** The rate of evaporation of the solvent is found to vary with the temperature by following an expression similar to the Clausius-Clapeyron law, the rate of evaporation being proportional to the vapor pressure of the solvent (Figure 12–4). It is expressed by the relationship:

$$F = 1.41 \times \exp\left(-\frac{7610}{RT}\right) \quad (12\text{-}14)$$

where $T$ is expressed in $K$.

**Validity of the model.** The model is tested by comparing the kinetics of drying obtained experimentally with those determined by calculation, for various temperatures ranging from 30°C to 46°C and various concentrations of solvent: Figure 12–5 when the temperature is 30°C; Figure 12–

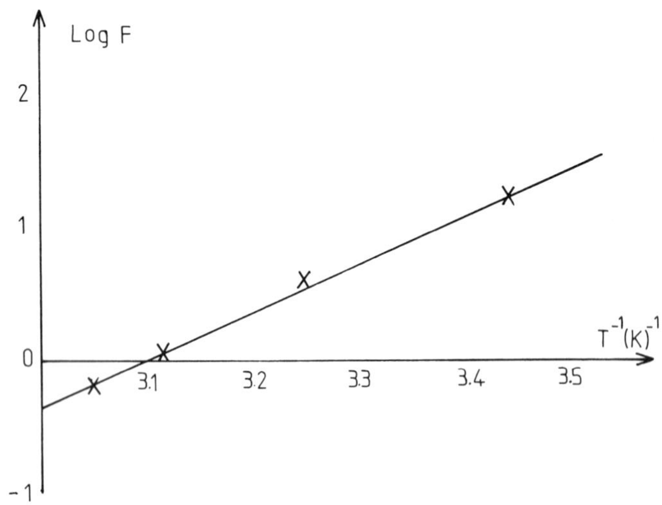


**Figure 12–4** Variation of the rate of evaporation with temperature. (Reprinted from *Prog. Organ. Coatings*, 15, (1987), 163–172, H. P. Blandin, J. C. David, and J. M. Vergnaud, ref. 7, by courtesy of Elsevier Sequoia Publ.)

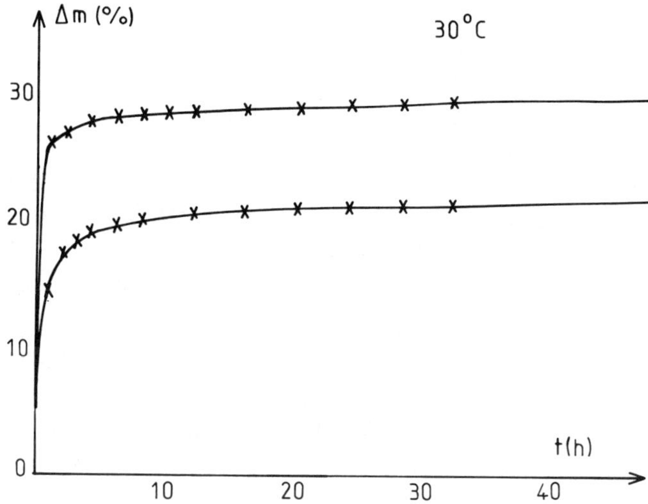


**Figure 12–5** Validity of the model of drying: Kinetics of drying at 30°C for two solvent concentrations. – : calculated. × : experiments. (Reprinted from *Prog. Organ. Coatings*, 15, (1987), 163–172, H. P. Blandin, J. C. David, and J. M. Vergnaud, ref. 7, by courtesy of Elsevier Sequoia Publ.)

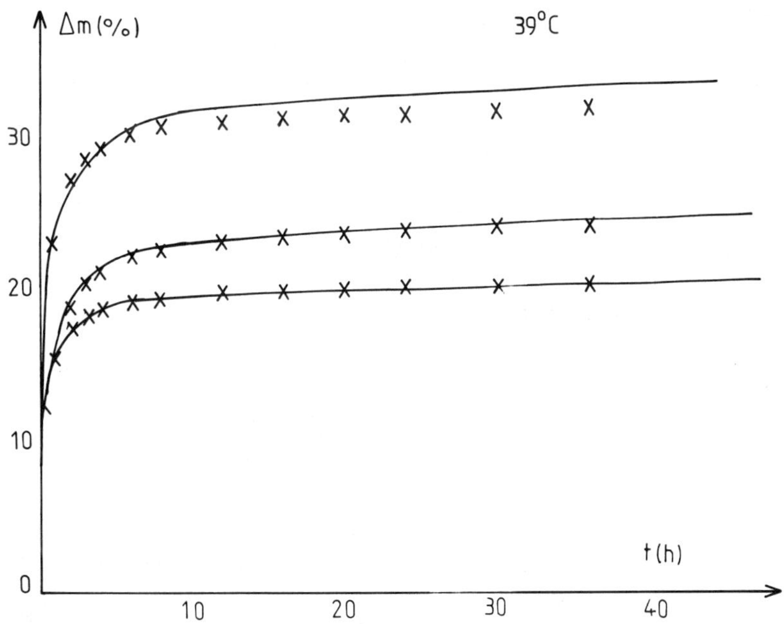


**Figure 12–6** Validity of the model of drying: Kinetics of drying at 39°C for three solvent concentrations. – : calculated. × : experiments. (Reprinted from *Prog. Organ. Coatings*, 15, (1987), 163–172, H. P. Blandin, J. C. David, and J. M. Vergnaud, ref. 7, by courtesy of Elsevier Sequoia Publ.)

6 at 39°C, for various solvent concentrations; and Figure 12–7 at 46°C, for various solvent concentrations.

In all these preceding cases, the theoretical kinetics of drying calculated with the model using various data (diffusivity and rate of evaporation) are in good agreement, proving the validity of the model.

The model is also tested for various thicknesses of the coating ranging from 0.45 mm to 0.67 mm, dried under the same conditions of temperature (20°C) and solvent concentration: Figure 12–8 with thickness = 0.45 mm; Figure 12–9 with thickness = 0.53 mm; and Figure 12–10 with thickness = 0.67 mm.

A good agreement is also obtained between the theoretical and experimental kinetics of drying, for the paints of various thicknesses.

**Effect of parameters on the kinetics of drying.** Four parameters are of interest for the process of drying of one-layer coatings.

- the temperature at which drying is conducted
- the concentration of solvent in the paint

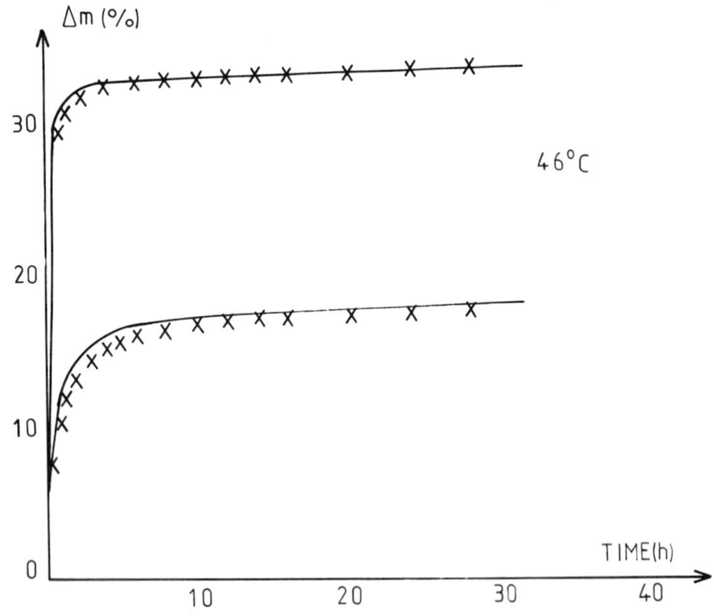


**Figure 12–7** Validity of the model of drying: Kinetics of drying at 46°C for two solvent concentrations. – :calculated. × :experimental. (Reprinted from *Prog. Organ. Coatings*, 15, (1987), 163–172, H. P. Blandin, J. C. David, and J. M. Vergnaud, ref. 7, by courtesy of Elsevier Sequoia Publ.)

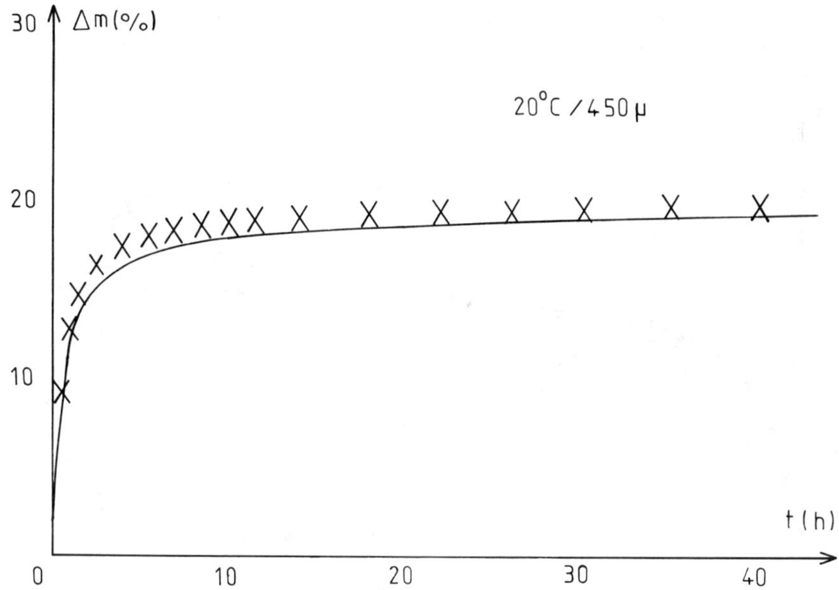


**Figure 12–8** Validity of the model: Kinetics of drying at 20°C for a coating of thickness = 0.45 mm. (Reprinted from *Prog. Organ. Coatings*, 15, (1987), 163–172, H. P. Blandin, J. C. David, and J. M. Vergnaud, ref. 7, by courtesy of Elsevier Sequoia Publ.)

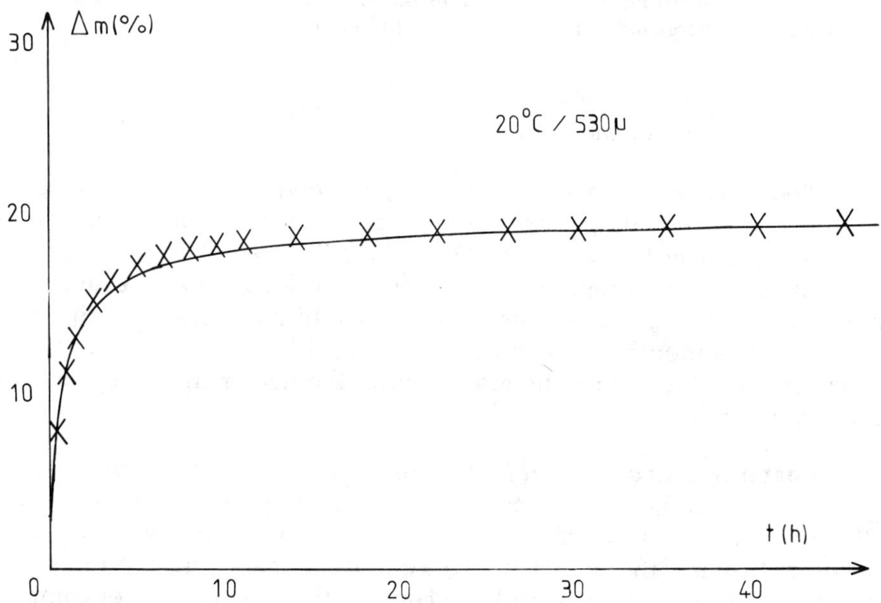


**Figure 12–9** Validity of the model: Kinetics of drying at 20°C for a coating of thickness = 0.53 mm. (Reprinted from *Prog. Organ. Coatings*, 15, (1987), 163–172, H. P. Blandin, J. C. David, and J. M. Vergnaud, ref. 7, by courtesy of Elsevier Sequoia Publ.)

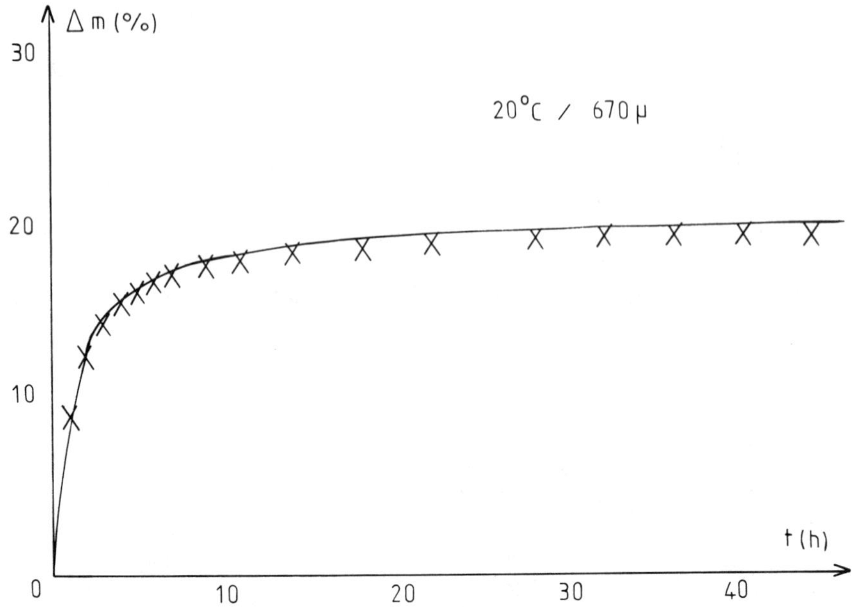


**Figure 12–10** Validity of the model: Kinetics of drying at 20°C for a coating of thickness = 0.67 mm. (Reprinted from *Prog. Organ. Coatings*, 15, (1987), 163–172, H. P. Blandin, J. C. David, and J. M. Vergnaud, ref. 7, by courtesy of Elsevier Sequoia Publ.)

- the nature of the solvent
- the thickness of the paint

***Temperature.*** The effect of temperature on the process of drying can be appreciated by comparing the kinetics of drying obtained at various temperatures over the range of 20°C to 46°C (Figures 12–5 to 12–7).

As shown in equations expressing the diffusivity and evaporation rate in terms of the temperature, the temperature acts not only upon the rate of evaporation but also on the value of diffusivity, with the simple statement that the higher the temperature, the higher the diffusivity and evaporation rate.

***Concentration of Solvent in the Paint.*** The effect of the initial solvent concentration in the paint on the kinetics of drying is illustrated in Figures 12–5 to 12–7 for various temperatures. As also expressed by the equation of the diffusivity, the higher the concentration of solvent, the higher the diffusivity. This result is of high concern for the achievement of complete dryness. It explains the great difficulty in drying a coating perfectly.

***Nature of the Solvent.*** The nature of the solvent is of interest for the drying, and it plays a role in the drying process not only with the evaporation rate but also with the diffusivity. It is easy to determine the effect of the solvent on the rate of evaporation because the rate of evaporation is proportional to the vapor pressure. It is rather difficult to predict the value of diffusivity of a solvent, and experiments are necessary for determining its value.

***Thickness of the Paint.*** The effect of the thickness of coatings on the kinetics of drying is of high significance, as shown in Figures 12–8 to 12–10 obtained with thicknesses ranging from 0.45 mm to 0.57 mm. The time of drying is proportional to the square of the thickness, as shown in the expression of the modulus $M$.

### 12-2-4 Conclusions

The process of drying of coatings is described by diffusion of the solvent through the coating and evaporation from the surface. The process of drying, in the case of a one-layer coating, can be predicted with the help of a model based on a numerical method with finite differences. The model not only takes into account the evaporation, but also the diffusion of the solvent.

The diffusivity, determined experimentally in the case of a fire-retardant coating, is found to be concentration dependent. The diffusivity and rate of evaporation vary with temperature according to relationships similar to the Clausius-Clapeyron law.

Various parameters are important: the temperature at which the drying is performed; the nature of the solvent and the solvent concentration in the paint; and the thickness of the paint.

As the validity of the model is demonstrated by varying each of these parameters in turn, it is possible to predict the effect of these parameters on the drying process with the help of the model and data such as the diffusivity and evaporation rate of the solvent.

Moreover, the model is capable of providing a fuller insight into the nature of the process, for example, by enabling calculation of the profiles of solvent concentration developed through the thickness of the coating and the amount of solvent remaining within the coating at any given time.

## 12-3 PROCESS OF DRYING OF COATINGS WITH VARIOUS LAYERS

As the thickness of a one-layer coating is rather thin, thick coatings are obtained by superimposing several layers (or at least one) upon each other.

The process of drying in the case of a one-layer coating is already complex in its nature, because of the following facts:

1. The resin dries by evaporation of the solvent, without chemical reaction, the solvent playing the role of diluent.
2. The rate of solvent loss depends on two factors: the rate of evaporation of the solvent from the surface, and the rate of solvent transport through the coating. During the drying operation, the solvent, which is first uniformly distributed through the paint sheet, diffuses through the thickness and evaporates from the surface, as described previously [7]. Thus, the diffusion of the solvent and its evaporation provide two areas of interest for controlling the drying process of the paint [1, 8]. However, the evaporation process is controlled by diffusion in the sense that the rate of evaporation depends largely on the rate at which the solvent is supplied to the surface by diffusion.
3. The solvent transport through the coating is controlled by transient diffusion, with diffusivity being dependent on concentration of solvent and temperature.
4. The rate of evaporation depends on the temperature and agitation of air, and on the nature of the solvent with its volatility.
5. The drying by solvent evaporation involves a shrinkage of the coating. At some stages of drying, the coating solidifies, and further shrinkage is constrained by adhesion to the substrate.

In the case of a two-layer coating, the process is more complicated, especially because of the following facts:

1. The first layer is already tack-free after a drying stage when the layers have quite different properties.
2. Moreover, a migration of the solvent from the second layer to the previous dry one is observed, according to a definite law.
3. A good knowledge of the profiles of concentration of the solvent through the previous dry coating is of high interest for studying the process of drying of the two-layer coating.

This problem is a good application for modeling by using a numerical method with finite differences.

### 12-3-1 Experimental

**Materials.** The coating used is a fire-retardant coating (PI 248—JIVAL), with the following properties:

- Density = 1.25
- Dry extract = 72.4 wt %
- Resins (pliolite): 10.8 wt %

- Pigment ($T_i$ $O_2$): 6.4
- Fire-retarding agents 55.2
- Solvents (Xylene + Benzine F) 27.6
- The coating dries tack-free within 8 h.
- The second layer can be applied after 24 h.

**Coating preparation and determination of weight loss.** Coatings of various thicknesses are applied to glass plates, and exposed to motionless air at room temperature (20°C). The process of drying is followed by recording the loss in weight of the coating on a balance.

**Preparation of two-layer coatings.** The operational conditions for preparing two-layer coatings are described in Table 12–1.

### 12-3-2 Theoretical

**Assumptions.** The following assumptions are made:

1. A one-dimensional transport of solvent is considered, because thin plane sheets of paint are used.
2. At the beginning of the drying process of the first layer, the concentration of solvent throughout the sheet is uniform.
3. The solvent transport through the coating is controlled by transient diffusion, with concentration-dependent diffusivity.
4. The rate of evaporation of solvent from the surface is proportional to the surface concentration.
5. Despite the shrinkage of the coating during the drying process, a frame of reference is fixed with respect to the paint, which is the thickness of the paint when it is deposited.

**TABLE 12–1** Conditions of Preparation of Two-Layer Coatings

| First Layer | | | Second Layer | |
|---|---|---|---|---|
| Time of Drying (h) | Thickness (mm) | $\Delta Ms^a$ (g) | Thickness (mm) | $\Delta Ms^b$ |
| 1 | 0.5 | 0.136 | 0.61 | 0.336 |
| 14 | 0.50 | 0.043 | 0.49 | 0.272 |
| 24 | 0.51 | 0.034 | 0.60 | 0.331 |

[a] $\Delta Ms$ is the amount of solvent remaining in the first layer when the second layer is applied.

[b] $\Delta Ms$ is the total amount of solvent in the second layer when it is applied over the previous one.

6. Of course, there is a profile of concentration of solvent through the first layer at the time of deposition of the second layer.
7. A convective transport of solvent takes place from the second layer to the first one, the rate of this matter transport being very fast.

**Mathematical treatment.** The transport of solvent through the thickness of the coating follows the one-dimensional diffusion equation:

$$\frac{\partial C}{\partial t} = \frac{\partial}{\partial x}\left(D\cdot\frac{\partial C}{\partial x}\right) \tag{12-1}$$

*For the first layer,* the initial and boundary conditions are:

| | | | | | |
|---|---|---|---|---|---|
| Initial | $t = 0$ | $0 < x < L$ | $C = C(n,o) = C_{in}$ | | (12-2) |
| Boundary | $t > 0$ | $x = L$ | $C = C(o,i)$ | surface | (12-3) |

The rate of evaporation of the solvent from the surface on the paint is proportional to the concentration of the solvent on the surface [7].

$$F = \frac{F_o\cdot C(o,i)}{C(o,o)} \tag{12-4}$$

where $F_o$ is the rate of evaporation of the solvent from the surface of the paint at the beginning of the drying process (at time $t = 0$), when the solvent concentration is $C(o,o)$.

*For the second layer,* the initial and boundary conditions are:

| | | | | | |
|---|---|---|---|---|---|
| Initial | $t = 0$ | $0 < x < L'$ | $C = C(n,o) = C_{in}$ | | (12-2′) |
| Boundary | $t > 0$ | second layer | $C = C(o,i)$ | surface | (12-3′) |
| | | first layer | $C = C(r,i)$ | surface | |

When the second layer is deposited, there is a discontinuinity in the solvent concentration at the interface separating these two layers. A convective transport of solvent is assumed to take place, which is able to make the concentration uniform through the two layers. A particular gradient of concentration of solvent is obtained at the end of this fast transport.

**Numerical analysis.** No analytical solution can be obtained for the problem for the following two reasons:

1. The diffusivity is concentration dependent.
2. A profile of concentration of solvent through the two-layer coating exists at the beginning of the drying after depositing the second layer.

An explicit numerical method with finite differences is used for the problem.

As shown in Figure 12–1, the thickness of the coating is divided into $N$ equal increments of space $\Delta x$ and the time into increments $\Delta t$, by concentration-reference planes ($n$: space, $i$: time).

The balance for the solvent is written on plane $n$ and enables us to express the new concentration after elapse of time $\Delta t$, $CN_n$, in terms of the previous concentrations obtained at the same plane $n$, and at the adjacent two planes $(n - 1)$ and $(n + 1)$.

$$CN(n) = \frac{1}{M_n} [C(n - 1) + (M_n - 2)\cdot C(n) + C(n + 1)] \qquad (12\text{-}6)$$

where $C_n$ and $CN_n$ are the concentrations at the plane $n$ at the previous time, and after the elapse of time $\Delta t$, respectively.

The dimensionless number is expressed as a function of the increments of space and time, and of the diffusivity. This dimensionless number varies with position and time because of the concentration dependence of the diffusivity.

$$M(n,i) = \frac{(\Delta x)^2}{\Delta t} \cdot \frac{1}{D(n,i)} \qquad (12\text{-}7)$$

For the plane adjacent to the glass plate, a thickness of $\Delta x/2$ is taken, and the new concentration is:

$$CN(N) = \frac{1}{M(N)} [2C(N - 1) + (M - 2)\cdot C(N)] \qquad (12\text{-}8)$$

For the plane in contact with the external atmosphere, a thickness $\Delta x/2$ is taken, and the matter balance calculated by considering the diffusion and evaporation rate on the surface gives:

$$CN(o) = C(o) + \frac{1}{3} [C(1) - CN(1)] + \frac{8}{3M} [C(1) - C(o)] - \frac{8}{3} \frac{F_o\cdot C(o,i)}{C(o,o)} \cdot \frac{\Delta t}{\Delta x} \qquad (12\text{-}9)$$

The amount of solvent remaining in the coating at the time $i\cdot\Delta t$ is obtained by integrating the solvent concentration with respect to space.

$$M_t = \left[\frac{3}{8}\cdot C(o) + \frac{9}{8} C(1) + \sum_{2}^{N-1} + C(n) + \frac{1}{2} C(N)\right] \cdot\Delta x \qquad (12\text{-}10)$$

where the concentrations are shown in various places through the thickness of the coating, at the same time $i\cdot\Delta t$.

### 12-3-3 Results

The following three points are considered successively:

- the determination of the diffusivity and evaporation rate of the solvent;
- the drying of the first layer, and determination of the profile of concentration at this end of this step;
- the deposition of the second layer, and drying of the two-layer coating.

#### Determination of diffusivity and evaporation rate

***Diffusivity.*** The diffusivity is determined in the same way as for the first layer. Short tests have been used [4] with various values of the concentration of solvent. The diffusivity is calculated from the slope of the straight line obtained by plotting the amount of solvent transported as a function of the square root of time (Figure 12–11).

$$\frac{M_t}{M_\infty} = \frac{2}{L}\left(\frac{D \cdot t}{\pi}\right)^{0.5} \tag{12-11}$$

At the constant temperature of 20°C, the diffusivity is found to vary

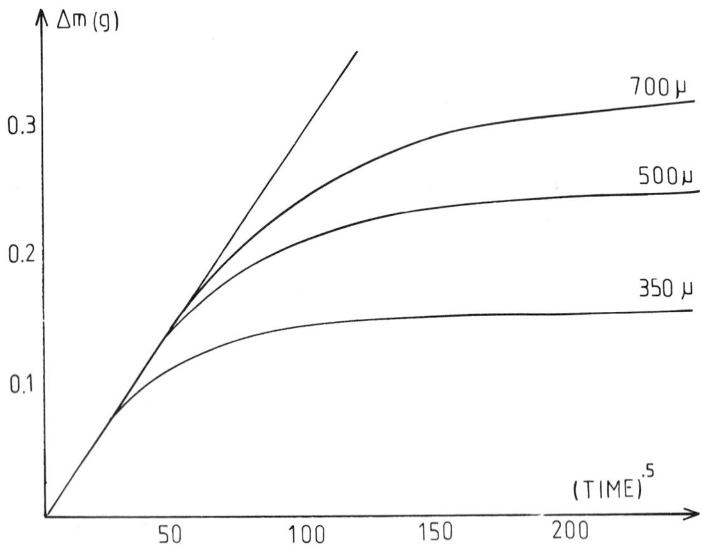


**Figure 12–11** Determination of diffusivity by using short test. Loss in weight of solvent as a function of the square root of time. (Reprinted from *J. Coat. Technol.*, 59, (1987), 27–32, H. P. Blandin, J. C. David, and J. M. Vergnaud, ref. 9, by courtesy of Fed. Soc. for Coatings Technology.)

with the solvent concentration by the equation:

$$D = 4.4 \times 10^{-7} \times \exp\left(-\frac{0.173}{C}\right) \quad (\text{cm}^2/\text{s}) \tag{12-15}$$

***Evaporation Rate.*** The value of the evaporation rate is obtained from the slope of the kinetic curve of the loss in weight at the beginning of the process, when this process is controlled by evaporation. The value obtained is:

$$F = 1.1 \times 10^{-5} \ \text{g/cm}^2\text{s} \quad \text{when} \quad C(o,o) = 22 \text{ wt } \% \tag{12-16}$$

**Drying process for the first layer.** Two results are of interest: the one concerned with the kinetics of the loss in weight following the evaporation of the solvent, the other with the profiles of concentration of solvent through the thickness of the paint at the end of the process.

The numerical model previously described gives information on the kinetics of the loss in weight for various thicknesses of paint ranging from 0.35 mm to 0.70 mm, as illustrated in Figure 12–12. A good agreement between theoretical and experimental values can be appreciated.

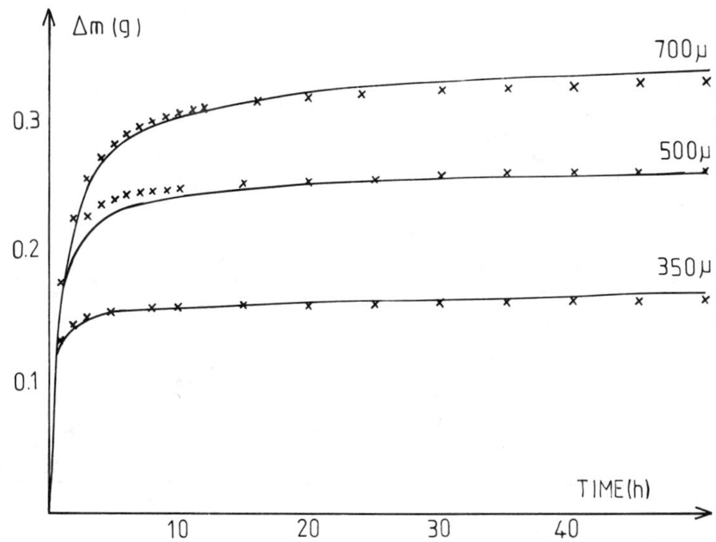


**Figure 12–12** Kinetics of loss in weight of solvent, for various thicknesses of the first layer of the coating. + :experiments. − :calculated. (Reprinted from *J. Coat. Technol.*, 59, (1987), 27–32, H. P. Blandin, J. C. David, and J. M. Vergnaud, ref. 9, by courtesy of Fed. Soc. for Coatings Technology.)

The profiles of concentration of solvent developed through the thickness of the paint are calculated at various times, and drawn in Figure 12–13 when the initial value of the thickness is 0.50 mm, and initial concentration of solvent is 0.27.

**Drying process for the second layer.** Very often thick coatings are obtained by superimposing several layers. Although the presence of the second layer presented more difficulty, the model is able to describe the case at hand by considering the evaporation from the surface and diffusion of the solvent through the two layers of the coating. Because of the second layer, two new important parameters are introduced: the thickness of these layers, and the time of drying selected for the first layer, this second parameter being responsible for the development of the profile of concentration of solvent.

From observation done during the deposition of the second layer, it has been concluded that a fast transport of solvent from the second layer to the first one takes place as soon as the second layer is deposited [9]. This transport is assumed to be obtained so that a linear gradient of solvent concentration is expanded through the thickness. The resulting solvent concentration takes into account the amount of solvent remaining in the

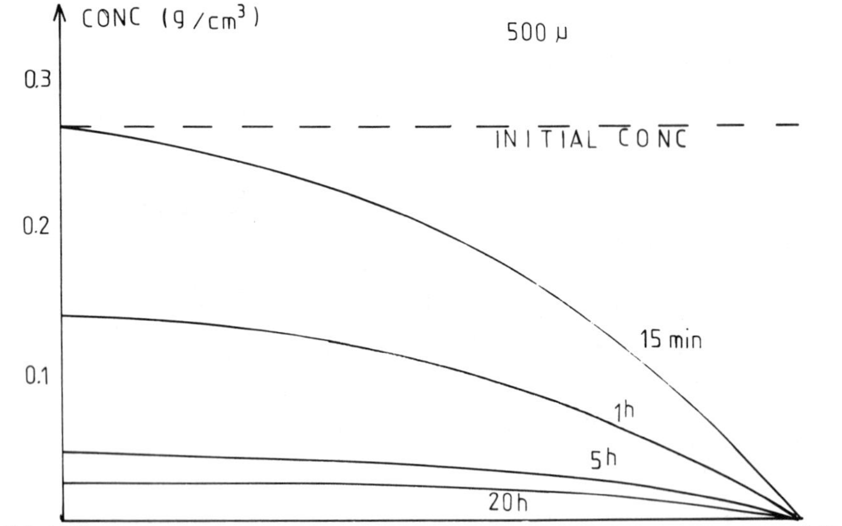


**Figure 12–13** Profiles of concentration of solvent developed through the thickness of the coating for various times. Initial thickness = 0.50 mm. (Reprinted from *J. Coat. Technol.*, 59, (1987), 27–32, H. P. Blandin, J. C. David, and J. M. Vergnaud, ref. 9, by courtesy of Fed. Soc. for Coatings Technology.)

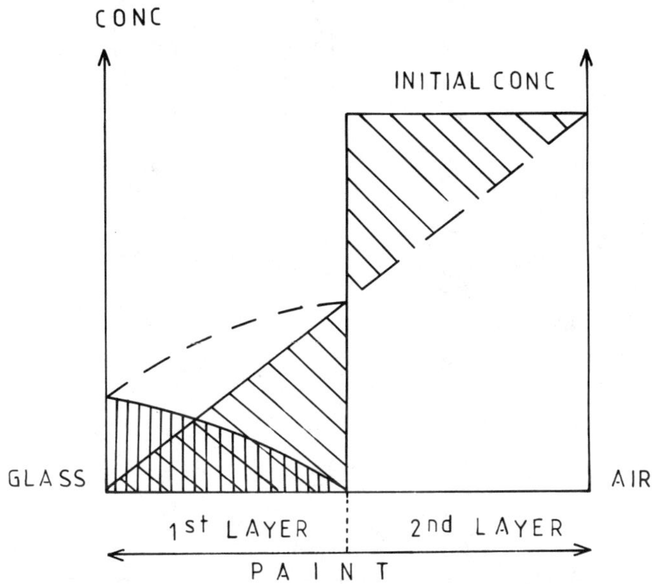


**Figure 12–14** Scheme of the transport of solvent between the two layers during the application of the second layer. Profiles of concentration of solvent: remaining in the first layer before the application: ||||; resulting from the solvent transfer: \\\\; resulting from these two contributions: ---. (Reprinted from *J. Coat. Technol.*, 59, (1987), 27–32, H. P. Blandin, J. C. David, and J. M. Vergnaud, ref. 9, by courtesy of Fed. Soc. for Coatings Technology.)

first layer at the time of deposition of the second layer, as shown in Figure 12–14.

A good agreement is found between experimental and calculated results for the kinetics of loss in weight obtained with the three coatings described in Table 12–1, from the beginning to the end of the drying process (Figure 12–15). The effect of the time of drying of the first layer on the process of drying of the second layer can also been appreciated.

The profiles of concentration of solvent grown through the two-layer coating are drawn in Figures 12–16 to 12–18 for each of the three samples, as they are calculated with the help of the model: Figure 12–16 when the time of drying of the first layer is 1 h; Figure 12–17 when the time of drying of the first layer is 14 h; and Figure 12–18 when the time of drying of the first layer is 24 h.

It is difficult to make experiments able to determine the profiles of concentration of solvent through the paint. However, integrating the concentration with respect to space at various times gives the theoretical kinetics of loss in weight which were in good agreement with the experimental kinetics.

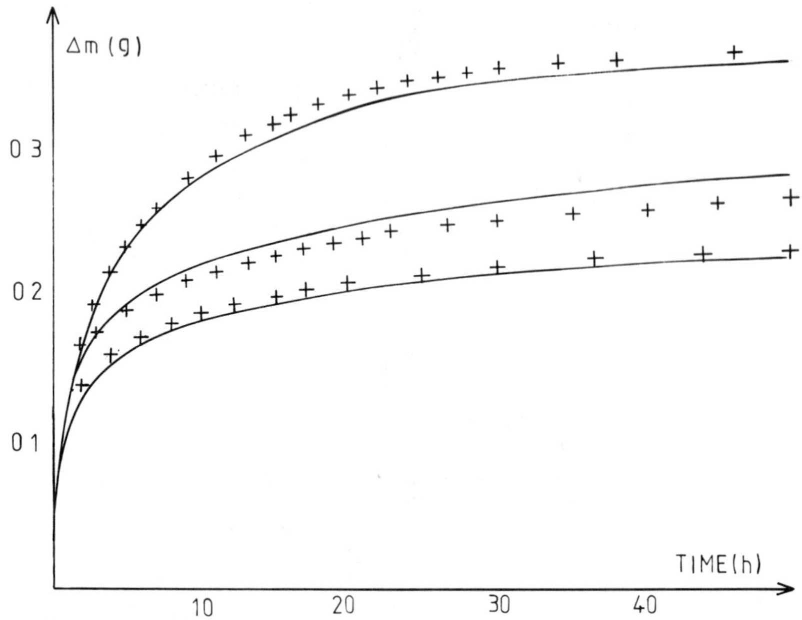


**Figure 12–15** Kinetics of drying of the two-layer coating, after deposition of the second layer, for various time of drying of the first layer: (1) 1 h (2) 14 h (3) 24 h; – : calculated. + : experiments. (Reprinted from *J. Coat. Technol.*, 59, (1987), 27–32, H. P. Blandin, J. C. David, and J. M. Vergnaud, ref. 9, by courtesy of Fed. Soc. for Coatings Technology.)

### 12-3-4 Conclusions

The drying process of paint is rather complicated, especially in the case of a two-layer paint. The process of drying is controlled either by evaporation of the solvent from the surface or by transport of solvent through the thickness of the paint.

The second transport through the solid is done by transient diffusion, with a concentration-dependent diffusivity. The value of the evaporation rate is proportional to the actual concentration of solvent on the paint surface.

Modeling by using a numerical method with finite differences is of great help for describing the process and giving a fuller insight on the process. In this case, the model is able to give three kinds of information:

1. The profiles of concentration of solvent expanded through the thickness of the paint, and this knowledge is important for the drying of the first layer.

2. The model led to the idea of carrying out experiments during the deposition of the second layer, in order to get a knowledge on the change in the concentration of solvent.
3. The kinetics of drying are obtained.

Of course, simulation of the process of drying is possible, and as fast as allowed by the computer. Thus, the effect of each parameter on the process of drying can be determined, and the operational conditions can be optimized.

***Remark 1.*** Equation expressing the concentration in the paint.

Equation (12-6) is written by considering a constant diffusivity [7].

As the diffusivity was found to depend on the concentration of the solvent [7, 9], a better equation is:

$$CN(n) = C_n + J_{n-1/2} - J_{n+1/2}$$

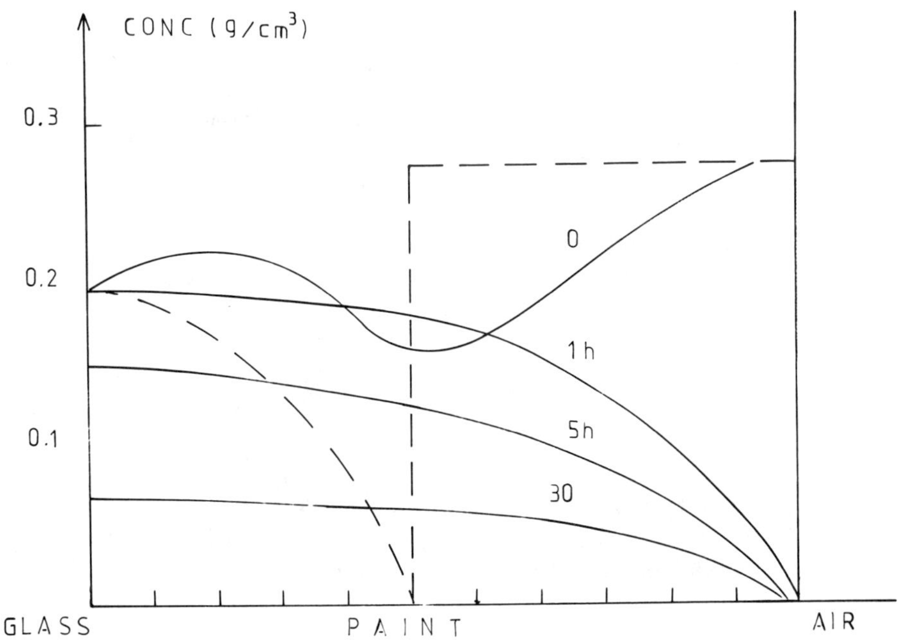


**Figure 12–16** Profiles of concentration of solvent calculated through the two-layer coating, at various times, with the first layer dried for 1 h. (Reprinted from *J. Coat. Technol.*, 59, (1987), 27–32, H. P. Blandin, J. C. David, and J. M. Vergnaud, ref. 9, by courtesy of Fed. Soc. for Coatings Technology.)

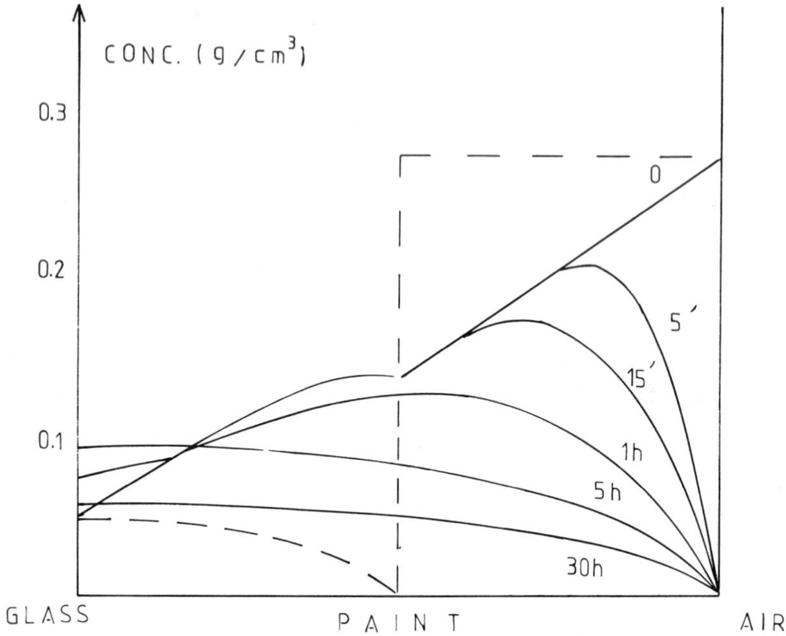


**Figure 12–17** Profiles of concentration of solvent calculated through the two-layer coating, at various times, with the first layer dried for 14 h. (Reprinted from *J. Coat. Technol.*, 59, (1987), 27–32, H. P. Blandin, J. C. David, and J. M. Vergnaud, ref. 9, by courtesy of Fed. Soc. for Coatings Technology.)

with the functions $J$ defined as follows:

$$J_{n-1/2} = \frac{\Delta t}{(\Delta x)^2} \cdot D_{n-1/2} \cdot (C_{n-1} - C_n)$$

$$J_{n+1/2} = \frac{\Delta t}{(\Delta x)^2} \cdot D_{n+1/2} \cdot (C_n - C_{n+1})$$

The concentration of the solvent at the position $n + \frac{1}{2}$ (and $n - \frac{1}{2}$) can easily be obtained by considering, for instance, a linear variation of the concentration between the planes $n$ and $n + 1$:

$$C_{n+1/2} = \frac{1}{2}(C_n + C_{n+1})$$

The diffusivity $D_{n+1/2}$ is then calculated in terms of this concentration.

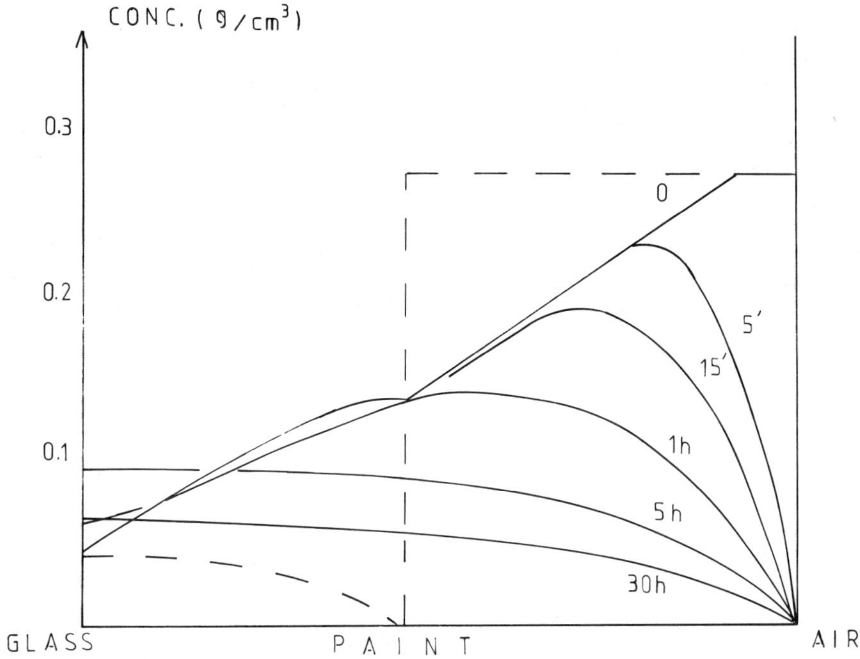


**Figure 12–18** Profiles of concentration of solvent calculated through the two-layer coating, at various times, with the first layer dried for 24 h. (Reprinted from *J. Coat. Technol.*, 59, (1987), 27–32, H. P. Blandin, J. C. David, and J. M. Vergnaud, ref. 9, by courtesy of Fed. Soc. for Coatings Technology.)

## REFERENCES

1. Y. Khatir, J. Bouzon, and J. M. Vergnaud, "Liquid sorption by rubber sheets and evaporation. Models and experiments," *J. Polymer Testing*, 6, (1986), 253–267.
2. D. Messadi and J. M. Vergnaud, "Simultaneous diffusion of benzyl alcohol into plasticized PVC and of plasticizer from polymer into liquid," *J. Appl. Polym. Sci.*, 26, (1981), 2315–2324.
3. J. M. Vergnaud, "Scientific aspects of plasticizer migration from plasticized PVC into liquids," *Polym-Plast. Technol. Eng.*, 20, (1983), 1–22.
4. J. L. Taverdet and J. M. Vergnaud, "Study of transfer process of liquid into and plasticizer out of plasticized PVC by using short tests," *J. Appl. Polym. Sci.*, 29, (1984), 3391–3400.
5. H. Liu, J. Y. Armand, J. Bouzon, and J. M. Vergnaud, "Effect of sample size and heating rate on the DSC process for reactions of high enthalpy," *Thermochim. Acta*, 126, (1988), 81–92.
6. J. Y. Armand and J. M. Vergnaud, "Effect of the value of heating rate in DSC

on the kinetic parameters, when there is high enthalpy of reaction," *Thermochim. Acta*, 131, (1988), 15–27.
7. H. P. Blandin, J. C. David, and J. M. Vergnaud, "Modeling of drying of coatings: effect of the thickness, temperature and concentration of solvent," *Progr. in Organ. Coatings*, 15, (1987), 163–172.
8. E. G. Shur, "Comparative evaporation rates of paint solvent (Part 1)," *Official Digest*, 27, (1955), 813–817.
9. H. P. Blandin, J. C. David, J. P. Illien, M. Malizewicz, and J. M. Vergnaud, "Modeling the drying process of coatings with various layers," *J. of Coating Technology*, 59, (1987), 27–32.

# Chapter Thirteen

# Absorption and Desorption of Moisture by Wood

## 13-1 INTRODUCTION

In the same way as living trees, pieces of wood used for houses exchange moisture with the surrounding atmosphere. In fact, wood is hygroscopic material, and its various physical properties such as dimensions, mass and density, as well as its mechanical properties are affected by the moisture content. Moreover, the problem of absorption of moisture is of high concern, because a moisture content in the wood higher than around 20 wt % is responsible for fungal attacks [1, 2]. As a result, it is of interest to have a good knowledge of the moisture content in wood, as well as moisture transport in relation to the environment.

Wood gains or loses moisture to the surrounding atmosphere, approaching the equilibrium moisture content (EMC). EMC is a function of various parameters characterizing the atmosphere such as the relative humidity of atmosphere (RH) and temperature as well as others concerned with the kind of wood and its previous history.

The rate of the transport of moisture is related to some of these preceding parameters, especially RH, temperature, and kinds of wood. It depends also on other parameters such as the velocity of atmosphere surrounding the sample, the value of moisture content and the presence of gradients of concentration within the wood.

Two resistances to the transport of moisture exist, the external resistance and the internal resistance, the sum of them being called total resistance [3]. Thus, two diffusivities are considered: the true diffusivity which characterizes the sole internal resistance, and the apparent diffusivity due to the presence of these two resistances.

Between the wood surface and surrounding atmosphere, the rate of transport is controlled by a convective matter transport called the surface emission coefficient [5]. This external resistance, resulting from the presence of a boundary layer, is proportional to the difference between the moisture content on the surface of wood and the value at the same position which is at equilibrium with the atmosphere [3, 4].

The rate of transport of moisture within the wood results from the superimposition of two movements [5]: (1) the vapor diffusion through the void structure, and (2) the diffusion of bound water through the cell walls.

The movements of moisture in wood at moisture contents below the fiber saturation point (FSP) have been described by transient diffusion, and many attempts have been made to measure the diffusivity either for the bound-water transport [6] or for the combined transport of bound water and water vapor [5, 7–10].

In spite of the fact that the variation of diffusivity with the moisture content in wood depends on the technique of determination [8, 11], some results of interest have been found.

1. Three main directions exist in wood: longitudinal, radial, and tangential. Very often, the sum of radial and tangential transports is called transverse transport.
2. The longitudinal diffusion is controlled by the vapor flow through the cell cavities [9], and the longitudinal diffusivity decreases with increasing moisture content at all temperatures [3, 9, 12].
3. The transverse diffusion is limited by the rate of bound-water flow through the cell walls, and the transverse diffusivity increases with the wood moisture content at all temperatures [3, 9, 12].
4. The radial diffusion increases largely with the moisture content, while the tangential diffusivity remains constant with moisture content [5].
5. Exponential relationships have been found to describe the dependency of longitudinal and radial diffusivities on moisture content during absorption [3, 5, 10, 13].
6. Temperature is also an important parameter for diffusion, and generally the diffusivity is increased with increasing temperature.

The main purpose in this chapter is to build mathematical models able to describe the kinetics of absorption and desorption of moisture by pieces of wood of various shapes. These models based on numerical methods with finite differences take all the known facts into account: the flux of moisture transport onto the wood surface and the diffusion of moisture within the wood with concentration-dependent diffusivities.

Experiments are also carried out with rather thin sheets of wood under definite conditions of temperature, RH, and moisture content, in order to obtain data such as diffusivities in terms of the moisture content. These experiments are performed by considering the transport of moisture in one direction, by preventing the sample from the other two transports with the help of impermeable paper.

Consequently, the models must be able to predict the process for rather big pieces of wood, by using all data determined with small aliquot samples taken from the real large piece, when the moisture is transported along one, two, or three directions.

## 13-2 MODELING OF MOISTURE EXCHANGE IN ONE DIMENSION (WITH CONSTANT AND PROGRAMMED RELATIVE HUMIDITY) BY WOOD

This chapter is concerned with the study of the kinetics of absorption and desorption of moisture (vapor) by pieces of wood, in order to be able to build a model for these processes.

Small pieces of wood are used for this study, and each main axis of diffusion within the wood sample is studied separately, by maintaining only the two opposite faces in contact with atmosphere.

The model, based on a numerical method with finite differences, takes into account the moisture transport by diffusion through the wood and the moisture transport between the atmosphere and the surface of wood. Two examples are given in this chapter:

1. One with the absorption and desorption, by considering a constant relative humidity (R.H.) of the surrounding atmosphere.
2. Another with the absorption of moisture, by considering a programmed relative humidity of the atmosphere.

### 13-2-1 Experimental

**Materials.** Cubes of a side of 2 cm of Scotspine sapwood (Pinus sylvestris) are used. Diffusion along one main axis is studied, by contacting the sample with the surrounding atmosphere (with the desired RH). The two opposite faces of the sample can exchange moisture, while the other four faces are kept away from moisture with an impermeable film. Experiments have been made for checking the validity of this fact. As a result from these preliminary studies, the uptake of moisture occurs by unidirectional diffusion through the sample.

**Apparatus and experiment.** An oven (Couprie SP4) with a temperature control (relative error of 0.5°C) and for relative humidity of air (relative error of 2 percent), with an air stirring system, is used.

The wood sample, previously equilibrated in air at a given RH at 30°C, is in contact with air at a higher RH for absorption and at a lower RH for desorption. The sample is weighed at intervals. A reference weight for the sample is determined when the sample is dried to a constant weight at 103°C.

In the case of programmed relative humidity of air, the RH is programmed with a constant rate.

### 13-2-2 Theoretical

**Assumptions.** The following assumptions are made in order to describe the process:

1. The sample is homogeneous, and the initial concentration of moisture is uniform through the wood.
2. The flux of moisture through the air-wood interface is proportional to the difference of concentrations on the wood surface: the actual one at the time of experiment, the other which is at equilibrium with the surrounding atmosphere.
3. The moisture is transported through the wood by transient diffusion, below the fiber saturation point (FSP).
4. The diffusivity obtained from experiments is concentration dependent.
5. Changes in dimensions during absorption and desorption are neglected.
6. The relative humidity of air is either kept constant at the desired value or programmed with a constant rate.

**Mathematical treatment.** The flux of moisture through the air-wood interface is defined by:

$$F = K \cdot (C_{eq} - C_{t,o}) \tag{13-1}$$

where $K$ is the coefficient of moisture transfer, $C_{t,o}$ is the concentration on the wood surface at time $t$, and $C_{eq}$ is the concentration on the wood surface at equilibrium with the surrounding atmosphere. Of course, $C_{eq}$ is obtained on the surface after an infinite time of contact.

The diffusion of moisture in one dimension in the wood is governed by Fick's law.

$$\frac{\partial C}{\partial t} = \frac{\partial}{\partial x}\left(D \cdot \frac{\partial C}{\partial x}\right) \tag{13-2}$$

where $D$ is the diffusivity.

**Numerical analysis.** No analytical solution can be found for the preceding equations with concentration-dependent diffusivity, and the problem is solved by using a numerical method with finite differences.

The cross section of the wood sample is divided into $N$ slices of constant thickness $\Delta x$, by concentration-reference planes (Figure 13–1).

The matter balance is calculated in the plane $n$, and the new concentration $CN_n$ after elapse of time $\Delta t$ is obtained in terms of the concentrations at the same plane $n$ and adjacent planes.

Inside the wood:

$$CN_n = \frac{1}{M_n}\left[C_{n-1} + (M - 2) \cdot C_n + C_{n+1}\right] \tag{13-3}$$

On the wood surface:

$$CN_o = \frac{1}{M_o}\left[2NC_{eq} + (M - 2N - 2) \cdot C_o + 2C_1\right] \tag{13-4}$$

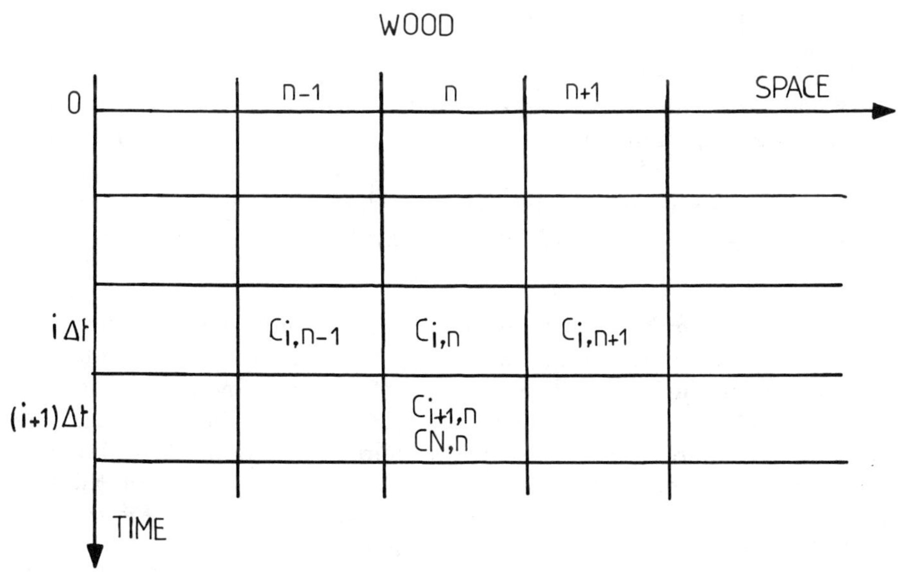


**Figure 13–1** Space-time diagram for calculating moisture concentration.

where $C_n$ and $C_o$ are the concentrations at position $n$ and on the surface, respectively.

The dimensionless numbers $M$ and $N$ are expressed by:

$$M_n = \frac{(\Delta x)^2}{\Delta t} \cdot \frac{1}{D_n} \tag{13-5}$$

$$N = \frac{K \cdot \Delta x}{D} \tag{13-6}$$

where $\Delta x$ and $\Delta t$ are the increments of space and time.

The amount of moisture located in the wood is obtained by integrating the concentration of moisture with respect to space.

$$M_t = \sum_{n=0}^{N} C_n \tag{13-7}$$

In the case of programmed relative humidity of air, the value of moisture concentration on the sample which is at equilibrium with the actual value of RH, $C_{eq}$, is programmed with time in the same way.

### 13-2-3 Results with Constant RH of Air

**Diffusivity and coefficient of moisture transport.** The rate of transport of moisture through the air-wood interface is controlled by the surface emission coefficient (or coefficient of moisture transport). This can be determined by two ways:

- Either with the method [14] based on Newman's solutions to the diffusion equation in which the external resistance is expressed in terms of the coefficient of moisture transport.
- Or with the method which has been successfully applied to the drying of coatings [15] and the desorption of rubber previously saturated by toluene [16].

The second method is more simple, since only one experiment is needed. The flux of moisture $F$ transported through the air-wood interface is obtained from the kinetics of absorption of moisture, by determining the rate of transport at the beginning of the process. The coefficient of moisture transport $K$ is then calculated with the help of equation (13-1). For this very short absorption, the value of the moisture concentration on the surface $C_o$ is known, while $C_{eq}$, the potential concentration of moisture at equilibrium, can be determined by experiments.

The values of the coefficient of moisture transport $K$ are given in Table 13–1. These values, expressed in cm/s, are about constant for various values of the relative humidity of air ranging from 55 percent to 80 percent.

**TABLE 13–1** Diffusivity and Coefficient of Matter Transport

| RH (%) | MC (%) | $M_\infty$ (%) | $D_1 10^7$ ($cm^2/s$) | $K$ (cm/s) |
|---|---|---|---|---|
| 55–60 | 11.20 | 0.39 | 7.85 | 0.57 |
| 60–65 | 11.61 | 0.43 | 7.05 | 0.53 |
| 65–70 | 12.03 | 0.41 | 5.8 | 0.57 |
| 70–75 | 12.44 | 0.42 | 5.0 | 0.60 |
| 75–80 | 12.85 | 0.40 | 4.3 | 0.64 |

***Diffusivity.*** The rate of moisture transport within the wood is quite different according to the main axis chosen for the wood. The rate of transport through the longitudinal direction is much higher than the rate through the transversal one. The transversal transport is practically negligible compared with the longitudinal one in the case of cubic samples. This fact appears clearly in Figure 13–2 where the kinetics of the longitudinal absorption are nearly equal to the kinetics of the total absorption (longitudinal + transversal). In this figure, the wood previously equilibrated with air at RH = 55%, is then immersed into air at RH = 80%, the temperature being maintained at 30°C.

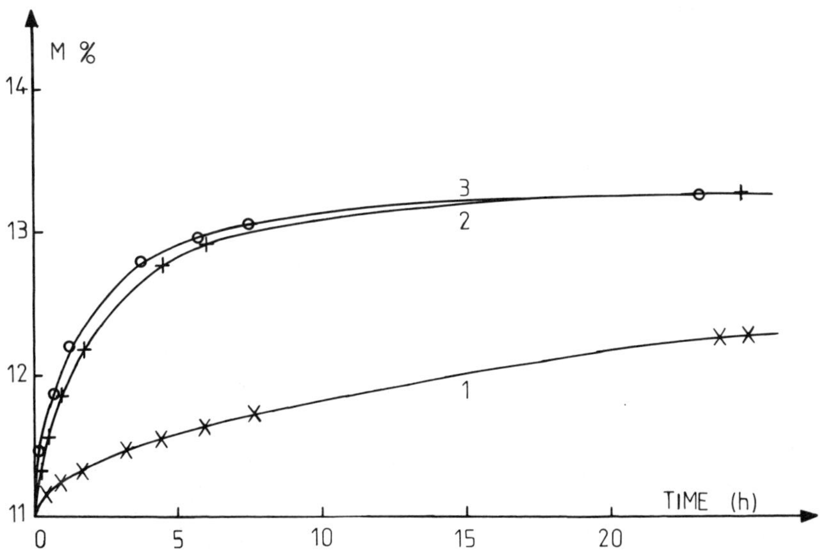


**Figure 13–2** Kinetics of absorption by a cube of 2 cm, by considering various diffusions: 1 = transversal; 2 = longitudinal; 3 = longitudinal + transversal. RH of air = 55 and 80%. (Reprinted from *Wood Sci. Technol.*, 22, (1988), 11–20, A. Droin, J. L. Taverdet, and J. M. Vergnaud, ref. 17, by courtesy of Springer-Verlag.)

Straight lines are obtained by plotting the amount of moisture absorbed as a function of the square root of time (Figure 13–3). These results can be explained by Fickian diffusion, and they are of interest to determine diffusivity with the help of equation (13-8).

$$\frac{M_t}{M_\infty} = \frac{4}{L}\left(\frac{D \cdot t}{\pi}\right)^{0.5} \tag{13-8}$$

where $M_t$ and $M_\infty$ are the amounts of moisture transported at time $t$ and infinite time, and $L$ is the thickness of the cube.

As it is desirable to know the diffusivities at various moisture contents, successive absorptions of moisture have been conducted with the same sample. From one step to the following one, the RH of air is increased by 5 percent in a discontinuous way. Between two steps, the time is chosen so that the equilibrium of absorption is attained (Figure 13–4). The sample used at the beginning of the process is previously equilibrated with air at RH = 55% at 30°C.

The values of longitudinal diffusivity decrease when the moisture content is increased from 11 percent to 12.85 percent. These values are of the same order of magnitude as those obtained previously under similar conditions (RH ranging from 39 percent to 47 percent) [12].

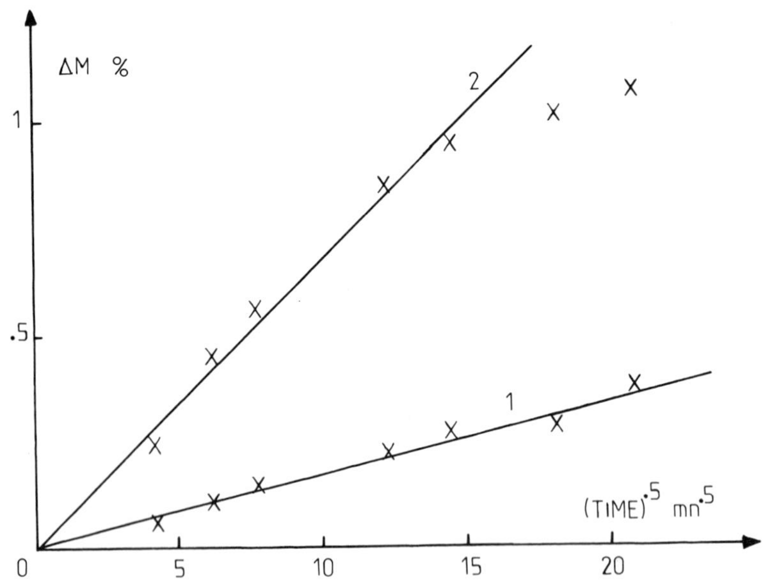


**Figure 13–3** Moisture absorbed (in percent of the dried wood) versus (time)$^{0.5}$, at 30°C, 1 = transversal; 2 = longitudinal. (Reprinted from *Wood Sci. Technol.*, 22, (1988), 11–20, A. Droin, J. L. Taverdet, and J. M. Vergnaud, ref. 17, by courtesy of Springer-Verlag.)

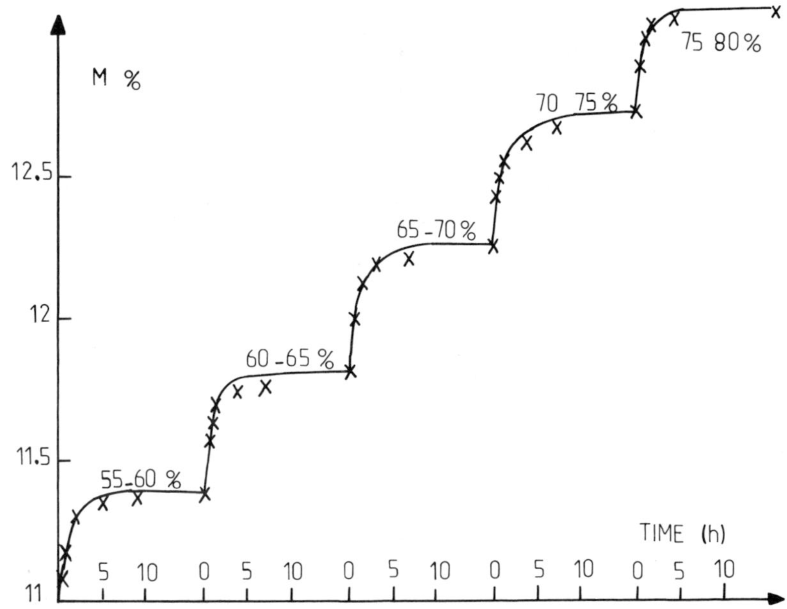


**Figure 13–4** Longitudinal absorption of moisture step by step, at 30°C, x: experiments. –: calculated. (Reprinted from *Wood Sci. Technol.*, 22, (1988), 11–20, A. Droin, J. L. Taverdet, and J. M. Vergnaud, ref. 17, by courtesy of Springer-Verlag.)

**Validity of the model and mathematical simulation.** The amount of moisture absorbed by the wood as a function of time is calculated when the RH of air is increased by 5 percent step by step, by using the model and the values of diffusivity and coefficient of matter transport [17]. As show in Figure 13–4, a rather good agreement is obtained between the experimental and calculated results.

Moreover, the model is able to calculate the profiles of concentration of moisture developed through the thickness of the sample. It is very difficult to determine these gradients by experiments, as shown previously in the case of plasticized PVC [18]. However, a proof for their validity exists, because the kinetics of absorption of moisture are obtained by integrating these concentrations with respect to the thickness of the sample. These profiles are calculated in the case of longitudinal absorption, when the RH of air is increased from 55 percent to 60 percent (Figure 13–5).

### 13-2-4 Results with Programming of the RH of Air

**Diffusivity and moisture content.** As the relative humidity of air is varied continuously during the process, it is necessary to obtain a rela-

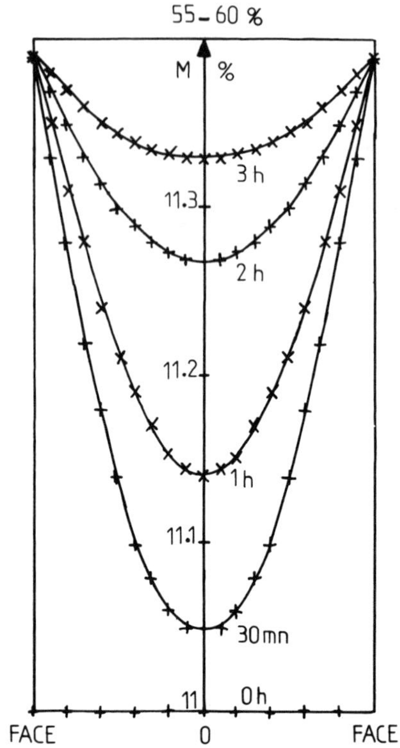


**Figure 13–5** Profiles of moisture concentration inside the wood in the case of longitudinal absorption when RH is increased from 55 to 60%. (Reprinted from *Wood Sci. Technol.*, 22, (1988), 11–20, A. Droin, J. L. Taverdet, and J. M. Vergnaud, ref. 17, by courtesy of Springer-Verlag.)

tionship between the RH of air and the moisture content at equilibrium with this RH.

The moisture content at equilibrium is measured successively for various RH of air at constant temperature (30°C). The isotherm of absorption is obtained from these measurements.

Within the 55 percent to 80 percent range of RH, the moisture content at equilibrium with air can then be expressed in terms of the RH by a linear function:

$$\text{EMC} = 0.088 \times \text{RH} + 6.15 \tag{13-9}$$

The diffusivity is determined by using the short-test technique with plane sheets [17, 19] under transient conditions. For each experiment (RH increasing by 5 percent), the diffusivity is calculated during a short time, and the moisture content in wood may be considered constant. So the moisture content corresponding to each value of diffusivity may be taken as the initial value of moisture content.

The variation of the transverse diffusivity with moisture content is drawn in Figure 13–6, showing a linear relation between the logarithm of diffusivity and moisture content (in percent).

$$D_t = 1.5 \times 10^{-6} \times \exp(0.0228 \times \text{MC}) \quad (\text{cm}^2/\text{s}) \tag{13-10}$$

**Validity of the model in classical absorption.** Two wood samples of various thicknesses (0.5 cm and 2 cm), previously equilibrated with air of RH = 55% at 30°C, are in contact with air of RH = 80%. The kinetics of transverse absorption of moisture are drawn (Figure 13–7) for these two samples.

Two results are worth noting for the kinetics of absorption:

1. A good agreement is obtained between experiments and theoretical values, proving the validity of the model.
2. The effect of the thickness is clearly shown. The equilibrium is almost reached at 13 h for the thin sample, and a time of 52 h is necessary

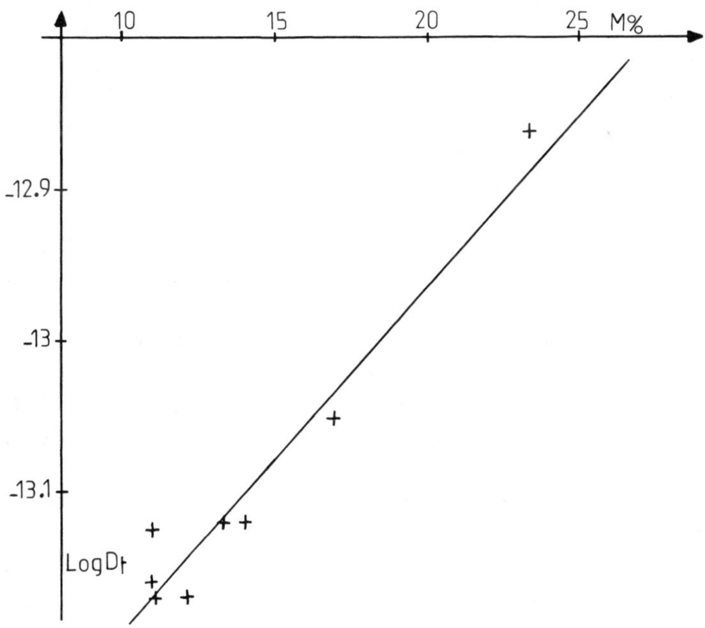


**Figure 13–6** Logarithm of transverse diffusivity versus moisture content (percent of dried wood) at 30°C. (Reprinted from *Wood Sci. Technol.*, 22, (1988), 299–310, A. Droin, J. L. Taverdet, and J. M. Vergnaud, ref. 20, by courtesy of Springer-Verlag.)

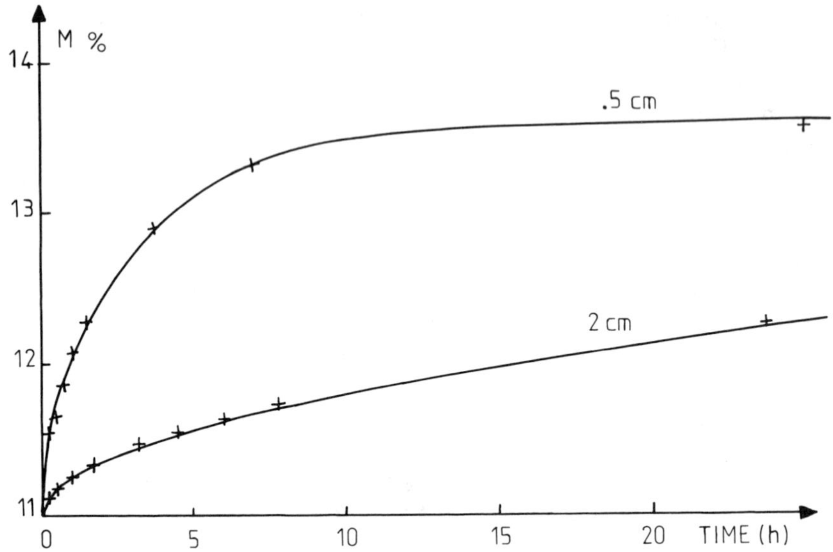


**Figure 13–7** Kinetics of transverse absorption of moisture by sample of various thicknesses (0.5 and 2 cm) at 30°C. Wood equilibrated with RH = 55%. Air at RH = 80%. (Reprinted from *Wood Sci. Technol.*, 22, (1988), 299–310, A. Droin, J. L. Taverdet, and J. M. Vergnaud, ref. 20, by courtesy of Springer-Verlag.)

for the thick sample to attain equilibrium. A proportionality exists between the square of the thickness of the sample and time, as shown in equation (13–8).

The profiles of moisture concentration are drawn (Figure 13–8), as they are calculated through the thickness of the two wood samples (0.5 cm and 2 cm thick).

**Validity of the model with programming of RH of air.** The relative humidity of air is varied with a constant rate (5 percent RH per hour), thus increased from 55 percent to 80 percent and decreased from 80 percent to 55 percent, at a constant temperature (30°C). The two wood samples of thickness (0.5 cm and 2 cm), previously equilibrated with air of RH = 55%, are contacted with this air of varying RH, and the transverse absorption and desorption are determined [20].

Experimental and theoretical kinetics of moisture transport are shown in Figure 13–9, with the following results:

1. A rather good agreement is obtained, especially for the thicker sample during the whole process of absorption and desorption, and for the thinner sample at the beginning of the process during the absorption process.

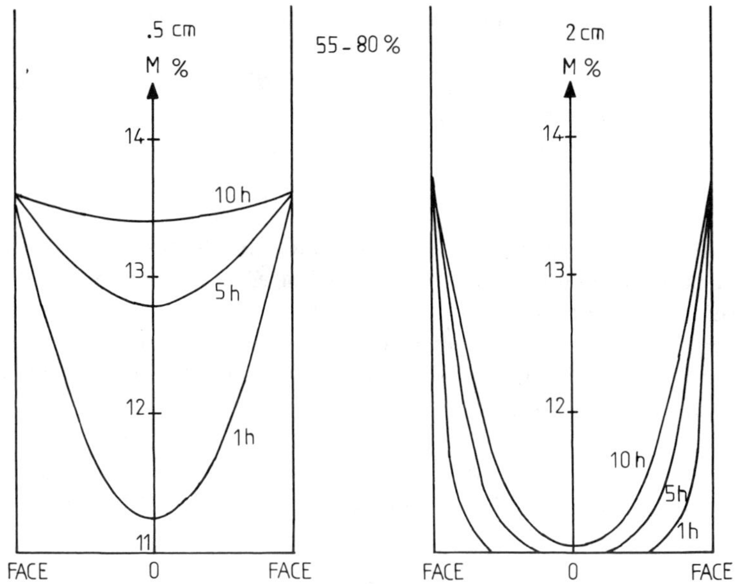


**Figure 13–8** Profiles of moisture concentration through the thickness of samples at 30°C (0.5 and 2 cm thick). (Reprinted from *Wood Sci. Technol.*, 22, (1988), 299–310, A. Droin, J. L. Taverdet, and J. M. Vergnaud, ref. 20, by courtesy of Springer-Verlag.)

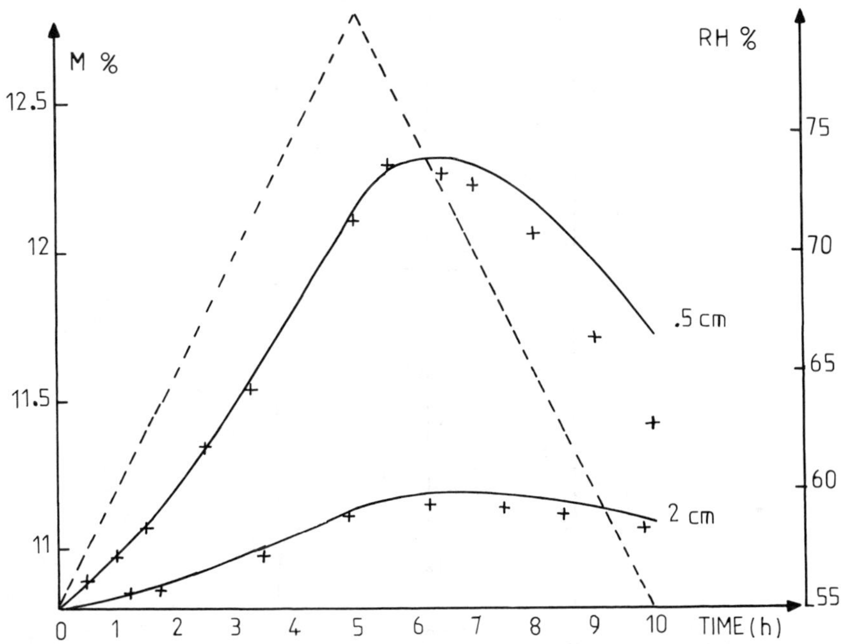


**Figure 13–9** Variation of MC in wood versus time, when RH of air is increased (55–80%) and decreased (80–55%), at 30°C (0.5 and 2 cm thick). (Reprinted from *Wood Sci. Technol.*, 22, (1988), 299–310, A. Droin, J. L. Taverdet, and J. M. Vergnaud, ref. 20, by courtesy of Springer-Verlag.)

2. The curves (experimental and calculated) are shifted to longer time, with respect to the linear RH history with time. The retardation in the absorption is also observed for the maximum of the moisture content which does not correspond with the maximum of the RH of air.
3. A rather poor agreement is obtained for the desorption with the thinner sample. (This is perhaps due to intempestive additional desorption done when the sample is weighed, this loss in weight being proportional to the moisture content of wood and to the surface of wood in contact with air of course, the relative loss in weight is higher for the thinner sheet).

The profiles of concentration of moisture developed through the thin sample (0.5 cm) (Figure 13–10) and the thick sample (2 cm) (Figure 13–11)

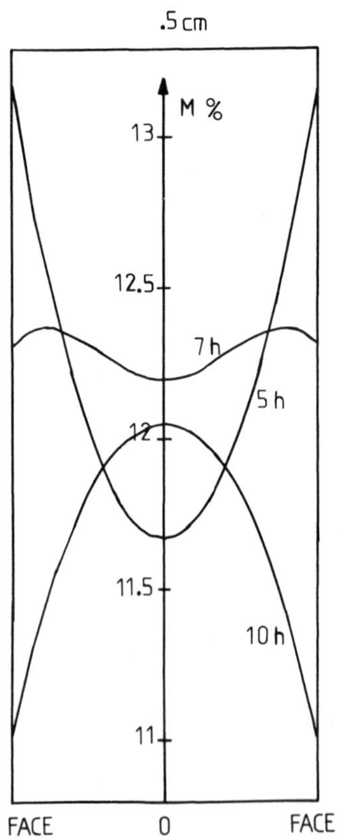


**Figure 13–10** Profiles of concentration of moisture through the thickness (0.5 cm) of wood with programmed RH (5 RH per h). (Reprinted from *Wood Sci. Technol.*, 22, (1988), 299–310, A. Droin, J. L. Taverdet, and J. M. Vergnaud, ref. 20, by courtesy of Springer-Verlag.)

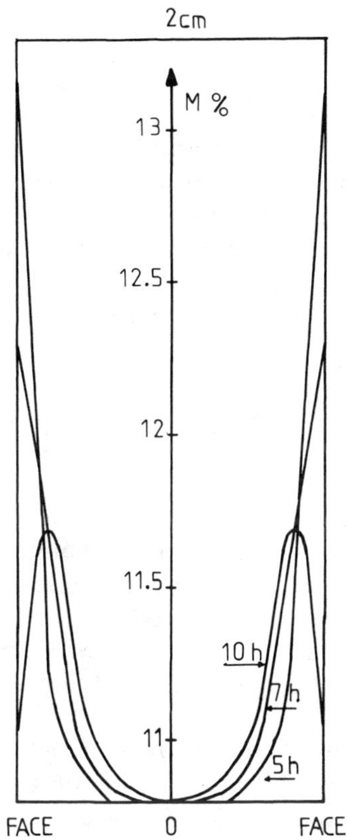


**Figure 13–11** Profiles of concentration of moisture through the thickness (2 cm) of wood with programmed RH (5 RH per h). (Reprinted from *Wood Sci. Technol.*, 22, (1988), 299–310, A. Droin, J. L. Taverdet, and J. M. Vergnaud, ref. 20, by courtesy of Springer-Verlag.)

are calculated during the process of absorption-desorption. The following facts can be drawn from these complex profiles:

1. The absorption is far from equilibrium after 7 h when the maximum RH is attained in the wood, even in the thin sheet. The equilibrium is nearly obtained in the case of classical absorption with constant RH.
2. The rate of variation for the RH is low enough to provoke a large change in the concentration of moisture throughout the thin sample. It is too high to provoke a pertubation within the thick sample. From calculation with the modulus *M*, it is shown that a rate of 5/16 RH per hour can be responsible for the same profiles of concentration throughout the thick sample as those obtained within the thin sample with the high rate of 5 RH per hour. The value of 16 corresponds with the square of the ratio of thicknesses of these two samples.

### 13-2-5 Conclusions

The mathematical model based on numerical methods with finite differences is able to describe the process of absorption and desorption of moisture when the RH of surrounding air is constant, and, when the RH of air is changed. In all these cases, a one-dimensional transport of moisture occurs through the thickness of the wood.

The diffusivity depends generally on the local moisture content, and the concentration dependence of diffusivity is obtained by using short tests at various RH of surrounding atmosphere.

## 13-3 MODELING OF MOISTURE EXCHANGE BY WOOD IN TWO DIMENSIONS

The first purpose of this chapter is to show that mathematical models are able to describe the process of moisture exchange in more complex cases, for example, when the moisture transports by longitudinal and transverse diffusion in the wood are superimposed.

The problem considered is that of a two-dimensional transport obtained with parallelepipedic boards in the following two cases:

1. The moisture transport in the board is obtained only along two directions: the longitudinal and one of the two transverse directions, the other third being stopped or at least hindered by any ways.
2. One of the two transverse directions of the board is very long with regard to the other, and it is necessary to know the profiles of moisture concentration developed through the cross section perpendicular to this transverse direction (Figure 13–12).

It is also interesting to know the contribution of these two kinds of transport of moisture at various places in the cross section of the board. The effect of the ratio of the length of the board in the two dimensions: length of board with longitudinal diffusion, thickness of board with transverse direction is also of concern [21].

### 13-3-1 Experimental

**Materials.** Parallelepipedic samples of Scotspine sapwood (Pinus sylvestris) with various dimensions are used:

- Cubic samples of 2 cm per side, for determining the diffusivity of moisture.

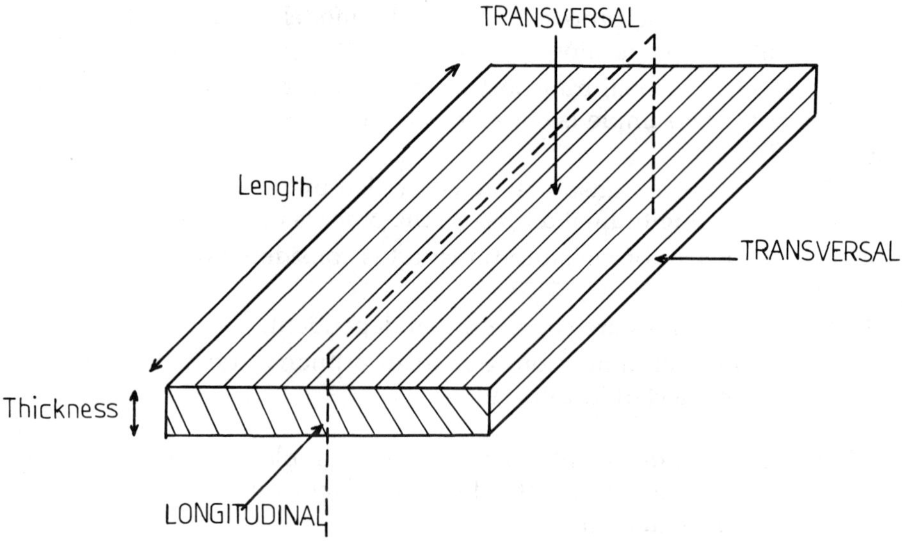


**Figure 13–12** Scheme of the two directions of moisture transport through the wood.

- Samples of thickness 1 cm and length of 1 cm, 4 cm, and 10 cm for studying the bidimensional transport of moisture.

These samples are cut in such a way that longitudinal diffusion takes place along the longer dimension (1 cm, 4 cm, and 10 cm) and that transversal diffusion is composed of an equal contribution of radial and tangential diffusion.

**Apparatus and experiment.** An oven (Couprie SP4) with a temperature control (error of 0.5°C) and of relative humidity of atmosphere (error of 2 percent for RH) is used.

After equilibration in air at RH = 55% at 30°C, the samples are contacted with air kept at a higher RH for absorption. The change in weight of samples is measured at intervals. The reference weight of each sample is determined after drying to constant weight at 103°C. The moisture content (MC) in wood is calculated with regard to this reference weight.

### 13-3-2 Theoretical

**Assumptions.** Some assumptions are made in order to clarify the problem:

1. The initial concentration of moisture is uniform in each sample at the

beginning of the process. (In fact, the model can be used whatever the initial concentration, provided that this profile is known.)

2. A two-dimensional transport of moisture through the wood is considered. The sample is cut along the main axes of the wood (Figure 13–12).
3. These moisture transports are controlled by transient diffusion. The one is the longitudinal diffusion, the other is the transverse transfer.
4. The coefficient of moisture transport is considered on the wood surface.
5. Both diffusivities are concentration dependent.
6. The changes of dimensions due to absorption or desorption of moisture are neglected in calculation.

**Mathematical treatment.** As the wood sample is cut along the principal axes of the original tree, the diffusion of moisture is described by the two-dimensional equation:

$$\frac{\partial C}{\partial t} = \frac{\partial}{\partial L}\left(D_1 \cdot \frac{\partial C}{\partial L}\right) + \frac{\partial}{\partial T}\left(D_t \cdot \frac{\partial M}{\partial T}\right) \tag{13-11}$$

$D_1$ and $D_t$ being the principal diffusivity along the principal axes of diffusion.

On the surface of the wood, the flux of moisture transfer $F$ is expressed in terms of the coefficient of moisture transfer $h$.

$$F = h \cdot (C_{eq} - C_o) \tag{13-12}$$

where $C_o$ and $C_{eq}$ are the moisture concentrations on the surface of wood at any time, and after infinite time at equilibrium, respectively.

**Numerical analysis.** As no analytical solution can be found for the problem, an explicit numerical method with finite differences is used.

The cross section of the sample considered (Figure 13–13) is divided into a number of equal finite slices of thickness $\Delta L$ and $\Delta T$ along the principal axes of diffusion. The moisture balance is written on the slice centered on the position of coordinates $i$, $j$, by using Fick's first law.

$$F_{i,j+1/2} = -D_{i,j+1/2} \cdot \left(\frac{\partial C}{\partial L}\right)_{i,j+1/2} \tag{13-13}$$

The gradient of concentration at the position $(i,j + \frac{1}{2})$ is approximated by the chord slope between position $(i,j)$ and $(i,j + 1)$.

$$\left(\frac{\partial C}{\partial L}\right)_{i,j+1/2} = \frac{C_{i,j} - C_{i,j+1}}{\Delta L} \tag{13-14}$$

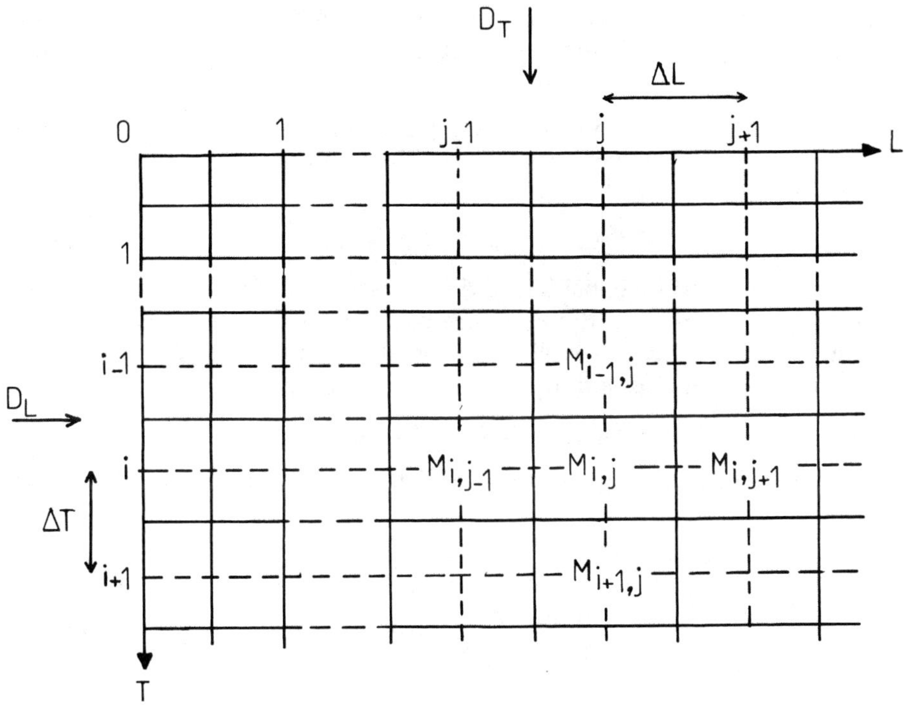


**Figure 13–13** Diagram for calculation of moisture concentration with two-dimensional transport: longitudinal $L$ and transversal $T$.

From these calculations, the new concentration after elapse of time $\Delta t$ is obtained at any place in terms of the previous concentrations at the same place and adjacent places.

***Inside the Wood.***

$$CN_{i,j} = C_{i,j} + \frac{C_{i+1,j} - C_{i,j}}{G_{T,i+1/2,j}} - \frac{C_{i,j} - C_{i-1,j}}{G_{T,i-1/2,j}} + \frac{C_{i,j+1} - C_{i,j}}{G_{L,i,j+1/2}} - \frac{C_{i,j} - C_{i,j-1}}{G_{L,i,j-1/2}} \tag{13-15}$$

with the function $G$ expressed in terms of diffusivity

$$G_{L,i,j+1/2} = \frac{(\Delta L)^2}{\Delta t} \cdot \frac{1}{D_{L,i,j+1/2}} \tag{13-16}$$

the diffusivity at position $(i,j + \frac{1}{2})$ being calculated as follows:

$$D_{L,i,j+1/2} = \frac{1}{2}\,[D_{L,i,j+1} + D_{L,i,j}] \tag{13-17}$$

where $C_{i,j}$ and $CN_{i,j}$ are the concentrations at position $i,j$ and at time $t$, and at time $t + \Delta t$, $\Delta t$, $\Delta L$, and $\Delta T$ being the increments of time, and space along $j$ and $i$ axes.

***On the Surface of Wood.*** Two equations are used for the surface. The first is used when equilibrium is reached on the surface as soon as the process starts.

$$C_{i,o} = C_{o,j} = C_{o,o} = C_{eq} \tag{13-18}$$

The second is used by considering the transport of moisture onto the surface. In the second case, the moisture balance is calculated on the slice of thickness $\Delta L/2$ and $\Delta T$ as well as on the slice of thickness $\Delta T/2$ and $\Delta L$.

After the approximation

$$CN_{i,1/4} - C_{i,1/4} = CN_{i,o.} - C_{i,o} \tag{13-19}$$

the moisture balance is given for the slice centered on the coordinates ($i$, 0.25):

$$CN_{i,o} = C_{i,o} + \frac{2h\cdot\Delta t}{\Delta L}(C_{eq} - C_{i,o}) - \frac{2(C_{i,o} - C_{i,1})}{G_{L,i,1/2}} + \frac{C_{i-1,o} - C_{i,o}}{G_{T,i-1/2,o}} - \frac{C_{i,o} - C_{i+1,o}}{G_{T,i+1/2,o}} \tag{13-20}$$

***On the Corner.*** The matter balance is done on the slice of thickness $\Delta L/2$ and $\Delta T/2$ in contact with each corner. The new concentration at the corner $C_{o,o}$ is:

$$CN_{o,o} = C_{o,o} + \frac{2h\cdot\Delta t}{\Delta L}(C_{eq} - C_{o,o}) - \frac{2(C_{o,o} - C_{o,1})}{G_{L,o,1/2}} + \frac{2h\cdot\Delta t}{\Delta T}(C_{eq} - C_{o,o}) - \frac{2(C_{o,o} - C_{1,o})}{G_{T,1/2,o}} \tag{13-21}$$

***Amount of Moisture Located in Wood.*** The amount of moisture located in the wood sample is calculated by integrating the moisture concentration with respect to $L$ and $T$ axes.

### 13-3-2 Results

#### Diffusivity, EMC, and coefficient of moisture transport

***Diffusivity.*** The diffusivity is obtained by using the short-test technique [19] performed in transient conditions. The RH of air is increased by small increments of 5 percent and the kinetics of absorption are followed up to equilibrium for each step (Figure 13–4). The diffusivity, calculated with the help of the equation:

$$\frac{M_t}{M_\infty} = \frac{4}{L}\left[\frac{D \cdot t}{\pi}\right]^{0.5} \tag{13-22}$$

is then determined in terms of the moisture concentration.

The longitudinal diffusivity is then expressed by [21]:

$$\text{Log } D_t = -0.177 \times \text{EMC} - 8.1 \tag{13-23}$$

while the transversal diffusivity is:

$$\text{Log } D_T = 0.0228 \times \text{EMC} - 13.4 \tag{13-24}$$

the $D$s being in cm²/s.

***Moisture Content at Equilibrium (EMC or $C_{eq}$).*** The value of EMC, determined from the weight of the sample, is expressed as a linear function of the RH of air, within the 55 percent to 80 percent range of RH [20].

$$\text{EMC} = C_{eq} = 0.088 \times \text{RH} + 6.15 \tag{13-25}$$

***Coefficient of Moisture Transfer ($h$).*** As shown previously, the coefficient of moisture transport onto the surface can be determined by two ways [14, 15]. The value obtained is $h = 10^{-4}$ cm/s.

**Validity of the model**

***Kinetics of Transport.*** The wood samples with following dimensions

| | | | |
|---|---|---|---|
| longitudinal (cm) | 1 | 4 | 10 |
| transverse (cm) | 1 | 1 | 1 |

are previously equilibrated at 30°C in air of RH = 55%, and then in contact with air of RH = 80%.

The experimental and theoretical kinetics of absorption of moisture drawn in Figure 13–14 illustrate the validity of the model.

Comparisons between these four curves enable us to draw two conclusions concerned with the rate of moisture onto the surface:

1. The coefficient of moisture transport plays the role of a limiting step for the whole process, when it is coupled with the longitudinal diffusion.
2. This coefficient $h$ does not alter the whole process in the case of transverse diffusion. This is due to the relative value of the coefficient $h$ with regard to the values of longitudinal and transverse diffusivities, and also to the values of the thickness of the sample.

***Profiles of Concentrations.*** The profiles of moisture concentration are easily obtained by calculation. Two ways are used for drawing these profiles:

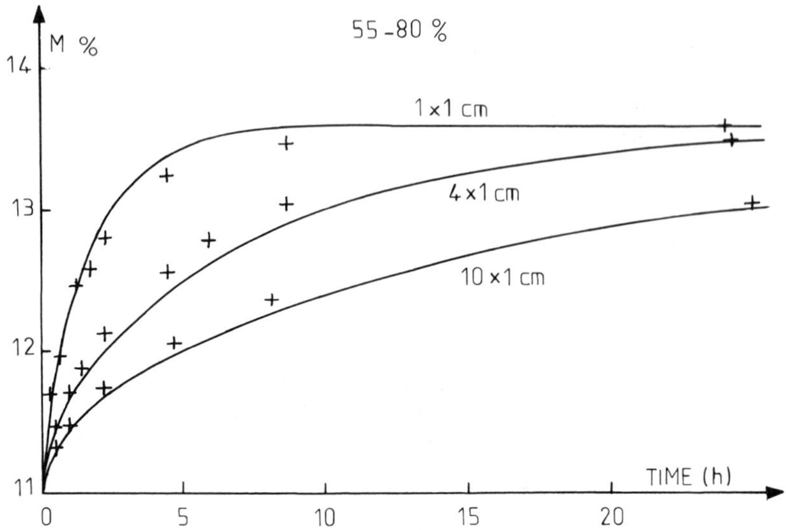


**Figure 13–14** Kinetics of moisture absorption at 30°C (55–80% RH) for three samples. +: experimental. −: calculated. (Reprinted from *Holzforschung*, 43, (1989), 297–302, A. Droin-Josserand, J. L. Taverdet, and J. M. Vergnaud, ref. 21, by courtesy of Walter de Grugter and Co.)

1. The first uses a three-axes diagram, one of these axes being concerned with the value of moisture content (Figure 13–15). In this figure, the profiles are obtained at 10 h for the sample of dimension 4 × 1 cm. It is clear that at this time the moisture content (MC) on both surfaces (in contact with longitudinal and transversal transport) has reached the value at equilibrium.
2. Another attempt for illustrating these results is made by drawing throughout the section considered in the sample the line along which there is the same value of MC, at a definite time. These profiles (iso-concentration) indicated at various times (2, 5, and 10 h), show clearly the progress of moisture within the midplane of the sample (Figure 13–16). Other profiles are drawn in the midplane of the various samples (1 × 1 cm; 4 × 1 cm; 10 × 1 cm) at the same time, when the wood is submitted to the same conditions (Figure 13-17).

These profiles of moisture concentration can give a good idea of the contribution of each diffusional transport at any place and time. Some results are drawn:

1. When these lines are parallel to each other and parallel to the surface, the transport perpindicular to this surface is preponderant.
2. The distance between two lines indicate the value of the gradient of

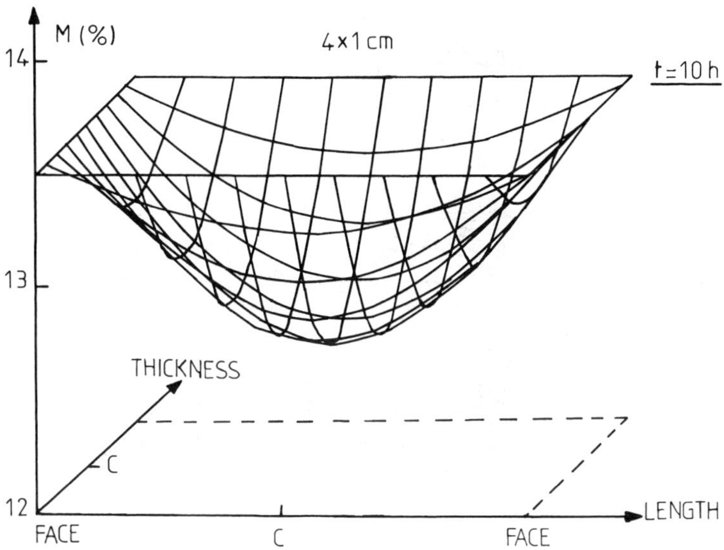


**Figure 13–15** Profiles of moisture concentration with three-axes diagram, at 10 h for the sample 4 × 1 cm. (Reprinted from *Holzforschung*, 43, (1989), 297–302, A. Droin-Josserand, J. L. Taverdet, and J. M. Vergnaud, ref. 21, by courtesy of Walter de Grugter and Co.)

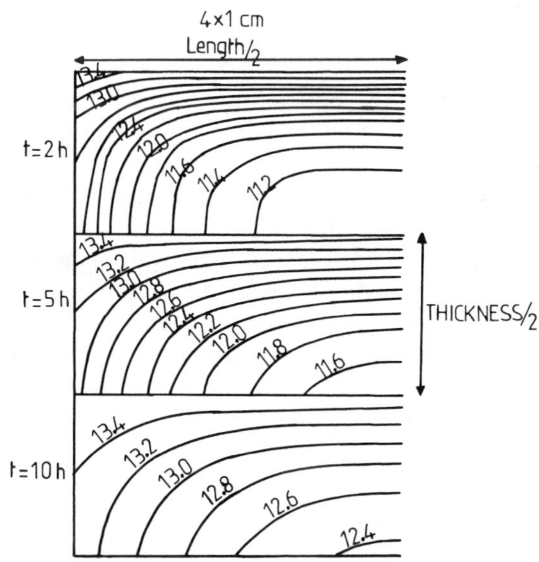


**Figure 13–16** Curves of same moisture concentration, at various times (2, 4, and 5 h) within the midplane of the sample of dimensions 4 × 1 cm. (Reprinted from *Holzforschung*, 43, (1989), 297–302, A. Droin-Josserand, J. L. Taverdet, and J. M. Vergnaud, ref. 21, by courtesy of Walter de Grugter and Co.)

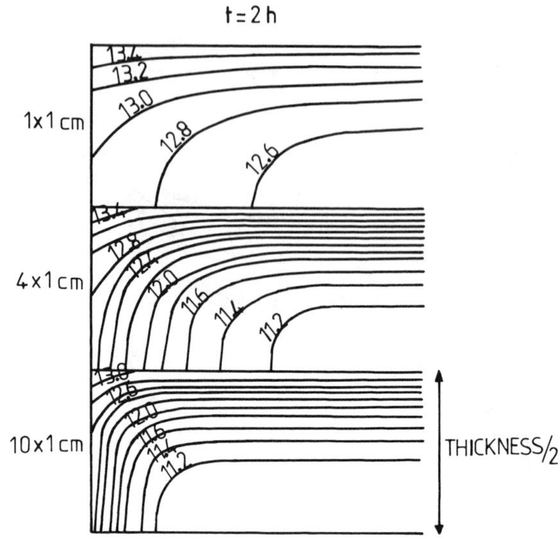


**Figure 13–17** Curves of same moisture concentration, within the midplane of various samples, at the same time (2 h). (Reprinted from *Holzforschung*, 43, (1989), 297–302, A. Droin-Josserand, J. L. Taverdet, and J. M. Vergnaud, ref. 21, by courtesy of Walter de Grugter and Co.)

moisture concentration, with the simple statement that the smaller the distance, the higher the gradients of concentration.

**Conclusions.** The mathematical model based on a numerical method with finite differences is able to describe the process of absorption of moisture by a two-dimensional system.

The model takes into account the two types of moisture transport: the external one with the coefficient of moisture transport through the air-wood interface; the internal one with diffusion of moisture along two principal longitudinal and transverse axes of diffusion.

Moreover, the profiles of moisture concentrations can be calculated, as they are developed in various cross sections of the sample. Their presentation in the way of isoconcentration curves shows clearly the contribution of each diffusional transport to the whole process.

## 13-4 MODELING OF MOISTURE EXCHANGE BY WOOD IN THREE DIMENSIONS

A first objective is concerned with the determination of the kinetics of exchange (absorption, desorption) of moisture by pieces of wood having various shapes and dimensions (sheets, boards, cubes, parallelepipeds).

Generally, pieces of wood are cut from the tree along its principal directions. Still, the problem is difficult because of the moisture transport which follows the three principal directions and of the concentration dependence of the three principal diffusivities. As no analytical solution is given, models with numerical methods are built to describe the process.

Of course, it is of interest to be able to predict the kinetics of exchange of moisture for real pieces of wood without having to do experiments on these materials, either because of time-consuming experiments with thick samples or because of the importance of the testing apparatus, without speaking of some destructive effect. A method exists [16] to attain this purpose by doing experiments on small samples and building models able to calculate the kinetics of exchange with the real sample. Two problems appear: the one with the question of how representative is a small sample taken of a large piece of wood (if it can be), the other with the complexity of the process.

Mathematical models are built by using explicit numerical methods with finite differences, and they take into account all the facts [22, 23]:

- moisture transport through the air-wood interface with the coefficient of moisture transport.
- simultaneous transport of moisture within the wood along the three principal directions
- the presence of concentration-dependent diffusivity along each of these principal directions

### 13-4-1 Experimental

**Materials.** Scots pine sapwood (Pinus sylvestris) with three shapes and various dimensions are cut out from the same tree (Table 13–2).

**Apparatus and experimentation.** Experiments are performed in an oven (Couprie SP4) with a temperature control (0.2°C) and of relative humidity (2% RH).

The wood samples, previously equilibrated in air at an RH of 55 percent and 30°C, are placed in air at a higher RH of 80 percent for studying

**TABLE 13–2** Shapes and Dimensions of Three Dimensional Samples

| Shape | Longitudinal (cm) | Radial (cm) | Tangential (cm) |
|---|---|---|---|
| Cube | 2 | 2 | 2 |
| Parallelepipede | 10 | 1 | 1 |
| Parallelepipede | 4 | 1 | 4 |

the absorption process. The samples are weighed at intervals, while their reference weight is obtained by drying them at 103°C to constant weight.

Each of the three principal diffusions (longitudinal, radial, tangential) are studied separately, in turn, by protecting the adequate surface from moisture exchange.

The coefficient of moisture transport onto the surface of wood is determined by one of the two ways described before [14, 15].

### 13-4-2 Theoretical

**Assumptions.** Some assumptions are made in order to explore the process and simplify the problem:

1. The flux of moisture transport through the air-wood interface is proportional to the difference of the moisture concentration on the surface and the concentration at equilibrium with the RH of air.
2. The moisture is transported within the wood by transient diffusion, below the fiber saturation point.
3. Three principal axes of diffusion exist with three principal diffusivities: longitudinal, tangential, and radial. The samples are cut by following these main directions.
4. Diffusivities depend on the concentration of moisture.
5. Variations in dimensions due to absorption and desorption are neglected in calculation.

**Mathematical treatment.** The transport of moisture in the wood is described by the transient Fick's law with the three principal axes of diffusion:

$$\frac{\partial C}{\partial t} = \frac{\partial}{\partial x}\left(D_t \cdot \frac{\partial C}{\partial x}\right) + \frac{\partial}{\partial y}\left(D_1 \cdot \frac{\partial C}{\partial z}\right) + \frac{\partial}{\partial z}\left(D_r \cdot \frac{\partial C}{\partial z}\right) \qquad (13\text{-}26)$$

where the $D_s$ are the principal diffusivity along the principal axes: $x$ for tangential, $y$ for longitudinal, and $z$ for radial.

The flux of moisture through the air-wood interface is given by:

$$F = h \cdot (C_{eq} - C_o) \qquad (13\text{-}12)$$

where $C_o$ and $C_{eq}$ are the actual concentration on the surface and the concentration at equilibrium with the surrounding atmosphere, and $h$ is the coefficient of moisture transport.

**Numerical analysis.** No analytical solution can be found for the preceding equations, and the problem is solved by using an explicit numerical method with finite differences.

The wood sample is divided into parallelepipedic pieces of dimensions $\Delta x$, $\Delta y$, and $\Delta z$. The small piece considered in Figure 13–18 is centered on the position defined by the integers $i$, $j$, $k$ with:

$$x = i \cdot \Delta x \qquad y = j \cdot \Delta y \qquad z = h \cdot \Delta z \tag{13-27}$$

***Within the Sample.*** The matter balance is calculated in the parallelepipedic piece centered on $i$, $j$, $k$, during the increment of time $\Delta t$. The new concentration of moisture at this position after elapse of time $\Delta t$ is thus obtained as a function of concentrations at the previous time in the same place and at the adjacent six places:

$$\begin{aligned} CN(i,j,k) = {} & C(i,j,k) + M_t\left(i + \frac{1}{2}, j,k\right) \times [C(i+1, j,k) - C(i,j,k)] \\ & - M_t\left(i - \frac{1}{2}, j,k\right) \times [C(i,j,k) - C(i-1, j, k)] \\ & + M_l\left(i, j + \frac{1}{2}, k\right) \times [C(i,j+1, k) - C(i,j,k)] \\ & - M_l\left(i,j - \frac{1}{2}, k\right) \times [C(i,j,k) - C(i,j-1, k)] \\ & + M_r\left(i,j,k + \frac{1}{2}\right) \times [C(i,j,k+1) - C(i,j,k)] \\ & - M_r\left(i,j,k - \frac{1}{2}\right) \times [C(i,j,k) - C(i,j,k-1)] \end{aligned} \tag{13-28}$$

with the three dimensionless numbers defined as follows:

$$M_t\left(i + \frac{1}{2}, j,k\right) = \frac{\Delta t}{(\Delta x)^2} \cdot D_t\left(i + \frac{1}{2}, j,k\right) \tag{13-29}$$

Diffusivities on each axis vary with their coordinates, following the change in moisture concentration. Two ways at least can be used. When the gradient of concentration is low, the mean diffusivity is calculated at position $(i + \frac{1}{2}, j,k)$.

$$D_t\left(i + \frac{1}{2}, j,k\right) = \frac{1}{2}\,[D_t(i,j,k) + D_t(i+1, j,k)]. \tag{13-30}$$

When the gradient of concentration is high, the mean concentration is calculated at position $(i + \frac{1}{2}, j,k)$:

$$D_t\left(i + \frac{1}{2}, j, k\right) = D_{ot} \times \exp A_t \cdot \left[\frac{C(i,j,k) + C(i+1, j,k)}{2}\right] \tag{13-31}$$

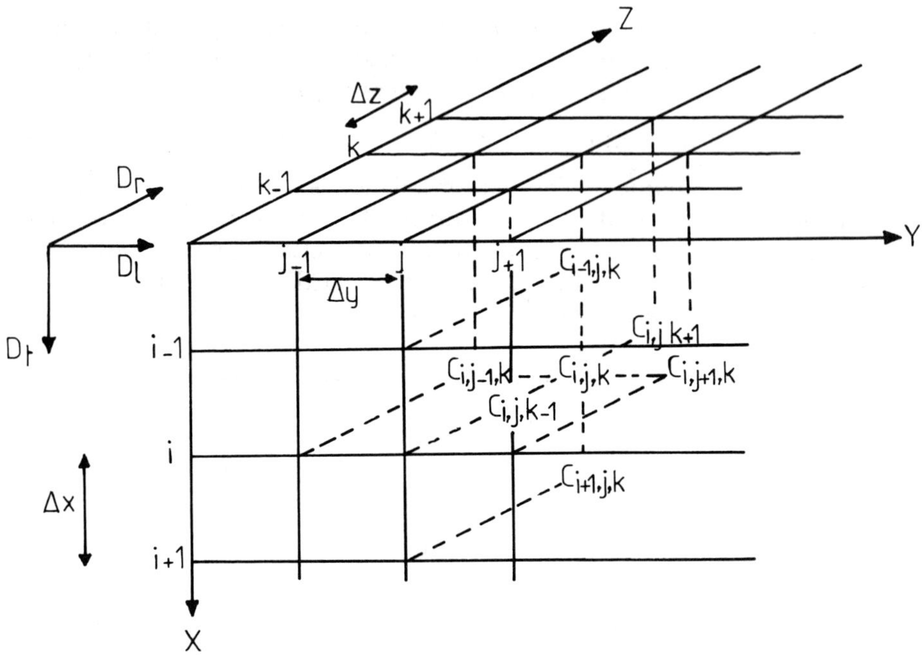


**Figure 13–18** Scheme for numerical analysis in case of three-dimensional transfer: $D_l$ = longitudinal; $D_t$ = tangential and $D_r$ = radial.

***On Each Face of the Sample* (z = o, k = o).** The calculation is done on the face ($z = o$, $k = o$) by considering the parallelepiped of dimensions $\Delta z/2$, $\Delta x$, and $\Delta y$ at position $(i, j, \frac{1}{4})$.

$$CN(i,j,o) = C(i,j,o) + \frac{2h \cdot \Delta t}{\Delta z} [C_{eq} - C(i,j,o)] - 2\, M_r \left(i,\, j,\, \frac{1}{2}\right) \times [C(i,j,o) - C(i,j,\, 1)] + M_l \left(i,\, j - \frac{1}{2},\, o\right) \times [C(i,j - 1,\, o) - C(i,j,o)] - M_l \left(i,j + \frac{1}{2},\, o\right) \times [C(i,j,o) - C(i,j + 1,\, o)] + M_t \left(i - \frac{1}{2},\, j,o\right) \times [C(i - 1,\, j,o) - C(i,j,o)] - M_t \left(i + \frac{1}{2},\, j,o\right) \times [C(i,j,o) - C(i + 1,\, j,o)] \tag{13-32}$$

The expressions for the other faces with $x$ = o and $t$ = o are obtained by circular permutation.

***On Each Line at the Intersection of Two Faces.*** The calculation is done at the position $(x,o,o)$ by considering the parallelepiped of dimensions $\Delta y/2$, $\Delta z/2$, and $\Delta x$:

$$CN(i,o,o) = C(i,o,o) + \frac{2K \cdot \Delta t}{\Delta y} [C_{eq} - C(i,o,o)] - 2M_1 \left(i, \frac{1}{2}, o\right) \times [C(i,o,o) - C(i,1,o)] + M_t \left(i - \frac{1}{2}, o,o\right) \times [C(i-1,o,o) - C(i,o,o)] - M_t \left(i + \frac{1}{2}, o,o\right) \times [C(i,o,o) - C(i+1,o,o)] + \frac{2K \cdot \Delta t}{\Delta z} \times [C_{eq} - C(i,o,o)] - 2M_r \left(i,o, \frac{1}{2}\right) \times [C(i,o,o) - C(i,o,1)] \tag{13-33}$$

The concentrations at the other intersections are also obtained by permutation.

***Corners (o,o,o).*** For the corner $(o,o,o)$, the calculation is done by considering the parallelepipede of dimensions $\Delta x/2$, $\Delta y/2$, and $\Delta z/2$ near the corner.

$$CN(o,o,o) = C(o,o,o) + 2h \cdot \Delta t [C_{eq} - C(o,o,o)] \times \left[\frac{1}{\Delta x} + \frac{1}{\Delta y} + \frac{1}{\Delta z}\right] - 2M_1 \left(o, \frac{1}{2}, o\right) \times [C(o,o,o) - C(o,1,o)] - 2M_r \left(o,o, \frac{1}{2}\right) \times [C(o,o,o) - C(o,o,1)] - 2M_t \left(\frac{1}{2}, o,o\right) \times [C(o,o,o) - C(1,o,o)] \tag{13-34}$$

***Amount of Moisture Located in the Sample.*** The amount of moisture located in the sample is obtained at any time by integrating the concentration of moisture with respect to the three dimensions.

### 13-4-3 Results

#### Coefficient of moisture transport, EMC, and diffusivities

***EMC: Moisture Concentration at Equilibrium.*** As shown previously, EMC is determined at 30°C for various values of relative humidity (RH) ranging from 55 percent to 80 percent. The isotherm of absorption being linear within this range, there is:

$$\text{EMC} = 0.088 \times \text{RH} + 6.15 \tag{13-25}$$

***Coefficient of Moisture Transport h.*** As shown previously, the coefficient of moisture transport through the air-wood interface, $h$, is:

$$h = 10^{-4} \quad \text{(cm/s)} \tag{13-35}$$

Of course, $h$, is the same for all the surfaces of the sample. This value is of the same order of magnitude as the values reported in the literature (24).

***Diffusivities.*** The diffusivities are determined by following the short-test method [19] under transient conditions. The kinetics of absorption are measured by increasing the RH of air by a small value (5 percent). The wide range of RH between 55 percent and 80 percent is thus covered. For each value of RH of air, the diffusivity is obtained from the straight line drawn by plotting the amount of moisture transported ($M_t$ or MC) in terms of the square root of time, by using the equation:

$$\frac{M_t}{M_{eq}} = \frac{4}{L}\left[\frac{D\cdot t}{\pi}\right]^{0.5} \tag{13-22}$$

where $L$ is the thickness of the sample.

The diffusivities are expressed as a function of the moisture content (MC) in the wood.

As shown previously, some results are worth noting:

1. Diffusivities along tangential and radial directions are about the same.
2. The diffusivity along the longitudinal direction is considerably higher than the other two diffusivities.
3. The longitudinal diffusivity decreases when the moisture content in the wood increases.
4. Diffusivities along tangential and radial directions increase slowly with the moisture content in the wood.

### Validity of the model

***Kinetics of Absorption of Moisture.*** As shown earlier, the best way for testing the validity of the model is to compare the kinetics of moisture exchange (absorption or desorption) obtained either from experiment or calculation.

Experiments are performed by using the three samples previously stabilized in air at RH of 55 percent and contacting them with a new atmosphere of RH = 80%, at 30°C.

The kinetics of absorption of moisture determined by experiments (+) and calculation (full line) are drawn for the cubic sample (2 × 2 × 2

**TABLE 13–3** Diffusivities (cm$^2$/s) in Three Principal Directions

| | |
|---|---|
| Longitudinal: | $D_l = 304 \times 10^{-6} \times \exp(-0.177 \times MC)$ |
| Tangential: | $D_t = 1.5 \times 10^{-6} \times \exp(0.0228 \times MC)$ |
| Radial: | $D_r = 1.5 \times 10^{-6} \times \exp(0.0228 \times MC)$ |

cm) in Figure 13–19, for parallelepipedic samples (4 × 1 × 4 cm) in Figure 13–20 and (10 × 1 × 1 cm) in Figure 13–21.

Two different kinetics of absorption (curves 1 and 2 in Figs. 13–19—13–21) are calculated, according to the boundary conditions chosen on the wood surface and then for the coefficient of moisture transport:

1. By assuming that the concentration of moisture on the wood surface is at equilibrium as soon as the process starts. This case is obtained for a high value of the coefficient of moisture transport $h$. It can be observed when the air is stirred very strongly [16] (curve 1 in Figs. 13–19 to 13–21).
2. By using the coefficient of moisture transport determined experimentally in motionless air ($h = 10^{-4}$ cm/s).

From these figures, the following conclusions are drawn:

1. Experimental values are located between the theoretical curves. This agreement is a proof for the validity of the model as well as for the accuracy of the data.
2. The motion of air surrounding the wood surface is of high importance,

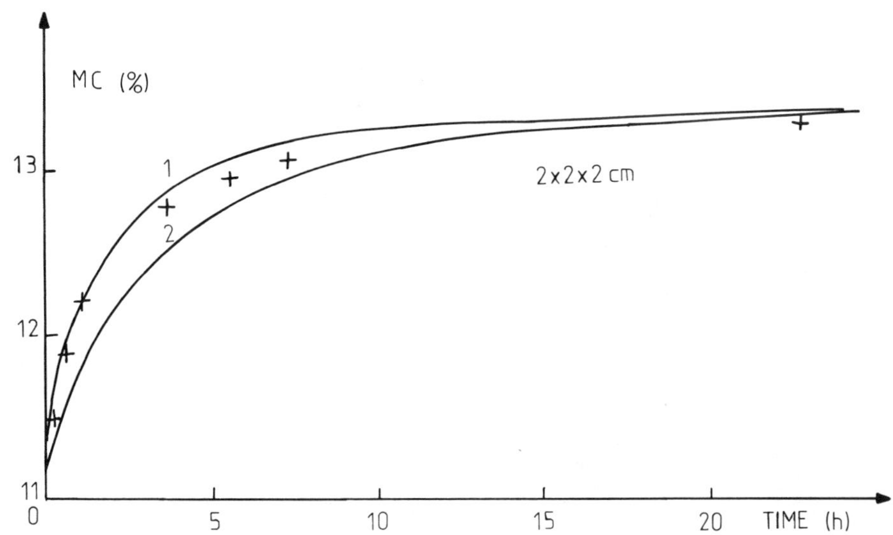


**Figure 13–19** Kinetics of moisture absorption by cubic sample (2 × 2 × 2 cm) at 30°C, previously stabilized with air of RH = 55%, and contacted with air of RH = 80%. (1) theoretical with high stirring of air. (2) theoretical without motion of air. (Reprinted from *Wood Sci. Technol.*, 23, (1989), 259–271, A. Droin-Josserand, J. L. Taverdet, and J. M. Vergnaud, ref. 22, by courtesy of Springer-Verlag.)

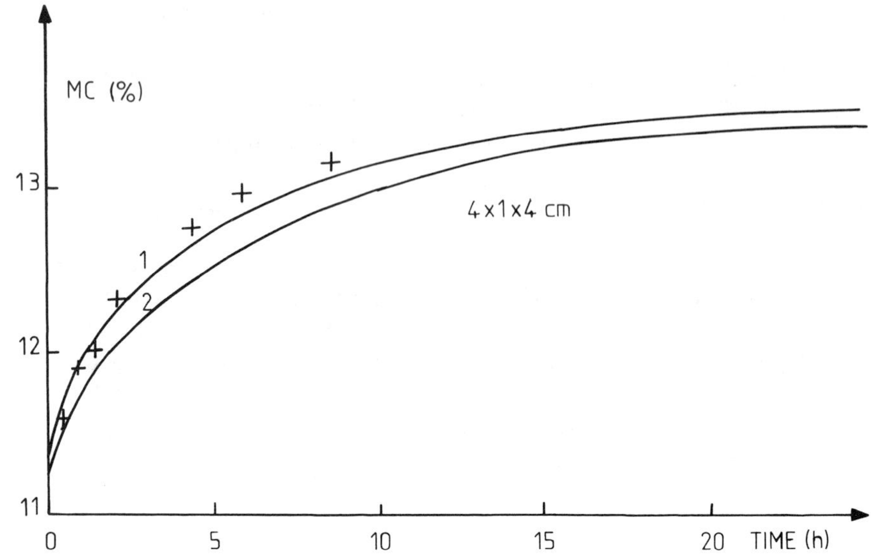


**Figure 13–20** Kinetics of moisture absorption by parallelepipedic sample (4 × 1 × 4 cm) at 30°C, stabilized at RH = 55%, and contacted with air of RH = 80%. (1) theoretical with high stirring of air. (2) theoretical without motion of air. (Reprinted from *Wood Sci. Technol.*, 23, (1989), 259–271, A. Droin-Josserand, J. L. Taverdet, and J. M. Vergnaud, ref. 22, by courtesy of Springer-Verlag.)

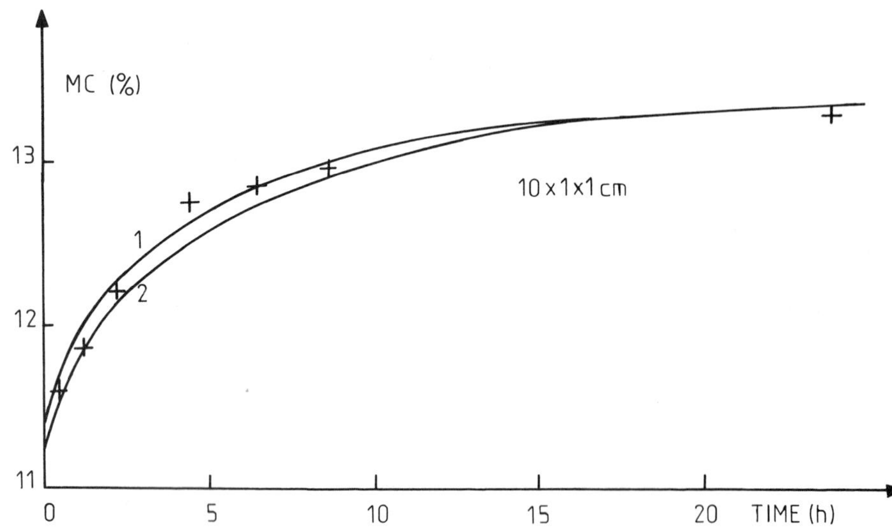


**Figure 13–21** Kinetics of moisture absorption by parallelepipedic sample (10 × 1 × 1 cm) at 30°C, stabilized at RH = 55 %, and contacted with air of RH = 80%. (1) theoretical with high stirring of air. (2) theoretical without motion of air. (Reprinted from *Wood Sci. Technol.*, 23, (1989), 259–271, A. Droin-Josserand, J. L. Taverdet, and J. M. Vergnaud, ref. 22, by courtesy of Springer-Verlag.)

especially at the beginning of the process: the higher the stirring, the higher the rate of absorption.

3. These curves correspond to the fastest and the lowest kinetics of absorption.
4. The kinetics of absorption depend on the shape and dimensions of the samples. The higher rate is obtained for the cube, for which the relative contribution of the longitudinal diffusion is the most important.
5. The effect of the motion of air on the kinetics of absorption is higher for the cube. As for the cube, the effect of longitudinal transport is the most effective, the limiting effect of low coefficient of moisture transport is more significant. There is an obvious statement that the higher the diffusivity, the more effective the limiting effect of a low coefficient of moisture transport on the kinetics.

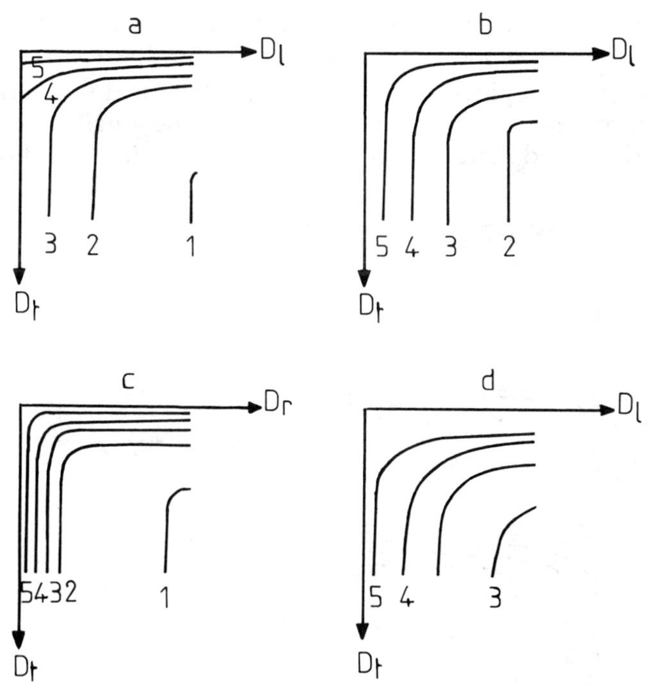


**Figure 13–22** Profiles of concentration through midplanes of the cube (2 × 2 × 2 cm). MC = 12 (1), 13 (2). (a) upper left: midplane parallel with longitudinal direction at 2 h, and motionless air. (b) upper right: midplane parallel with longitudinal direction at 2 h, and strong stirred air. (c) lower left: midplane perpendicular to longitudinal direction at 2 h, and motionless air. (d) lower right: midplane parallel with longitudinal direction at 5 h, and motionless air. (Reprinted from *Wood Sci. Technol.*, 23, (1989), 259–271, A. Droin-Josserand, J. L. Taverdet, and J. M. Vergnaud, ref. 22, by courtesy of Springer-Verlag.)

***Profiles of Concentration of Moisture.*** The profiles of concentration of moisture are drawn for the three samples; the cube in Figure 13–22, and parallelepipedes (4 × 1 × 4 cm) in Figure 13–23 and (10 × 1 × 1 cm) in Figure 13–24. These profiles are obtained during the same conditions of absorption as for the kinetics.

1. For the cube (Figure 13–22), two planes of symmetry are of interest: the midplane parallel with the longitudinal direction (*a*), the other midplane perpendicular to this longitudinal direction (*c*). The effect of time can be seen by comparing the profiles *a* and *d* obtained at 2 and 5 h. The effect of the motion of air is also seen in profiles *a* and *b*.
2. For the parallelepipede 4 × 1 × 4 cm (Figure 13–23), three planes of symmetry are considered: the midplane perpendicular to the longitudinal direction (*a*); the midplane parallel with the longitudinal and tangential direction (*b*); and the midplane parallel with the longitudinal and radial direction (*c*). The effect of each principal diffusivity on the whole transport can be appreciated in these curves.
3. For the parallelepipede 10 × 1 × 1 cm (Figure 13–24), two planes are considered: the midplane parallel with the longitudinal direction (*a*); and the midplane perpendicular to the longitudinal direction (*b*).

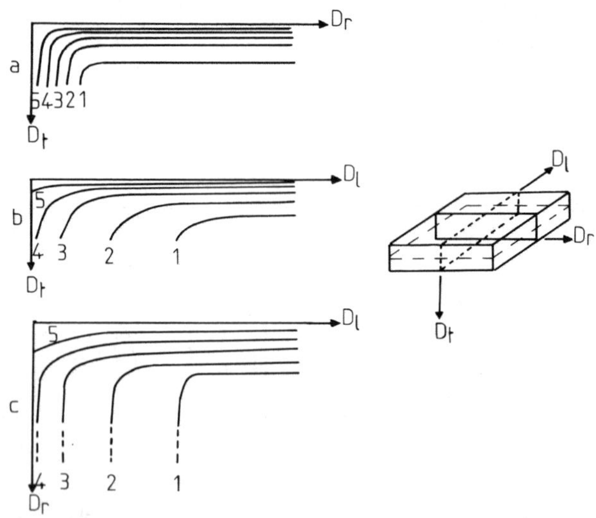


**Figure 13–23** Profiles of concentration through midplanes of the parallelepipede (4 × 1 × 4 cm). MC = 11.2 (1), 12.2 (2), 13.2 (3). (a) top: midplane perpendicular to longitudinal direction. (b) middle: midplane parallel with longitudinal and tangential directions. (c) bottom: midplane parallel with longitudinal and radial direction. (Reprinted from *Wood Sci. Technol.*, 23, (1989), 259–271, A. Droin-Josserand, J. L. Taverdet, and J. M. Vergnaud, ref. 22, by courtesy of Springer-Verlag.)

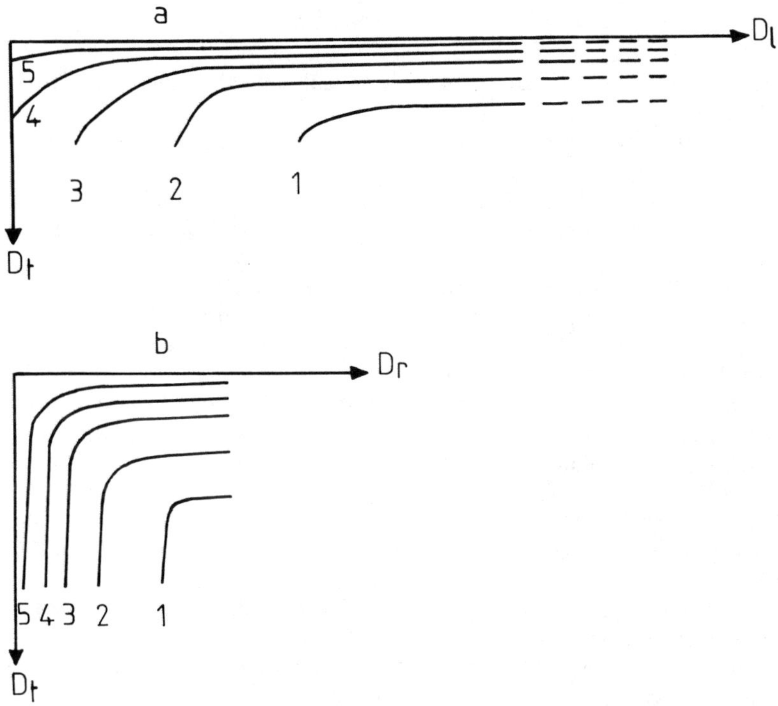


**Figure 13–24** Profiles of concentration through midplanes in parallelepipede (10 × 1 × 1 cm). MC = 12 (1), 13 (2). (a) top: midplane parallel with longitudinal and tangential direction. (b) bottom: midplane parallel with radial and tangential direction. (Reprinted from *Wood Sci. Technol.*, 23, (1989), 259–271, A. Droin-Josserand, J. L. Taverdet, and J. M. Vergnaud, ref. 22, by courtesy of Springer-Verlag.)

The effects of longitudinal and transversal diffusion on the whole transport can be compared in curves *a*.

**Conclusions.** The model is able to describe the process of absorption (and generally of moisture exchange) by wood samples of various sizes and shapes: cube and parallelepipede.

The model takes into account the external rate of transport with the coefficient of moisture transport and the internal rate of transport with the three principal axes of diffusion: longitudinal, tangential, and radial.

The kinetics of moisture absorption can then be predicted, as well as the profiles of moisture concentration, for wood boards of various shapes.

For calculation, some data are necessary:

- the coefficient of moisture transport
- the three principal diffusivities and their expression as a function of the moisture content

## REFERENCES


1. D. Dirol, "Evaluation d'une méthode d'essais mycologiques sur vermiculite," *Holzforschung*, 39, (1985), 11–16.
2. D. M. Griffin, "Water potential and wood decay fungi," *Ann. Rev. Phytopathol.*, 15, (1977), 319–329.
3. S. Avramidis and J. F. Siau, "An investigation of the external and internal resistance to moisture diffusion in wood," *Wood Sci. Technol.*, 21, (1987), 249–256.
4. C. Skaar, "Wood water relationships," in *The chemistry of solid wood*, ed. R. M. Rowell. (Washington, DC: Am. Chem. Soc., 1984) pp. 127–172.
5. C. Skaar, "Moisture movement in beech below the fiber saturation point," *Forest Prod. J.*, 8, (1958), 352–357.
6. A. J. Stamm, "Bound-water diffusion into wood in the fiber direction," *Forest Prod. J.*, 9, (1959), 27–32.
7. A. J. Stamm, "Combined bound water and water vapor diffusion into Sitka spruce," *Forest Prod. J.*, 12, (1960), 644–648.
8. G. L. Comstock, "Moisture diffusion coefficients in wood as calculated from absorption, desorption and steady-state data," *Forest Prod. J.*, 13, (1963), 97–103.
9. E. T. Choong, "Diffusion coefficients of softwoods by steady state and theoretical methods," *Forest Prod. J.*, 15, (1965), 21–27.
10. W. W. Moschler and R. E. Martin, "Diffusion equation solutions in experimental wood drying," *Wood. Sci.*, 1, (1968), 47–57.
11. J. C. Ashworth, "The mathematical simulation of the batch drying of softwoods timber," Ph.D. Thesis, Univ. of Canterbury, (1977).
12. J. F. Siau, "Flow in wood." (New York: Syracuse Univ. Press 1971).
13. H. N. Rosen, "Exponential dependency of the moisture diffusion coefficient on moisture content," *Wood Sci.*, 8, (1976), 174–179.
14. E. T. Choong and C. Skaar, "Diffusivity and surface emissivity in wood drying," *Wood and Fiber*, 4, (1972), 80–86.
15. H. B. Blandin, J. C. David, J. P. Illien, M. Malizewiez, and J. M. Vergnaud, "Modelling of drying of coatings: Effect of the thickness, temperature and concentration of solvent," *Progress in Organ. Coat.*, 15, (1987), 163–172.
16. Y. Khatir, J. Bouzon, and J. M. Vergnaud, "Liquid sorption by rubber sheets and evaporation; models and experiments," *Polym. Testing*, 6, (1986), 253–265.
17. A. Droin, J. L. Taverdet, and J. M. Vergnaud, "Modelling the kinetics of moisture absorption by wood," *Wood Sci. Technol.*, 22, (1988), 11–20.
18. D. Messadi, M. Hivert, and J. M. Vergnaud, "A new approach to the study of plasticizer migration from PVC into methanol," *J. Appl. Polym. Sci.*, 26, (1981), 667–677.
19. J. L. Taverdet and J. M. Vergnaud, "Study of transfer process of liquid into and plasticizer out of plasticized PVC by using short tests," *J. Appl. Polym. Sci.*, 29, (1984), 3391–3400.
20. A. Droin-Josserand, J. L. Taverdet, and J. M. Vergnaud, "Modelling the absorption and desorption of moisture by wood in an atmosphere of constant and programmed relative humidity," *Wood Sci. Technol.*, 22, (1988), 299–310.

21. A. Droin-Josserand, J. L. Taverdet, and J. M. Vergnaud, "Modelling of moisture absorption within a section of parallelepipedic sample of wood by considering longitudinal and transversal diffusion," *Holzforschung*, 43, (1989), 297–302.
22. A. Droin-Josserand, J. L. Taverdet, and J. M. Vergnaud, "Modelling the process of moisture absorption in three dimensions by wood samples of various shapes: cubic, parallelepipedic," *Wood. Sci. Technol.*, 23, (1989), 259–271.
23. A. Droin-Josserand, J. L. Taverdet, and J. M. Vergnaud, "Modelling the process of absorption and desorption of moisture by wood samples," in *Drying of Solids*, ed. Mujumdar, ed. Sakri, (1989).
24. E. T. Choong and C. Skaar," Separating internal and external resistance to moisture removal in wood drying," *Wood Sci.*, 1, (1969), 200–202.

# Author Index

# Subject Index